# Elements

| Element | Symbol | Atomic number | Relative atomic mass / g mol$^{-1}$ |
|---|---|---|---|
| Actinium | Ac | 89 | 227.03 |
| Aluminium | Al | 13 | 26.98 |
| Americium | Am | 95 | 241.06 |
| Antimony | Sb | 51 | 121.75 |
| Argon | Ar | 18 | 39.95 |
| Arsenic | As | 33 | 74.92 |
| Astatine | At | 85 | 210 |
| Barium | Ba | 56 | 137.34 |
| Berkelium | Bk | 97 | 249.08 |
| Beryllium | Be | 4 | 9.01 |
| Bismuth | Bi | 83 | 208.98 |
| Boron | B | 5 | 10.81 |
| Bromine | Br | 35 | 79.91 |
| Cadmium | Cd | 48 | 112.40 |
| Caesium | Cs | 55 | 132.91 |
| Calcium | Ca | 20 | 40.08 |
| Californium | Cf | 98 | 252.08 |
| Carbon | C | 6 | 12.01 |
| Cerium | Ce | 58 | 140.12 |
| Chlorine | Cl | 17 | 35.45 |
| Chromium | Cr | 24 | 52.01 |
| Cobalt | Co | 27 | 58.93 |
| Copper | Cu | 29 | 63.54 |
| Curium | Cm | 96 | 244.07 |
| Dysprosium | Dy | 66 | 162.50 |
| Einsteinium | Es | 99 | 252.09 |
| Erbium | Er | 68 | 167.26 |
| Europium | Eu | 63 | 151.96 |
| Fermium | Fm | 100 | 257.10 |
| Fluorine | F | 9 | 19.00 |
| Francium | Fr | 87 | 223 |
| Gadolinium | Gd | 64 | 157.25 |
| Gallium | Ga | 31 | 69.72 |
| Germanium | Ge | 32 | 72.59 |
| Gold | Au | 79 | 196.97 |
| Hafnium | Hf | 72 | 178.49 |
| Helium | He | 2 | 4.00 |
| Holmium | Ho | 67 | 164.93 |
| Hydrogen | H | 1 | 1.008 |
| Indium | In | 49 | 114.82 |
| Iodine | I | 53 | 126.90 |
| Iridium | Ir | 77 | 192.2 |
| Iron | Fe | 26 | 55.85 |
| Krypton | Kr | 36 | 83.80 |
| Lanthanum | La | 57 | 138.91 |
| Lawrencium | Lr | 103 | 262 |
| Lead | Pb | 82 | 207.19 |
| Lithium | Li | 3 | 6.94 |
| Lutetium | Lu | 71 | 174.97 |
| Magnesium | Mg | 12 | 24.31 |
| Manganese | Mn | 25 | 54.94 |
| Mendelevium | Md | 101 | 258.10 |
| Mercury | Hg | 80 | 200.59 |
| Molybdenum | Mo | 42 | 95.94 |
| Neodymium | Nd | 60 | 144.24 |
| Neon | Ne | 10 | 20.18 |
| Neptunium | Np | 93 | 237.05 |
| Nickel | Ni | 28 | 58.69 |
| Niobium | Nb | 41 | 92.91 |
| Nitrogen | N | 7 | 14.01 |
| Nobelium | No | 102 | 259 |
| Osmium | Os | 76 | 190.2 |
| Oxygen | O | 8 | 16.00 |
| Palladium | Pd | 46 | 106.4 |
| Phosphorus | P | 15 | 30.97 |
| Platinum | Pt | 78 | 195.08 |
| Plutonium | Pu | 94 | 239.05 |
| Polonium | Po | 84 | 210 |
| Potassium | K | 19 | 39.10 |
| Praseodymium | Pr | 59 | 140.91 |
| Promethium | Pm | 61 | 146.92 |
| Protactinium | Pa | 91 | 231.04 |
| Radium | Ra | 88 | 226.03 |
| Radon | Rn | 86 | 222 |
| Rhenium | Re | 75 | 186.2 |
| Rhodium | Rh | 45 | 102.91 |
| Rubidium | Rb | 37 | 85.47 |
| Ruthenium | Ru | 44 | 101.07 |
| Samarium | Sm | 62 | 150.35 |
| Scandium | Sc | 21 | 44.96 |
| Selenium | Se | 34 | 78.96 |
| Silicon | Si | 14 | 28.09 |
| Silver | Ag | 47 | 107.87 |
| Sodium | Na | 11 | 22.99 |
| Strontium | Sr | 38 | 87.62 |
| Sulfur | S | 16 | 32.06 |
| Tantalum | Ta | 73 | 180.95 |
| Technetium | Tc | 43 | 98.91 |
| Tellurium | Te | 52 | 127.60 |
| Terbium | Tb | 65 | 158.92 |
| Thallium | Tl | 81 | 204.37 |
| Thorium | Th | 90 | 232.04 |
| Thulium | Tm | 69 | 168.93 |
| Tin | Sn | 50 | 118.71 |
| Titanium | Ti | 22 | 47.90 |
| Tungsten | W | 74 | 183.85 |
| Uranium | U | 92 | 238.03 |
| Vanadium | V | 23 | 50.94 |
| Xenon | Xe | 54 | 131.30 |
| Ytterbium | Yb | 70 | 173.04 |
| Yttrium | Y | 39 | 88.91 |
| Zinc | Zn | 30 | 65.37 |
| Zirconium | Zr | 40 | 91.22 |

We work with leading authors to develop the
strongest educational materials in chemistry,
bringing cutting-edge thinking and best learning
practice to a global market.

Under a range of well-known imprints, including
Prentice Hall, we craft high quality print and
electronic publications which help
readers to understand and apply their content,
whether studying or at work.

To find out more about the complete range of our
publishing please visit us on the World Wide Web at:
www.pearsoneduc.com

A Companion Web Site accompanies

# INORGANIC CHEMISTRY

## by Catherine E. Housecroft and Alan G. Sharpe

Visit the *Inorganic Chemistry,* Companion Web Site at
*www.booksites.net/housecroft*
where you will find valuable teaching and learning material including:

For Students:
- Chapter-by-chapter annotated web links
- Multiple choice practice questions

For Lecturers:
- A secure, password protected site with teaching guidance
- All Chem 3D figures in the book are available for viewing in Chime

# INORGANIC
# CHEMISTRY

## SOLUTIONS MANUAL

Catherine E. Housecroft

Prentice
Hall

*An imprint of* **Pearson Education**

Harlow, England · London · New York · Reading, Massachusetts · San Francisco · Toronto · Don Mills, Ontario · Sydney
Tokyo · Singapore · Hong Kong · Seoul · Taipei · Cape Town · Madrid · Mexico City · Amsterdam · Munich · Paris · Milan

Pearson Education Ltd
Edinburgh Gate
Harlow
Essex CM20 2JE
England

and Associated Companies around the World.

*Visit us on the World Wide Web at:*
*www.pearsoneduc.com*

**First edition 2001**

ISBN 0582-31084-9

*British Library Cataloguing-in-Publication Data*
A catalogue record for this book can be obtained from the British Library

10 9 8 7 6 5 4 3 2 1
05 04 03 02 01

Typeset by Catherine Housecroft
Printed and bound in Great Britain by Henry Ling Ltd., at the Dorset Press, Dorchester, Dorset

# Preface

This *Solutions Manual* accompanies *Inorganic Chemistry* by Catherine E. Housecroft and Alan G. Sharpe, and provides answers to the end-of-chapter problems. *Inorganic Chemistry* builds upon the 3rd edition of Alan Sharpe's *Inorganic Chemistry* but in moving to the new text, we have introduced many more end-of-chapter problems and in-chapter 'worked examples' that give detailed methods of working for numerical problem-solving. We hope that providing both the worked examples and an accompanying *Solutions Manual* will give students, lecturers and tutors an invaluable resource. The *Solutions Manual* is organized to complement *Inorganic Chemistry* on a chapter-by-chapter basis. Cross references to the main text are used in order to provide interplay between the two books.

The problems in *Inorganic Chemistry* are of varying types and, given a limited page budget, it has not been possible to provide complete 'model' answers to every problem. Answers have therefore been tackled in several ways. *Numerical problems* have been given the fullest treatment with complete workings of all calculations. Where not supplied in the problems, data come from the Appendices in *Inorganic Chemistry*. This method of providing tabulated data is designed to give students practice in selecting appropriate data. *Descriptive answers* have generally been formulated in terms of a set of bullet points, providing the reader with an essay plan from which to develop a full answer. Cross references to relevant sections in the main text have been given to help the reader find additional, relevant material. Answers to problems asking for *suggested products in a reaction* include both the full reaction (with alternatives where appropriate) and comments on the type of reaction or how one might go about deducing possible products. Some problems ask for *rationalization of observables* and the answers have been planned to guide students through appropriate arguments illustrating, where possible, how to make use of tabulated data to quantify those arguments.

Students and lecturers are encouraged to make use of the accompanying website accessed via:

**http://www.booksites.net/housecroft**

which provides a range of additional information and links to appropriate websites.

I am deeply grateful to Dr. Alan Sharpe for reading through all the answers and providing criticism and suggestions for improvement. As always, Professor Edwin Constable has added his own criticisms to my writings and his contributions to the *Solutions Manual* have been invaluable. Many people within Pearson Education have played a role in the production of this book, but special thanks go to Bridget Allen, Kevin Ancient, Lynn Brandon, Julie Knight and Alex Seabrook. Finally, Philby and Isis have again played their part and must know just a little chemistry by now.

Catherine E. Housecroft
August 2000

***Important note to readers***: throughout the book, 'H&S' stands for *Inorganic Chemistry* by Catherine E. Housecroft and Alan G. Sharpe, Pearson Education, 2001.

# 1 Some basic concepts

1.1 (a) Al is monotopic, i.e. there is only one naturally occurring isotope.

$Z = 13$    Mass number = 27

Number of electrons = Number of protons = 13

Number of neutrons = 27 – 13 = 14

The nuclear data are summarized in the notation: $^{27}_{13}\text{Al}$

Notation:

Notation:
$^{54}_{26}\text{Fe}, \,^{56}_{26}\text{Fe}, \,^{57}_{26}\text{Fe}, \,^{58}_{26}\text{Fe}$

(b) Br ($Z = 35$) has 2 naturally occurring isotopes.

*Each* isotope has 35 electrons and 35 protons.

For the isotope with mass number 79:  number of neutrons = 79 – 35 = 44

For the isotope with mass number 81:  number of neutrons = 81 – 35 = 46

(c) Fe ($Z = 26$) has 4 naturally occurring isotopes.

*Each* isotope has 26 electrons and 26 protons.

For the isotope with mass number 54:  number of neutrons = 54 – 26 = 28

For the isotope with mass number 56:  number of neutrons = 56 – 26 = 30

For the isotope with mass number 57:  number of neutrons = 57 – 26 = 31

For the isotope with mass number 58:  number of neutrons = 58 – 26 = 32

1.2 Assume that $^3$H can be ignored since abundance is so low; error introduced by this assumption is negligible. The mass numbers of $^1$H and $^2$H are 1 and 2 respectively. Let % $^1$H = $x$, and % $^2$H = $100 - x$.

Then:

$$A_r = 1.008 = \frac{x \times 1}{100} + \frac{(100 - x) \times 2}{100}$$

$$100.8 = x + 200 - 2x$$

$$x = 99.20$$

This result gives 99.20% $^1$H and 0.80% $^2$H. The values do not agree with those in Appendix 5 (99.985 % $^1$H and 0.015% $^2$H) because we have used integral atomic masses for the isotopes. The accurate masses (to 5 sig. fig.) are 1.0078 and 2.0141, and if you work through the above calculation again, this gives 99.98% $^1$H and 0.02% $^2$H.

1.3 (a) Isotopic abundances: $^{32}$S 95.02 %, $^{33}$S 0.75 %, $^{34}$S 4.21 %, $^{36}$S 0.02 %. Relative intensities of peaks containing these isotopes must reflect their relative abundances.

$m/z = 256$ is assigned to $(^{32}\text{S})_8$ – the most abundant peak.

$m/z = 257$ is assigned to $(^{32}\text{S})_7(^{33}\text{S})$.

$m/z = 258$ is assigned to $(^{32}\text{S})_6(^{33}\text{S})_2$ and $(^{32}\text{S})_7(^{34}\text{S})$.

$m/z = 259$ is assigned to $(^{32}\text{S})_6(^{33}\text{S})(^{34}\text{S})$.

$m/z = 260$ is assigned to $(^{32}\text{S})_6(^{34}\text{S})_2$.

(b) Structure of $S_8$ is shown in **1.1**; the *parent ion* is due to $S_8$. Fragmentation by S–S bond cleavage produces $S_7$, $S_6$, $S_5$, $S_4$ ... and gives lower mass peaks.

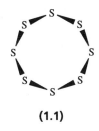

**(1.1)**

1.4 Equation 1.4 in H&S is:

$$\bar{v} = R\left(\frac{1}{2^2} - \frac{1}{n^2}\right) \qquad \text{where } R = 1.097 \times 10^7 \text{ m}^{-1}$$

For $n = 3$:

$$\bar{v} = 1.097 \times 10^7\left(\frac{1}{4} - \frac{1}{9}\right) = 1.524 \times 10^6 \text{ m}^{-1}$$

$$\bar{v} = \frac{1}{\lambda}$$

Convert $m^{-1}$ to m, and then to nm (1 nm = $10^{-9}$ m) to obtain wavelength in nm:

$$\lambda = \frac{1}{1.524 \times 10^6} \times 10^9 = 656.2 \text{ nm (to 4 sig. fig.)}$$

For $n = 4$:

$$\bar{v} = 1.097 \times 10^7 \left( \frac{1}{4} - \frac{1}{16} \right) = 2.057 \times 10^6 \text{ m}^{-1}$$

$$\lambda = \frac{1}{2.057 \times 10^6} \times 10^9 = 486.2 \text{ nm (to 4 sig. fig.)}$$

Use the same method for $n = 5$, and $n = 6$, to obtain calculated values of $\lambda$ of 434.1 nm and 410.2 nm respectively. Calculated values are consistent with observed lines in the Balmer series.

1.5    The radii, $r_n$, are found using equation 1.8 in H&S, with values of $n = 2$ and $n = 3$.

$$r_n = \frac{\varepsilon_0 h^2 n^2}{\pi m_e e^2}$$

where:  $\varepsilon_0$ = permittivity of a vacuum
$h$ = Planck constant
$m_e$ = electron rest mass
$e$ = charge on an electron

For $n = 2$:

$$r_n = \frac{8.854 \times 10^{-12} \times (6.626 \times 10^{-34})^2 \times 2^2}{\pi \times 9.109 \times 10^{-31} \times (1.602 \times 10^{-19})^2} = 2.117 \times 10^{-10} \text{ m}$$

or the radius may be quoted in pm: $r_n = 211.7$ pm      (1 pm = $10^{-12}$ m)
For $n = 3$:

$$r_n = \frac{8.854 \times 10^{-12} \times (6.626 \times 10^{-34})^2 \times 3^2}{\pi \times 9.109 \times 10^{-31} \times (1.602 \times 10^{-19})^2} = 4.764 \times 10^{-10} \text{ m}$$

or:    $r_n = 476.4$ pm

1.6    Each orbital is defined by a set of *three* quantum numbers: $n$, $l$ and $m_l$.
(a) 1s: $n = 1$;                       $l = 0 ... (n - 1)$, so only $l = 0$ is possible;
$m_l = +l, +(l - 1) ... 0 ... -(l - 1), -l$, but since $l = 0$, only $m_l = 0$ is possible.
Therefore, the set of quantum numbers that defines the 1s orbital is:
$n = 1$, $l = 0$, $m_l = 0$.
(b) 4s: $n = 4$;                       $l = 0, 1, 2, 3$, but for the s orbital, $l = 0$;
as for (a) above, if $l = 0$, $m_l = 0$.
Therefore, the set of quantum numbers that defines the 4s orbital is:
$n = 4$, $l = 0$, $m_l = 0$.
(c) 5s: $n = 5$;                       $l = 0, 1, 2, 3, 4$ but for the s orbital, $l = 0$;
as above, if $l = 0$, $m_l = 0$.
Therefore, the set of quantum numbers that defines the 5s orbital is:
$n = 5$, $l = 0$, $m_l = 0$.

1.7    For a 3p atomic orbital:      $n = 3$, $l = 1$, $m_l = +1$, 0 or $-1$.
Therefore, the three 3p orbitals are defined by the three sets of quantum numbers:

| $n$ | $l$ | $m_l$ |
|---|---|---|
| 3 | 1 | +1 |
| 3 | 1 | 0 |
| 3 | 1 | −1 |

1.8    The combination of $n = 4$ and $l = 3$ refers to $4f$ atomic orbitals. For $l = 3$, the possible values of $m_l$ are +3, +2, +1, 0, −1, −2, −3, and so there are 7 atomic orbitals in the $4f$ set. The sets of 3 quantum numbers that uniquely define each $4f$ orbital are:

| $n$ | $l$ | $m_l$ | $n$ | $l$ | $m_l$ |
|---|---|---|---|---|---|
| 4 | 3 | +3 | 4 | 3 | −1 |
| 4 | 3 | +2 | 4 | 3 | −2 |
| 4 | 3 | +1 | 4 | 3 | −3 |
| 4 | 3 | 0 | | | |

1.9    A *hydrogen-like* atom or *hydrogen-like* mononuclear ion contains one electron. To work out the answer to the problem, you need the atomic numbers ($Z$) of H (1), He (2) and Li (3). Thus, *atomic* H, He and Li contain 1, 2 and 3 electrons respectively.
(a) H⁺ has no electrons, and is therefore *not* hydrogen-like.
(b) He⁺ has 1 electron, and is therefore hydrogen-like.
(c) He⁻ has 3 electrons, and is therefore *not* hydrogen-like.
(d) Li⁺ has 2 electrons, and is therefore *not* hydrogen-like.
(e) Li²⁺ has 1 electron, and is therefore hydrogen-like.

1.10    (a) He⁺ is hydrogen-like, but it does not have the same nuclear charge as H. From footnote to Table 1.2 in H&S, radial part of the wavefunction $R(r)$ is of the form:

$$R(r) = 2\left(\frac{Z}{a_0}\right)^{\frac{3}{2}} e^{-\left(\frac{Zr}{a_0}\right)}$$

For H, $Z = 1$, but for He⁺, $Z = 2$. Thus, compared to the curve in Figure 1.5a in H&S for the H atom, a similar curve for He⁺ will be of a similar shape, but shifted to the left and drawn in towards the nucleus.
(b) The radial distribution functions for the $1s$ atomic orbitals in the H atom and He⁺ ion are shown in Figure 1.1. Note how (compared to the H atom) the increased nuclear charge on He⁺ results in there being a greater probability of finding the electron closer to the nucleus.

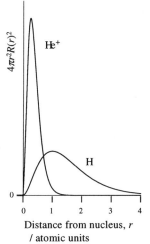

**Figure 1.1**  For answer 1.10.

1.11    Equation 1.16 in H&S is:

$$E = -\frac{k}{n^2} \qquad k = 1.312 \times 10^3 \text{ kJ mol}^{-1}$$

For $n = 1$:

$$E = -\frac{1.312 \times 10^3}{1} = -1312 \text{ kJ mol}^{-1}$$

For $n = 2$:

$$E = -\frac{1.312 \times 10^3}{4} = -328.0 \text{ kJ mol}^{-1}$$

For $n = 3$, $E = -145.8$ kJ mol⁻¹; for $n = 4$, $E = -82.00$ kJ mol⁻¹; for $n = 5$, $E = -52.50$ kJ mol⁻¹. The relative spacings of the energy levels are

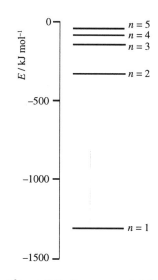

**Figure 1.2**  For answer 1.11.

shown in Figure 1.2 and decrease as $n$ increases. As one approaches the continuum (where $n = \infty$) the energy levels are extremely close together.

1.12    (a) Li:        $Z = 3$
Ground state electronic configuration is $1s^2 2s^1$.

(b) F:        $Z = 9$
Ground state electronic configuration is $1s^2 2s^2 2p^5$.

(c) S:        $Z = 16$
Ground state electronic configuration is $1s^2 2s^2 2p^6 3s^2 3p^4$.

(d) Ca:        $Z = 20$
Ground state electronic configuration is $1s^2 2s^2 2p^6 3s^2 3p^6 4s^2$.

(e) Ti:        $Z = 22$
Ground state electronic configuration is $1s^2 2s^2 2p^6 3s^2 3p^6 4s^2 3d^2$.

(f) Al:        $Z = 13$
Ground state electronic configuration is $1s^2 2s^2 2p^6 3s^2 3p^1$.

1.13    (a)  F:        $Z = 9$
The ground state electronic configuration is:

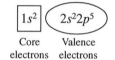

The energy level diagram
for only the valence electrons is:

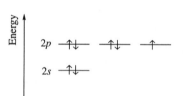

(b) Al:        $Z = 13$
The ground state electronic configuration is:

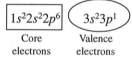

The energy level diagram
for only the valence electrons is:

(c)  Mg:        $Z = 12$
The ground state electronic configuration is:

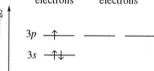

The energy level diagram
for only the valence electrons is:

1.14    (a) $IE_4$ of Sn = 4th ionization energy = energy required to remove the 4th electron from Sn in the gas phase. The equation that defines the process to which the value of $IE_4$ of Sn refers is:

$$Sn^{3+}(g) \longrightarrow Sn^{4+}(g) + e^-$$

Energy is needed, so the enthalpy change is positive – the reaction is endothermic. (b) ($IE_1 + IE_2 + IE_3$) is the sum of the first 3 ionization energies of Al(g), and corresponds to the overall change:

$$Al(g) \longrightarrow Al^{3+}(g) + 3e^-$$

This can be summarized in a thermochemical cycle as follows:

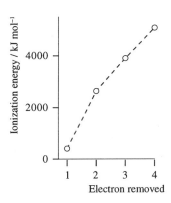

**Figure 1.3**  For answer 1.15.

1.15    Figure 1.3 shows the trend in the first four ionization energies of X. Value of $IE_1$ is much lower than that of $IE_2$, indicating that the first electron is removed relatively easily but that the second and subsequent electrons require significantly more energy. This trend is characteristic of a group 1 element, so X is likely to be an alkali metal. (X is actually Rb).

1.16    (a) Descending group 1: see Figure 1.4. All values are relatively low, indicating easy ionization of the first electron in M(g) to form $M^+$(g). The general decrease in values down the group reflects the reduced attraction between the nucleus and the valence electron that is being ionized, screening effects increasing as more shells of core electrons are added.
(b) Descending group 13: see Figure 1.5. Decrease from B to Al is as expected as core electrons in Al screen valence $3p$ electron from the nuclear charge. The lack of a continued decreasing trend can be explained in terms of the failure of the $d$- and $f$-electrons (which have a low screening power) to compensate for the increase in nuclear charge.

Screening: see Section 1.7 in H&S

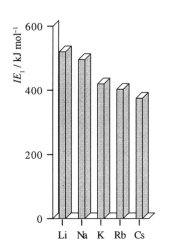

**Figure 1.4**  For answer 1.16a.

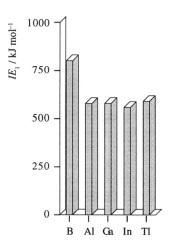

**Figure 1.5**  For answer 1.16b.

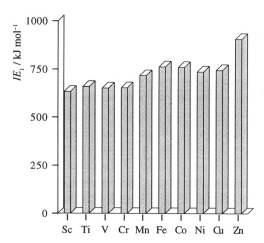

**Figure 1.6**  For answer 1.16c.

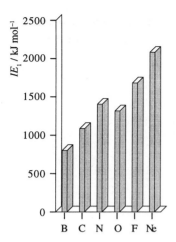

**Figure 1.7** For answer 1.16d.

(c) Crossing the first row of the *d*-block: see Figure 1.6. Overall trend is for a general increase in $IE_1$ but differences between values is small. For *each* metal, the ionization:

$$M(g) \rightarrow M^+(g) + e^-$$

corresponds to the removal of a 4*s* electron.

(d) Crossing the row of elements from B to Ne: see Figure 1.7. General increase in $IE_1$ is caused by an increase in $Z_{\text{eff}}$. Nitrogen has a ground state electronic configuration $2s^22p^3$ and the 2*p* level is half-occupied; there is a certain stability (see Box 1.7 in H&S) associated with this configuration and it is more difficult to ionize N than O.

(e) Going from Xe to Cs: Xe ($Z = 54$) is a noble gas with ground state electronic configuration $[Kr]5s^24d^{10}5p^6$, so removal of an electron involves removal from a fully occupied quantum shell and requires a very large amount of energy. Cs ($Z = 55$) is an alkali metal with ground state electronic configuration $[Xe]6s^1$. Removal of the valence electron from a *singly* occupied orbital in Cs requires a much smaller (although *still large*) amount of energy than removing the first electron from the full shell in Xe (375.7 versus 1170 kJ mol$^{-1}$).

(f) Going from P to S: P ($Z = 15$) has a ground state configuration $[Ne]3s^23p^3$, while that of S ($Z = 16$) is $[Ne]3s^23p^4$. The values of $IE_1$ are 1012 and 999.6 kJ mol$^{-1}$, and the decrease is explained as for that from N to O in part (d) above.

1.17   (a) See Figure 1.8.

(b) *Increase from H to He*: removing an electron from a singly occupied 1*s* orbital in H, and from a fully occupied (spin-paired electrons) orbital in He. *Decrease from He to Li*: ground state electronic configuration of Li is $1s^22s^1$, so first electron to be ionized is from a half-occupied 2*s* orbital; much less energy needed than for the removal of an electron from He ($1s^2$). *Increase from Li to Be*: in Be, the first electron to be removed comes from a filled 2*s* orbital (spin-paired electrons) compared to the removal of the single electron from the same orbital in Li. *Decrease from Be to B*: Ground state electronic configuration of B is $1s^22s^22p^1$, so the first electron to be removed is from a singly occupied 2*p* orbital. This requires less energy than removal of the electron from the $1s^22s^2$ configuration of Be. *Trend from B to Ne*: see answer 1.16d.

1.18   (a) For the reaction:        $O(g) + 2e^- \rightarrow O^{2-}(g)$

the energy change corresponds to the sum of the first two electron affinities. However, the use of energy terms labelled 'electron affinities' in tables of data requires particular care. Usually, *enthalpy changes* for the addition of an electron

**Figure 1.8** For answer 1.17.

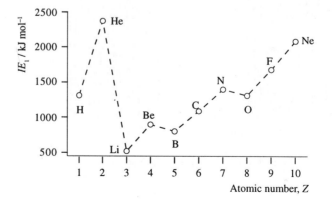

$(\Delta_{EA}H)$ are required (e.g. in a thermochemical cycle) rather than the actual electron affinity ($EA$). These quantities are related by:

See Box 1.6 in H&S

$$\Delta_{EA}H(298 \text{ K}) \approx \Delta_{EA}U(0 \text{ K}) = -EA$$

Let the enthalpy change for the above reaction be $\Delta_r H$:

$$\Delta_r H = \Delta_{EA}H_1 + \Delta_{EA}H_2$$

where $\Delta_{EA}H_1$ and $\Delta_{EA}H_2$ are the enthalpy changes for the processes:

$$O(g) + e^- \rightarrow O^-(g) \qquad \text{and} \qquad O^-(g) + e^- \rightarrow O^{2-}(g)$$

From Table 1.5 in H&S:

$$\Delta_r H = -141 + 798 = +657 \text{ kJ mol}^{-1}$$

(b) Formation of $O^{2-}(g)$ from $O(g)$ is highly *endothermic*, the contributing factor being the highly *endothermic* attachment of an electron to $O^-(g)$. This contrasts with *exothermic* attachment of an electron to neutral $O(g)$. Difference is due to repulsive forces felt by second electron as it approaches an anion. The fact that many metal oxides with ionic lattices are thermodynamically stable means that there must be sufficient energy liberated as the $M^{n+}(g)$ and $O^{2-}(g)$ ions come together (attractive forces) to more than offset the highly endothermic formation of $O^{2-}(g)$. Also, the endothermic formation of $M^{n+}(g)$ from $M(g)$, and the endothermic cleavage of O=O must be considered.

See the discussion of lattice energies in Chapter 5

1.19    (a) $F_2$

:F : F:

Remember that *no* stereochemistry is shown in a Lewis structure

(b) $BF_3$

:F:

B

F    F

(c) $NH_3$

H

H : N : H

(d) $H_2Se$

H

H : Se :

(e) $H_2O_2$

: O : H

H : O :

(f) $BeCl_2$

: Cl : Be : Cl :

(g) $SiH_4$

H

H : Si : H

H

(h) $PF_5$

1.20    The way to proceed is to allocate electrons to form N–H or N–F bonds, and then see how many unpaired electrons remain on each N atom.

(a) N$_2$H$_4$                                    (b) N$_2$F$_4$

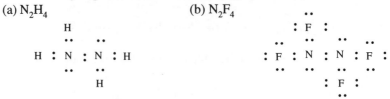

N–N single bond                          N–N single bond

➤ Isomers of N$_2$F$_2$ are possible: see answers 1.31f and g

(c) N$_2$F$_2$                                    (d) [N$_2$H$_5$]$^+$

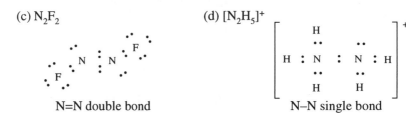

N=N double bond                          N–N single bond

1.21    Figure 1.9 shows the experimental structure of O$_3$. Resonance structures show possible covalent bonding descriptions. Overall bonding picture is a combination of resonance structures, although not all may contribute equally. *Resonance structures do not exist as separate species.* In O$_3$, each O atom has 6 valence electrons. Possible resonance structures (ignoring those with unreasonably high charge separation, e.g. O$^{2-}$ and O$^{2+}$ centres) are:

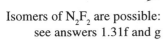

**Figure 1.9** The molecular structure of O$_3$.

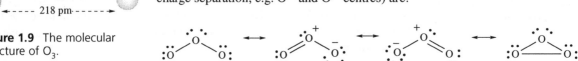

1.22    (a) First write down the ground state electronic configuration to determine the valence electrons available. Then draw out a set of resonance structures.

Li$_2$
Li:    $Z = 3$          $1s^2 2s^1$          1 valence electron available
Bonding in Li$_2$ is described in similar way to that in H$_2$; resonance structures are:

Li——Li    ⟷    Li$^+$  Li$^-$    ⟷    Li$^-$  Li$^+$

B$_2$
B:    $Z = 5$          $1s^2 2s^2 2p^1$          3 valence electrons available
Bonding in B$_2$ is described in terms of the following resonance structures (remember that in Lewis structures, odd electrons are paired wherever possible):

:B——B:    ⟷    B≡B    ⟷    B$^+$  B$^-$    ⟷    B$^-$  B$^+$

C$_2$
C:    $Z = 6$          $1s^2 2s^2 2p^2$          4 valence electrons available
The bonding in C$_2$ is described in terms of the following resonance structures:

:C=C:    ⟷    C̄≡C̄$^+$    ⟷    $^+$C̄≡C̄$^-$    ⟷    :C̄—C̄$^+$    ⟷    $^+$C̄—C̄$^-$:

(b) In each of the resonance structures drawn in part (a), all electrons are paired and so each molecule is predicted to be diamagnetic. There is therefore agreement with experiment for Li$_2$ and C$_2$, but not for B$_2$.

1.23  (a) $H_2$

H:    $Z = 1$        $1s^1$            1 valence electron available

$$H\text{——}H \quad\longleftrightarrow\quad H^+ \ \ H^- \quad\longleftrightarrow\quad H^- \ \ H^+$$

Na——Na

**(1.2)**

The left-hand resonance structure is consistent with an H–H single bond; bond order = 1. An ambiguity is not knowing to what extent the ionic resonance hybrids contribute. In the remaining parts of the question, only covalent resonance structures are considered; additional contributions may be made by ionic forms.

.. ..
S══S
.. ..

**(1.3)**

(b) $Na_2$

Na:    $Z = 11$        $1s^2 2s^2 2p^6 3s^1$        1 valence electron available

The bonding picture involves resonance structure **1.2** and this is consistent with a bond order of 1.

(c) $S_2$

S:    $Z = 16$        $1s^2 2s^2 2p^6 3s^2 3p^4$        6 valence electrons available

The bonding picture will be analogous to that in $O_2$ (see the end of Section 1.12 in H&S); it involves resonance structure **1.3**, and is consistent with a bond order of 2.

: N≡N :

**(1.4)**

(d) $N_2$

N:    $Z = 7$        $1s^2 2s^2 2p^3$        5 valence electrons available

Pairing of electrons, and obeying the octet rule, leads to **1.4** as the major contributing resonance structure. This is consistent with a bond order of 3. This conclusion assumes that resonance structure **1.5** does not contribute significantly.

.. ..
N——N
.. ..

**(1.5)**

(e) $Cl_2$

Cl:    $Z = 17$        $1s^2 2s^2 2p^6 3s^2 3p^5$        7 valence electrons available

The bonding is described in terms of resonance structure **1.6**, consistent with a bond order of 1.

.. ..
: Cl——Cl :
.. ..

**(1.6)**

1.24  Ground state electronic configuration of He ($Z = 2$) is $1s^2$. Within VB theory, the resonance structures that could be drawn (remembering that electrons are paired so far as possible) are:

$$He══He \quad\longleftrightarrow\quad He^{2+} \ \ He^{2-} \quad\longleftrightarrow\quad He^{2-} \ \ He^{2+}$$

The double bond formation is not possible with only the $1s$ orbital per He atom, and the ionic form is unreasonable (look at ionization energies for He). It is concluded that $He_2$ is not a viable species. (The real question is 'What is the stability of $He_2$ with respect to 2He?' and VB theory does not give an answer to this.)

1.25  (a) and (b) First, determine the available valence orbitals for each atom.

He:    $Z = 2$

Available valence orbital is the $1s$ atomic orbital. For the combination of two $1s$ atomic orbitals, the MO diagram is as shown on the right. Each He *atom* provides 2 electrons, so $He_2$ possesses 4 electrons which would fully occupy both the $\sigma(1s)$ and $\sigma^*(1s)$ MOs.

In $[He_2]^+$, there are 3 electrons; these occupy the MOs shown in the diagram to give an electronic configuration $\sigma(1s)^2\sigma^*(1s)^1$. In $[He_2]^{2+}$, there are 2 electrons and they occupy the MOs to give an electronic configuration of $\sigma(1s)^2$. The bond order in $[He_2]^+$ is $^1/_2(2-1) = ^1/_2$, and that in $[He_2]^{2+}$ is $^1/_2(2-0) = 1$. Both ions are therefore viable species. Note that $[He_2]^{2+}$ is isoelectronic with $H_2$.

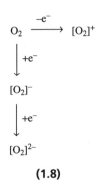

**(1.7)**

1.26    (a) O: $Z = 8$    Ground state electronic configuration = $1s^2 2s^2 2p^4$. The $1s$ electrons are core; the $2s$ and $2p$ orbitals and electrons are in the valence shell. Define the O–O internuclear axis to coincide with, for example, the $z$ axis (diagram **1.7**). For $O_2$, the MO diagram is constructed by allowing:

- $2s$ orbital of each O atom to interact to give $\sigma$-bonding combination and antibonding counterpart;
- $2p_z$ orbitals on the O atoms to interact to give $\sigma$-bonding MO and $\sigma^*$ MO;
- $2p_x$ orbitals on the O atoms to interact to give $\pi$-bonding MO and $\pi^*$ MO;
- $2p_y$ orbitals on the O atoms to interact to give $\pi$-bonding MO and $\pi^*$ MO.

The $\pi$-bonding MOs are degenerate and so are the two $\pi^*$ MOs:

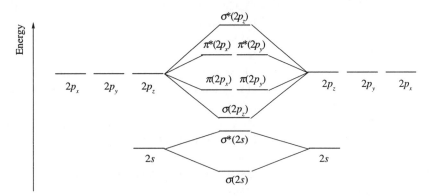

$O_2 \xrightarrow{-e^-} [O_2]^+$

$\downarrow +e^-$

$[O_2]^-$

$\downarrow +e^-$

$[O_2]^{2-}$

**(1.8)**

Now put in the electrons. Each O atom has 6 valence electrons, so there are 12 electrons in $O_2$ and these occupy the lowest lying MOs to give an electronic configuration of $\sigma(2s)^2\sigma^*(2s)^2\sigma(2p_z)^2\pi(2p_x)^2\pi(2p_y)^2\pi^*(2p_x)^1\pi^*(2p_y)^1$.

(b) Trend in O–O bond distances: $O_2$, 121 pm; $[O_2]^+$, 112 pm; $[O_2]^-$, 134 pm; $[O_2]^{2-}$, 149 pm, i.e. $[O_2]^{2-} > [O_2]^- > O_2 > [O_2]^+$.

Using scheme **1.8**, and assuming that the same qualitative MO diagram (above) is appropriate for each species, the electronic configurations of each species are:

$[O_2]^{2-}$    $\sigma(2s)^2\sigma^*(2s)^2\sigma(2p_z)^2\pi(2p_x)^2\pi(2p_y)^2\pi^*(2p_x)^2\pi^*(2p_y)^2$

$[O_2]^-$    $\sigma(2s)^2\sigma^*(2s)^2\sigma(2p_z)^2\pi(2p_x)^2\pi(2p_y)^2\pi^*(2p_x)^2\pi^*(2p_y)^1$

$O_2$    $\sigma(2s)^2\sigma^*(2s)^2\sigma(2p_z)^2\pi(2p_x)^2\pi(2p_y)^2\pi^*(2p_x)^1\pi^*(2p_y)^1$

$[O_2]^+$    $\sigma(2s)^2\sigma^*(2s)^2\sigma(2p_z)^2\pi(2p_y)^2\pi(2p_y)^2\pi^*(2p_x)^1\pi^*(2p_y)^0$

The bond orders are therefore:

| | | | |
|---|---|---|---|
| $[O_2]^{2-}$ | $\frac{1}{2}(8-6) = 1$ | $[O_2]^-$ : | $\frac{1}{2}(8-5) = 1\frac{1}{2}$ |
| $O_2$ | $\frac{1}{2}(8-4) = 2$ | $[O_2]^+$ : | $\frac{1}{2}(8-3) = 2\frac{1}{2}$ |

These bond orders are consistent with the sequence of bond lengths above, with the lowest bond order corresponding to the longest bond. This correlation can be made along a series of *closely related* species as here.

(c) A paramagnetic species contains one or more unpaired electrons: $[O_2]^-$, $O_2$ and $[O_2]^+$ are paramagnetic, but $[O_2]^{2-}$ is diamagnetic.

1.27    (a) N–H    $\chi^P(N) = 3.0$    $\chi^P(H) = 2.2$    Polar: $N^{\delta-}$–$H^{\delta+}$    See **1.9**.

(b) F–Br    $\chi^P(F) = 4.0$    $\chi^P(Br) = 3.0$    Polar: $F^{\delta-}$–$Br^{\delta+}$    See **1.10**.

(c) C–H    $\chi^P(C) = 2.6$    $\chi^P(H) = 2.2$    Very slightly polar: $C^{\delta-}$–$H^{\delta+}$    See **1.11**.

(d) P–Cl    $\chi^P(P) = 2.2$    $\chi^P(Cl) = 3.2$    Polar: $P^{\delta+}$–$Cl^{\delta-}$    See **1.12**.

(e) N–Br    $\chi^P(N) = 3.0$    $\chi^P(Br) = 3.0$    Non-polar

$\overset{\delta-}{N}\!\!-\!\!\overset{\delta+}{H}$    $\overset{\delta-}{F}\!\!-\!\!\overset{\delta+}{Br}$

**(1.9)**    **(1.10)**

$\overset{\delta-}{C}\!\!-\!\!\overset{\delta+}{H}$    $\overset{\delta+}{P}\!\!-\!\!\overset{\delta-}{Cl}$

**(1.11)**    **(1.12)**

| Group | | | | |
|---|---|---|---|---|
| 13 | 14 | 15 | 16 | 17 |
| B | C | N | O | F |
| Al | Si | P | S | Cl |

**1.28** Two species are isoelectronic if they possess the same *total* number of electrons. 'Isoelectronic' is also used to mean 'same number of *valence* electrons', although strictly this usage should always be qualified. In this problem, assume that 'isoelectronic' is strictly applied and that you are looking for pairs of species with the same *total* number of electrons. Rather than count the number of electrons, use the periodic table (see margin) to help you: moving one place to the right in the table means one more electron (e.g. going from O to F), and moving one place to the left means one less electron (e.g. going from S to P). A negative charge *adds* an electron; a positive charge takes one away, e.g. $N^+$ is isoelectronic with C. Using these 'rules', the following species form isoelectronic pairs:

HF and $[OH]^-$ NH$_3$ and $[H_3O]^+$
$CO_2$ and $[NO_2]^+$ SiCl$_4$ and $[AlCl_4]^-$

The remaining molecules do not have isoelectronic partners in the list.

**1.29** The VSEPR model is summarized in the following 'rules'.

➢
Use Table 1.8 ('Parent' shapes) in H&S

- Each valence-shell electron pair of central atom E in molecule EX$_n$ with E–X single bonds is stereochemically significant; electron pair-electron pair repulsions determine the shape of EX$_n$.
- Electron-electron repulsions decrease in the order:
   lone pair-lone pair > lone pair-bonding pair > bonding pair-bonding pair.
- Where E–X bonds are of different bond orders, electron-electron repulsions decrease in the order:
   triple bond-single bond > double bond-single bond > single bond-single bond.
- Electron-electron repulsions between bonding pairs in EX$_n$ depend on the difference between electronegativities of E and X; repulsions are less the more the E–X bonding electron density is drawn towards X.

(a) H$_2$Se

**(1.13)**

Central atom is Se
Se is in group 16, so number of valence electrons = 6
Number of bonding pairs (2 Se–H bonds) = 2
Number of lone pairs = 2
Total number of electron pairs = 4 = 2 bonding and 2 lone pairs
'Parent' shape = tetrahedral, see structure **1.13**
*Molecular* shape = bent.

(b) $[BH_4]^-$

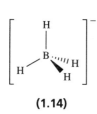

**(1.14)**

Central atom is B
B is in group 13, so number of valence electrons = 3
Add one extra electron for the negative charge
Number of bonding pairs (4 B–H bonds) = 4
No lone pairs
Total number of electron pairs = 4 = 4 bonding pairs
'Parent' shape = tetrahedral, see structure **1.14**
*Molecular* shape = tetrahedral.

(c) NF$_3$

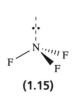

**(1.15)**

Central atom is N
N is in group 15, so number of valence electrons = 5
Number of bonding pairs (3 N–F bonds) = 3
Number of lone pairs = 1
Total number of electron pairs = 4 = 1 lone and 3 bonding pairs
'Parent' shape = tetrahedral, see structure **1.15**
*Molecular* shape = trigonal pyramidal.

(1.16)

(d) $SbF_5$

Central atom is Sb

Sb is in group 15, so number of valence electrons = 5

Number of bonding pairs (5 Sb–F bonds) = 5

Number of lone pairs = 0

Total number of electron pairs = 5 = 5 bonding pairs

'Parent' shape = trigonal bipyramidal,  see structure **1.16**

*Molecular* shape = trigonal bipyramidal.

(e) $[H_3O]^+$

Central atom is O

O is in group 16, so number of valence electrons = 6

Subtract one electron for the positive charge

Number of bonding pairs (3 O–H bonds) = 3

Number of lone pairs = 1

Total number of electron pairs = 4 = 1 lone and 3 bonding pairs

'Parent' shape = tetrahedral,  see structure **1.17**

*Molecular* shape = trigonal pyramidal.

(1.17)

(f) $IF_7$

Central atom is I

I is in group 17, so number of valence electrons = 7

Number of bonding pairs (7 I–F bonds) = 7

Number of lone pairs = 0

Total number of electron pairs = 7 = 7 bonding pairs

'Parent' shape = pentagonal bipyramidal,  see structure **1.18**

*Molecular* shape = pentagonal bipyramidal.

(1.18)

(g) $[I_3]^-$

Central atom is I

I is in group 17, so number of valence electrons = 7

Add one extra electron for the negative charge

Number of bonding pairs (2 I–I bonds) = 2

Number of lone pairs = 3

Total number of electron pairs = 5 = 2 bonding and 3 lone pairs

'Parent' shape = trigonal bipyramidal,  see structure **1.19**

*Molecular* shape = linear.

(1.19)

(h) $[I_3]^+$

Central atom is I

I is in group 17, so number of valence electrons = 7

Subtract one extra electron from the positive charge

Number of bonding pairs (2 I–I bonds) = 2

Number of lone pairs = 2

Total number of electron pairs = 4 = 2 bonding and 2 lone pairs

'Parent' shape = tetrahedral,  see structure **1.20**

*Molecular* shape = bent.

(1.20)

(i) $SO_3$

Central atom is S

S is in group 16, so number of valence electrons = 6

Each S–O bond is a double bond; number of S=O bonds = 3

For purposes of interelectronic repulsions, count *each* bond as 1 electron 'pair'

Number of lone pairs = 0

Total number of electron 'pairs' = 3 = 3 bonding 'pairs'

'Parent' shape = molecular shape = trigonal planar,  see structure **1.21**

(1.21)

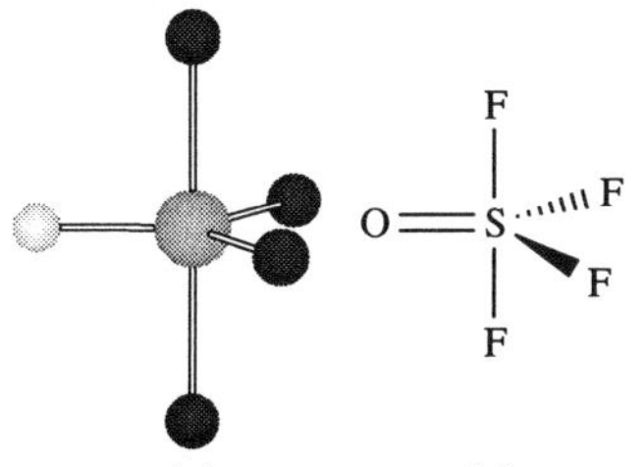

(a)          (b)

**Figure 1.10**  For answer 1.30.

1.30  Figure 1.10a shows the structure of $SOF_4$. It is trigonal bipyramidal, and since S has 6 valence electrons, there are no stereochemically active lone pairs.

Molecular and 'parent' shape = trigonal bipyramidal
Valence electrons from S = 6, from O = 2, from each F = 1
Total valence electrons = 12 = 6 pairs available for bonding
Total number of electron 'pairs' from Figure 1.10 = 5 = 5 bonding 'pairs'
Number of lone pairs = 0
One electron 'pair' in the VSEPR model arises from a double bond.

The structure can be rationalized if one considers the bond orders (see Figure 1.10b). Since double bond-single bond repulsions > single bond-single bond repulsions, electronic repulsions are minimized if the O atom lies in the equatorial plane.
(a) S–F single bonds
(b) S=O double bond

1.31  Shapes are found using VSEPR theory – method as in answer 1.29.
(a) $H_2S$
This is like $H_2Se$ (see answer 1.29a). First, consider each S–H bond:
S–H    $\chi^P(S) = 2.6$    $\chi^P(H) = 2.2$        Polar bond:    $S^{\delta-}-H^{\delta+}$
Now consider the molecule as a whole, taking into account lone pairs. The *bond* dipole moments reinforce each other, and the molecular dipole moment is enhanced by the 2 lone pairs. The direction of the resultant dipole moment is shown in **1.22**.

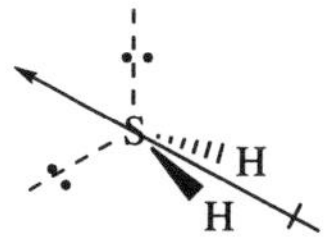

**(1.22)**

(b) $CO_2$
VSEPR theory is consistent with a linear molecule (2 bonding 'pairs'; C=O double bonds and no lone pairs), see structure **1.23**.
For the bond dipole moments:
C–O    $\chi^P(C) = 2.6$    $\chi^P(O) = 3.4$                Polar bond:    $O^{\delta-}-C^{\delta+}$
*But*, because the molecule is linear and symmetrical, the two bond dipole moments cancel one another (**1.24**), and the *molecule is non-polar.*

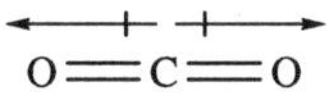

O=C=O    O=C=O
**(1.23)**        **(1.24)**

(c) $SO_2$
VSEPR theory is consistent with a bent molecule (2 bonding 'pairs' – S=O double bonds and 1 lone pair), see structure **1.25**.
In order to work out if the molecule is polar, look first at each bond dipole moment:
S–O    $\chi^P(S) = 2.6$    $\chi^P(O) = 3.4$        Polar bond:    $S^{\delta+}-O^{\delta-}$
The bond vectors reinforce each other (**1.25**), but resultant of these bond moments opposes the effect of the lone pair. Qualitatively difficult to assess direction of net molecular dipole moment. Experimental value for dipole moment in gas phase $SO_2$ is 1.63 D – diagram **1.26** shows direction in which this dipole moment acts.

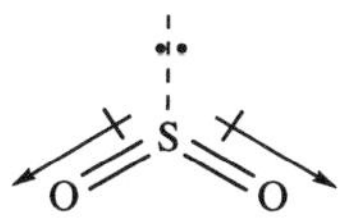

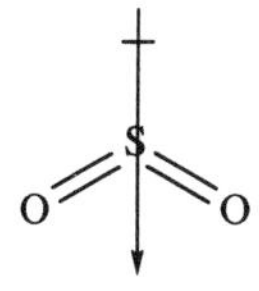

**(1.25)**

**(1.26)**

(d) $BF_3$
VSEPR theory consistent with trigonal planar $BF_3$ (3 bonding and no lone pairs), see structure **1.27**. Now consider each bond dipole moment:
B–F    $\chi^P(B) = 2.0$    $\chi^P(F) = 4.0$        Polar bond:    $B^{\delta+}-F^{\delta-}$
Each bond is polar, *but* molecule is non-polar because the 3 bond dipole moments (which are vectors) cancel out. Show this by resolving the vectors into two opposing directions. Let each vector be *V*. The vectors act as shown below:

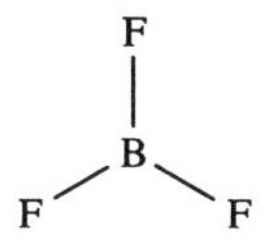

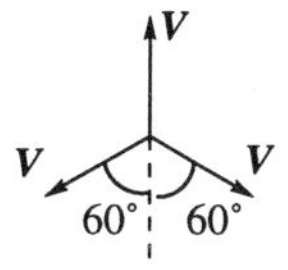

**(1.27)**

Resolving vectors into opposing upward and downward directions:
In an upward direction:    total vector = *V*
In a downward direction:    total vector = $2 \times V \cos 60$

$$= 2 \times \frac{V}{2} = V$$

∴ Two equal and opposite vectors act, and they cancel out.

**(1.28)**

**(1.29)**    **(1.30)**

**(1.31)**

$$H \!\!-\!\! C \!\equiv\! N$$

**(1.32)**

1.32

**(1.33)**

**(1.34)**

**(1.35)**

**(e) $PF_5$**

VSEPR theory consistent with trigonal bipyramidal molecule (**1.28**). Determine whether each *bond* is polar by considering electronegativity values:

P–F    $\chi^P(P) = 2.2$    $\chi^P(F) = 4.0$          Polar bond:    $P^{\delta+}$–$F^{\delta-}$

Each bond is polar, *but* the molecule is non-polar because:

- the 3 bond dipole moments in equatorial plane cancel out, and the reasoning for this is the same as for trigonal planar $BF_3$ in part (d);
- the 2 axial bond dipole moments are of equal magnitude but act in opposite directions, therefore the net dipole moment in the axial direction is zero.

**(f) *cis*-$N_2F_2$**

Answer 1.20c showed a Lewis structure for $N_2F_2$ although we ignored the possibility of isomers. Starting from this Lewis structure, apply VSEPR model to show that *cis*-$N_2F_2$ contains two trigonal planar N centres. The molecule is planar (**1.29**) because the N=N double bond constrains structure. Now consider each N–F bond:

$\chi^P(N) = 3.0$    $\chi^P(F) = 4.0$          Polar bond:    $N^{\delta+}$–$F^{\delta-}$

Each bond is polar, but resultant moment of the 2 bonds opposes resultant moment due to the two lone pairs (**1.30**). It is impossible at a qualitative level to decide whether the vector $V_2 > V_1$, or $V_2 < V_1$. The experimental value of the molecular dipole moment is 0.16 D, and theoretical calculations indicate that $V_2 < V_1$.

**(g) *trans*-$N_2F_2$**

As in part (f), start from Lewis structure in answer 1.20c; use VSEPR model to show that *trans*-$N_2F_2$ contains two trigonal planar N centres. The molecule is planar (**1.31**). Although each N–F bond is polar (as above), bond dipole moments cancel out, *and* dipole moments due to lone pairs cancel out. Therefore, *trans*-$N_2F_2$ is non-polar.

**(h) HCN**

Start from a Lewis structure and apply VSEPR model. HCN is linear (**1.32**). Now consider electronegativites of the atoms:

$\chi^P(H) = 2.2$          $\chi^P(C) = 2.6$          $\chi^P(N) = 3.0$

Each bond is polar bond:    $H^{\delta+}$–$C^{\delta-}$    $C^{\delta+}$–$N^{\delta-}$

The 2 bond dipole moments reinforce each other, and resultant molecular dipole moment acts in the direction shown in diagram **1.32**.

**(a) $BF_2Cl$**

VSEPR theory is consistent with trigonal planar molecule (3 bonding pairs, no lone pairs). No geometrical isomers are possible – any view that you draw can be converted into any other by a simple rotation, e.g. (i) to (ii) in **1.33**.

**(b) $POCl_3$**

Starting from a Lewis structure, followed by application of VSEPR theory gives tetrahedral structure with a P=O bond and P–Cl bonds. No isomers are possible.

**(c) $MePF_4$**

The structure of $MePF_4$ is related to $PF_5$ (trigonal bipyramidal, see answer 1.30e) but with one F atom replaced by a methyl group. The trigonal bipyramid contains 2 distinct sites: axial and equatorial. 2 isomers are possible (**1.34**).

**(d) $[PF_2Cl_4]^-$**

Start with Lewis structure, and apply VSEPR model: $[PF_2Cl_4]^-$ is octahedral. Two F atoms can be mutually adjacent or opposite giving *cis*- and *trans*-isomers respectively (**1.35** and **1.36**).

**(1.36)**

# 2 Nuclear properties

**Some general notes on notation**

For an atom of an element of atomic number $Z$:

$Z$ = number of electrons = number of protons

Number of neutrons = mass number $(A) - Z$

The mass and atomic numbers are stated in the full notation for the element symbol:

$$\xrightarrow{\text{Mass number}} \;\; {}^{19}_{\phantom{1}9}\text{F}$$
$$\xrightarrow{\text{Atomic number}}$$

The atomic number is fixed for a given element, but the mass number depends on the particular isotope, e.g. naturally occurring isotopes of Ba include:

$${}^{130}_{56}\text{Ba}, \quad {}^{134}_{56}\text{Ba}, \quad {}^{135}_{56}\text{Ba}, \quad {}^{137}_{56}\text{Ba}, \quad {}^{138}_{56}\text{Ba}$$

Sometimes, only the mass number is given, but $Z$ can be found in a periodic table.

---

2.1 (a) ${}^{19}_{9}\text{F}$ :     $Z = 9$ = number of electrons = number of protons

Number of neutrons = 19 − 9 = 10

(b) ${}^{59}_{27}\text{Co}$ :     $Z = 27$ = number of electrons = number of protons

Number of neutrons = 59 − 27 = 32

(c) ${}^{235}_{92}\text{U}$ :     $Z = 92$ = number of electrons = number of protons

Number of neutrons = 235 − 92 = 143

2.2 (a) and (b): see introductory notes above, and Section 1.3 in H&S.

(c) Mass of one atom of ${}^{1}\text{H}$ is *exactly* equal to the sum of masses of one proton and one electron, *but* mass of any other nuclide is *less* than sum of masses of protons, neutrons and electrons present. This difference in mass is the *mass defect* and gives measure of *binding energy* of protons and neutrons in the nucleus. Einstein's equation:

$$\Delta E = \Delta mc^2 \qquad c = \text{speed of light in a vacuum} = 2.998 \times 10^8 \text{ m s}^{-1}$$

relates the loss in mass ($\Delta m$) to the energy liberated ($\Delta E$) when the nuclear particles combine to form a nucleus. Huge amounts of energy are liberated.

(d) Binding energies can be expressed per nucleus, or as an average binding energy per nucleon. A nucleon is one particle in the nucleus. 'Binding energies per nucleon' allow comparisons between elements to be made:

$$\text{Binding energy per nucleon} = \frac{\text{Binding energy per nucleus}}{\text{Number of protons } + \text{ number of neutrons}}$$

2.3 Data from Appendix 5 in H&S: naturally occurring Ba contains 0.11% ${}^{130}\text{Ba}$, 0.10% ${}^{132}\text{Ba}$, 2.42% ${}^{134}\text{Ba}$, 6.59% ${}^{135}\text{Ba}$, 7.85% ${}^{136}\text{Ba}$, 11.23% ${}^{137}\text{Ba}$, 71.70% ${}^{138}\text{Ba}$. The mass spectrum of naturally occurring atomic Ba shows this distribution of isotopes, and is represented in Figure 2.1. The relatively low abundances of ${}^{130}\text{Ba}$ and ${}^{132}\text{Ba}$ mean that the intensities of these peaks are too low to be resolved on the scale of the chart in the figure. A vertical expansion is needed to see them; this is shown in the inset to Figure 2.1.

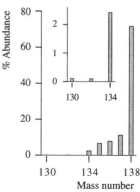

**Figure 2.1** For answer 2.3.

2.4    An α-particle is a helium nucleus, [$^4$He]$^{2+}$. In *nuclear reaction* equations, an α-particle is usually written as neutral He, since He$^{2+}$ ions readily pick up electrons. (a) For decay of radium-224, an α-particle is emitted:

$$^{224}_{88}\text{Ra} \longrightarrow {}^{A}_{Z}\text{X} + {}^{4}_{2}\text{He}$$

To work out the unknown product X, use the fact that atomic numbers and mass numbers on the left and right sides of the equation must balance. For atom X:

$$A = 224 - 4 = 220$$
$$Z = 88 - 2 = 86$$

The identity of X is found by looking up (e.g. periodic table) the element with atomic number 86: X = Rn. The final equation is:

$$^{224}_{88}\text{Ra} \longrightarrow {}^{220}_{86}\text{Rn} + {}^{4}_{2}\text{He}$$

(b) Helium is monatomic, so an atom of He gas corresponds to an α-particle. Data given:

α-Particles emitted at a rate of $7.65 \times 10^{12}$ s$^{-1}$ per mole of $^{224}$Ra
He production rate = $2.9 \times 10^{-10}$ dm$^3$ s$^{-1}$
1 mole He occupies 22.7 dm$^3$ at 273 K, 1 bar

$$\therefore \quad \text{Moles He produced per second} = \frac{2.9 \times 10^{-10}}{22.7} = 1.28 \times 10^{-11}$$

1 mole of He contains the Avogadro number of atoms of He. From above:

$1.28 \times 10^{-11}$ moles contains $7.65 \times 10^{12}$ atoms of He

$$\therefore \quad \text{1 mole contains} \ \frac{7.65 \times 10^{12}}{1.28 \times 10^{-11}} = 5.98 \times 10^{23} \text{ atoms}$$

This is an estimate of the Avogadro number (actual value = $6.022 \times 10^{23}$ mol$^{-1}$).

2.5    *Radioactive decay is always first order.* There are two ways to approach the problem.
**Method 1**    Plot moles of $^{218}$Po against time and measure several half-lives (Figure 2.2).

**Figure 2.2**    Decay curve for $^{218}$Po. The decay is exponential.

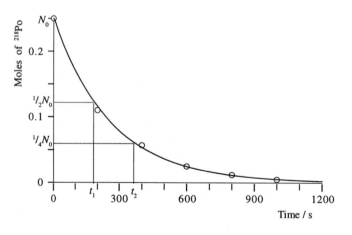

The data are plotted in Figure 2.2, and the time taken for the number of moles of $^{218}$Po to halve:

$$t_1 \approx 180 \text{ s}$$

and then halve again:

$$t_2 - t_1 \approx 180 \text{ s}$$

These are the first two half-lives. You should measure at least three half-lives, and take the average value ($\approx 180$ s). (Actual value of half-life is 182 s.) The rate constant, $k$, is then found from the equation:

$$k = \frac{\ln 2}{t_{\frac{1}{2}}} \approx \frac{\ln 2}{180} \approx 3.9 \times 10^{-3} \text{ s}^{-1}$$

***Method 2***   Since the decay is first order, a plot of ln (moles) against time will be linear. This comes from the integrated form of the first order rate equation:

$$\ln [^{218}\text{Po}] = \ln [^{218}\text{Po}]_0 - kt$$

where: $[^{218}\text{Po}]$ = concentration of $^{218}\text{Po}$ at time $t$, $[^{218}\text{Po}]_0$ is the initial concentration, and $k$ is the rate constant. You can use moles of $^{218}\text{Po}$ directly, and a plot of ln (moles $^{218}\text{Po}$) against time is shown in Figure 2.3. The points have been fitted with the best straight line.
The gradient of the line gives $k$ directly:

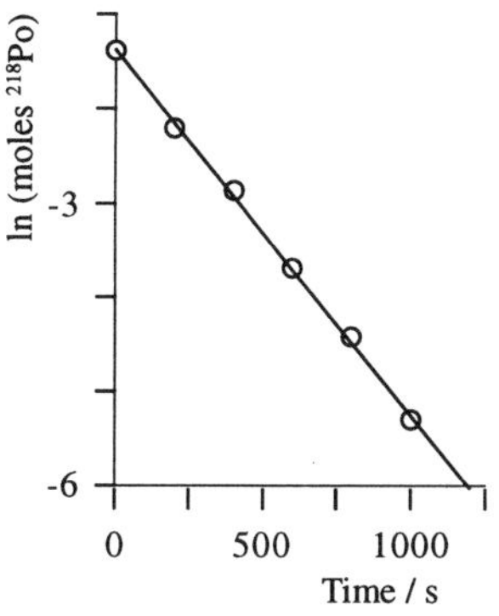

$$\text{Gradient} = -k = -(3.9 \times 10^{-3}) \text{ s}^{-1}$$
$$k = 3.9 \times 10^{-3} \text{ s}^{-1}$$

**Figure 2.3**  Plot for method 2 in answer 2.5.

The half-life is then found from the equation:

$$t_{\frac{1}{2}} = \frac{\ln 2}{k} = \frac{\ln 2}{3.9 \times 10^{-3}} \approx 1.8 \times 10^2 \text{ s}$$

2.6    For $^{90}\text{Sr}$:

$$t_{\frac{1}{2}} = 29.1 \text{ yr}$$

$$= 29.1 \times 365 \times 24 \times 60 \times 60 \text{ s} = 9.18 \times 10^8 \text{ s}$$

$$k = \frac{\ln 2}{t_{\frac{1}{2}}} = \frac{\ln 2}{9.18 \times 10^8} = 7.55 \times 10^{-10} \text{ s}^{-1}$$

2.7    The table is completed as follows:

| Reaction type | Change in number of protons | Change in number of neutrons | Change in mass number | Is a new element formed? |
|---|---|---|---|---|
| $\alpha$-Particle loss | –2 | –2 | –4 | Yes |
| $\beta$-Particle loss | +1 | –1 | 0 | Yes |
| Positron loss | –1 | +1 | 0 | Yes |
| (n,$\gamma$) reaction | 0 | +1 | +1 | No |

2.8    First, notice on the figure (reproduced here, in part, as Figure 2.4) that there are two types of decay step:
(i) mass number decreases by 4, and $Z$ decreases by 2;
(ii) mass number stays the same, and $Z$ increases by 1.
These correspond to (i) $\alpha$-particle loss, and (ii) $\beta$-particle loss. For example, the

**Figure 2.4** Decay series for answer 2.8. A more fully labelled version of this figure appears as Figure 2.3 in H&S.

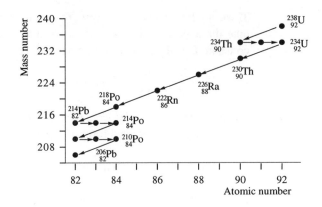

first decay step is loss of an α-particle:

$$^{238}_{92}\text{U} \rightarrow {}^{234}_{90}\text{Th} + {}^{4}_{2}\text{He}$$

and the next 2 steps each involve the loss of a β-particle:

$$^{234}_{90}\text{Th} \xrightarrow{-\beta^-} {}^{234}_{91}\text{Pa} \xrightarrow{-\beta^-} {}^{234}_{92}\text{U}$$

See Table 2.1 in H&S for details of all the decay steps.

2.9    The notation follows the convention:

$$\text{Initial nuclide}\left(\begin{array}{c}\text{incoming particles} \\ \text{or quanta}\end{array}, \begin{array}{c}\text{outgoing particles} \\ \text{or quanta}\end{array}\right)\text{final nuclide}$$

Expansion of each notational form gives the following nuclear reactions:

(a)   $^{58}_{26}\text{Fe}(2n, \beta)^{60}_{27}\text{Co}$          $^{58}_{26}\text{Fe} + 2n \longrightarrow {}^{60}_{27}\text{Co} + \beta^-$

(b)   $^{55}_{25}\text{Mn}(n, \gamma)^{56}_{25}\text{Mn}$          $^{55}_{25}\text{Mn} + n \longrightarrow {}^{56}_{25}\text{Mn} + \gamma$

(c)   $^{32}_{16}\text{S}(n, p)^{32}_{15}\text{P}$          $^{32}_{16}\text{S} + n \longrightarrow {}^{32}_{15}\text{P} + p$

(d)   $^{23}_{11}\text{Na}(\gamma, 3n)^{20}_{11}\text{Na}$          $^{23}_{11}\text{Na} + \gamma \longrightarrow {}^{20}_{11}\text{Na} + 3n$

2.10    To tackle these questions, use the facts that:
$\Sigma$(Mass numbers of reactants) = $\Sigma$(Mass numbers of products)
$\Sigma$(Atomic numbers of reactants) = $\Sigma$(Atomic numbers of products)

$\Sigma$ = summation of   ➤

(a) Let the unknown product be $^A_Z\text{X}$:

$$^{235}_{92}\text{U} + {}^{1}_{0}n \longrightarrow {}^{142}_{56}\text{Ba} + {}^{A}_{Z}\text{X} + 2{}^{1}_{0}n$$

Thus:   $235 + 1 = 142 + A + (2 \times 1)$     $\therefore$    $A = 92$
and:     $92 = 56 + Z$          $\therefore$    $Z = 36$

Use the value of $Z$ to identify the element from the periodic table: $^A_Z\text{X} = {}^{92}_{36}\text{Kr}$

(b) Let the unknown product be $^A_Z\text{X}$ :

$$^{235}_{92}\text{U} + {}^{1}_{0}n \longrightarrow {}^{137}_{52}\text{Te} + {}^{A}_{Z}\text{X} + 2{}^{1}_{0}n$$

Thus:   $235 + 1 = 137 + A + (2 \times 1)$      $\therefore$   $A = 97$
and:    $92 = 52 + Z$                         $\therefore$   $Z = 40$

Use the value of $Z$ to identify the element from the periodic table: $^A_Z X = \; ^{97}_{40}Zr$.

2.11   'Fast' neutrons are high-energy ($\approx 1$ MeV) neutrons, and are produced by the nuclear fission of uranium-235. 'Slow' (thermal) neutrons have energies of $\approx 0.05$ eV; they are involved in (n,$\gamma$) reactions, and in nuclear fission of uranium-235.
(a) This is a fast neutron reaction: conversion of nitrogen-14 into carbon-14.
(b) This is an (n,$\gamma$) reaction, caused by collision of uranium-238 with a slow neutron.
(c) This reaction is nuclear fission of uranium-235, brought about by collision with a slow neutron.

2.12   The graph described in question 2.12 is analogous to that in Figure 2.3 (p. 17) and so the gradient gives the rate constant, $k$, for the first order decay:

$$\text{Gradient} = -0.0023 \text{ day}^{-1} = -k$$
$$k = 0.0023 \text{ day}^{-1}$$

The half-life is related to $k$ by the equation:

$$t_{\frac{1}{2}} = \frac{\ln 2}{k} = \frac{\ln 2}{0.0023} \approx 300 \text{ days}$$

2.13   Naturally occurring CO is essentially $^{12}C^{16}O$, and the absorption at $2170 \text{ cm}^{-1}$ can be assigned to the stretching mode of this species. Enrichment to $^{13}C^{16}O$ would result in a shift in the absorption. First consider the relationship between the wavenumber and reduced mass of the molecule; the force constant can be ignored if we assume that its change is negligible upon enrichment:

$$\bar{v} \propto \sqrt{\frac{1}{\mu}}$$

The reduced mass is related to the masses (in kg) of the atoms by the equation:

$$\frac{1}{\mu} = \frac{1}{m_1} + \frac{1}{m_2}$$

For $^{12}C^{16}O$:

$$\frac{1}{\mu} = \frac{1}{(12 \times 1.67 \times 10^{-27})} + \frac{1}{(16 \times 1.67 \times 10^{-27})} \qquad \mu = 1.145 \times 10^{-26} \text{ kg}$$

For $^{13}C^{16}O$:

$$\frac{1}{\mu} = \frac{1}{(13 \times 1.67 \times 10^{-27})} + \frac{1}{(16 \times 1.67 \times 10^{-27})} \qquad \mu = 1.198 \times 10^{-26} \text{ kg}$$

The ratio of the values of $\bar{v}$ for $^{12}C^{16}O$ ($\bar{v}_2$) and $^{13}C^{16}O$ ($\bar{v}_1$) is given by:

$$\frac{\bar{v}_1}{\bar{v}_2} = \sqrt{\frac{\mu_2}{\mu_1}}$$

and so:

$$\bar{v}_1 = \bar{v}_2 \sqrt{\frac{\mu_2}{\mu_1}} = 2170 \sqrt{\frac{1.145 \times 10^{-26}}{1.198 \times 10^{-26}}} = 2120 \text{ cm}^{-1}$$

2.14    First, look at the ratio of the wavenumbers to see if the change is due to the incorporation of $^2$H (D) in place of $^1$H since the experimental conditions differed in this regard:

$$\frac{2300}{1630} = 1.411 \approx \sqrt{2}$$

Using the relationships in answer 2.13, the above ratio shows that the shift indeed arises from deuteration.

Compound **A** is Na$_2$HPO$_3$, and so compound **B** is Na$_2$DPO$_3$.

**A** contains [HPO$_3$]$^{2-}$, and **B** contains [DPO$_3$]$^{2-}$.

An O–H bond would exchange deuterium with D$_2$O:

$$\text{O–H} + \text{D}_2\text{O} \rightleftharpoons \text{O–D} + \text{DOH}$$

Since **A** does *not* exchange deuterium in D$_2$O, [HPO$_3$]$^{2-}$ must contain a P–H bond (not an O–H bond). Therefore, **A** has structure **2.1**.

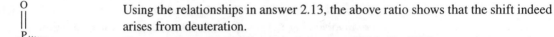

**(2.1)**

2.15    The amount of a sparingly soluble salt that dissolves in, say, 1 dm$^3$ is very small. Determining the solubility of such a salt by dissolution, evaporating the solution to dryness, followed by weighing the residue, involves significant error in the recorded mass. However, using the radioactivity of a sample as a means of determining amount present is accurate, provided that the salt contains an isotope of short half-life so that the radioactivity is sufficient for accurate determination. The next answer shows how the method of *isotope dilution* works.

2.16    This answer illustrates the method of isotope dilution. Assume that the mass of radioactive Pb is negligible compared to the 0.0100 g in the sample.

$$\text{Total amount of Pb} = \frac{0.0100}{207} = 4.83 \times 10^{-5} \text{ moles}$$

This number of moles of Pb is initially in solution. Now, the sparingly soluble salt PbCrO$_4$ is precipitated, leaving a *small number of moles* of PbCrO$_4$ still dissolved in solution. A 10 cm$^3$ aliquot of this solution is now taken, and from the radioactivity of the residue obtained from this aliquot, we can say:

*Ratio* dissolved Pb(II) : total (i.e. initial) Pb(II) = 4.17 × 10$^{-5}$

Let the number of moles of Pb(II) dissolved in the 10 cm$^3$ aliquot = $y$

Therefore:

$$\frac{y}{4.83 \times 10^{-5}} = 4.17 \times 10^{-5}$$

$$y = 4.17 \times 10^{-5} \times 4.83 \times 10^{-5} = 2.01 \times 10^{-9} \text{ moles}$$

The solubility, expressed in mol dm$^{-3}$, is therefore 2.01 × 10$^{-7}$ mol dm$^{-3}$.

2.17    Measurement in Hz gives a value of the coupling constant $J$ which is *independent* of the magnetic field. If it were recorded as a chemical shift difference ($\delta_1 - \delta_2$), the value would depend on the field strength. For example, consider a coupling constant $J_{HH} = 10$ Hz. In a 100 MHz $^1$H NMR spectrum, this corresponds to a chemical shift difference ($\delta_1 - \delta_2$) = 0.1, but in a 250 MHz spectrum, $J_{HH} = 10$ Hz corresponds to a value of ($\delta_1 - \delta_2$) = 0.04. Reporting coupling constants in Hz makes comparisons between spectra easier and more reliable. (Read the section entitled *'Resonance frequencies and chemical shifts'* in Box 2.4 in H&S.)

**Questions 2.18-2.23: some general notes**

Refer to the discussion in Box 2.4 in H&S, and the discussion and examples in Section 2.11 in H&S if you need further explanations about these NMR spectroscopic questions and answers.

2.18    Values of spin-spin coupling constants decrease with the bond separation of the nuclei that are spin-coupled. The fact that it is possible to observe long range couplings between $^{31}P$ and $^{19}F$, and between $^{31}P$ and $^{1}H$, but not between non-equivalent and remote $^{1}H$ nuclei, suggests that for a pair of directly attached nuclei, values of $J_{PF}$ and $J_{PH}$ are much greater than $J_{HH}$. [Typical values for directly attached nuclei are: $J_{PF} \leq 1500$ Hz, $J_{PH} \leq 800$ Hz, $J_{HH} \leq 10$ Hz.]

2.19    See structure **2.2**. There are 2 $^{13}C$ environments, labelled *a* and *b* in structure **2.2**. Both $^{1}H$ and $^{19}F$ are NMR active – $^{1}H$: 100%, $I = \frac{1}{2}$; $^{19}F$: 100%, $I = \frac{1}{2}$. Nucleus $^{13}C(a)$ couples to 3 adjacent $^{19}F$ to give a binomial quartet. Because $J(^{13}C-^{19}F)$ values are large for directly attached nuclei, long range coupling is *also* seen. Therefore, $^{13}C(b)$ couples to the 3 $^{19}F$ nuclei (a 2-bond coupling) to give another binomial quartet. The quartet with $J = 284$ Hz is assigned to $^{13}C(a)$, and that with $J = 44$ Hz to $^{13}C(b)$. No coupling of $^{13}C(b)$ to the $^{1}H$ nucleus is observed – they are not directly attached.

F—$\overset{F}{\underset{F}{\overset{|}{\underset{|}{C}}}}$$\overset{a}{}$—$\overset{\overset{O}{\parallel}}{C}$$\overset{b}{\underset{O—H}{}}$

**(2.2)**

2.20    Consider structures of $Ph_3P$ and $Ph_2PH$ (**2.3**). In $^{31}P$ NMR spectrum, $Ph_2PH$ exhibits a doublet with a large $J_{PH}$ (directly attached $^{1}H$ and $^{31}P$ nuclei). In the spectrum of $Ph_3P$, a singlet is expected. (This ignores any small couplings to the *ortho*-H atoms of the phenyl rings which might be resolved). It would also be instructive to run the proton decoupled $^{31}P$ NMR spectrum ($^{31}P\{^{1}H\}$ NMR spectrum). This instrumentally removes $^{31}P-^{1}H$ coupling and in the spectrum of $Ph_2PH$, the doublet will collapse to a singlet on going from the $^{31}P$ NMR to $^{31}P\{^{1}H\}$ NMR spectrum.

Trigonal pyramidal

R = H or $C_6H_5$

**(2.3)**

2.21    (a) The binomial decet (10 line pattern) arises from coupling of the $^{31}P$ nucleus to 9 equivalent $^{1}H$ nuclei, each with $I = \frac{1}{2}$; remember there is free rotation about each P–C and C–H single bond. The value of 2.7 Hz corresponds to $J_{PH}$.
(b) In the $^{1}H$ NMR spectrum, one signal is observed (all protons are equivalent) and it is a doublet due to coupling to the $^{31}P$ nucleus. The magnitude of $J_{PH} = 2.7$ Hz *must* be the same as is observed in the $^{31}P$ NMR spectrum.

$H_3C$—$\overset{P^{\cdots\cdots}CH_3}{\underset{CH_3}{}}$

**(2.4)**

2.22    (a) See structure **2.5**. The $^{29}Si$ nucleus couples to 2 directly attached, equivalent $^{1}H$ nuclei, H(*c*), to give a triplet, $J_{SiH} = 194$ Hz.
(b) In $^{1}H$ NMR spectrum, there are 3 proton environments to consider (*a*, *b* and *c* in structure **2.5**), but question only asks about protons H(*c*). Critical point is that although $^{29}Si$ is spin active with $I = \frac{1}{2}$, it is only present in 4.7% abundance, and so 95.3% of protons attached to Si do *not* couple to $^{29}Si$; these protons give a singlet in the $^{1}H$ NMR spectrum. 4.7% of protons H(*c*) *do* couple to $^{29}Si$, and these protons give a doublet, the centre of which coincides with the singlet due to 95.3% of the protons. The spectrum is shown in Figure 2.5, which also shows where the coupling constant (194 Hz) is measured.

**(2.5)**

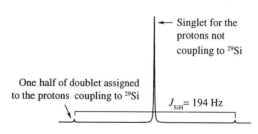

**Figure 2.5** Simulation of part of the $^{1}H$ NMR spectrum of compound **2.5**.

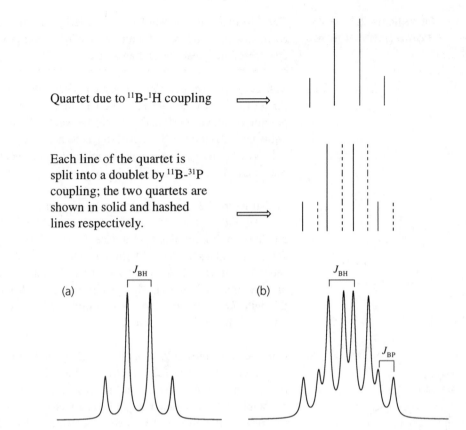

Quartet due to $^{11}$B-$^{1}$H coupling ⟹

Each line of the quartet is split into a doublet by $^{11}$B-$^{31}$P coupling; the two quartets are shown in solid and hashed lines respectively. ⟹

(a)    $J_{BH}$

(b)    $J_{BH}$    $J_{BP}$

**(2.6)**

**(2.7)**

**Figure 2.6**    Simulations of the $^{11}$B NMR spectra of (a) THF·BH$_3$ and (b) PhMe$_2$P·BH$_3$.

2.23    Figure 2.6 shows the $^{11}$B NMR spectra of THF·BH$_3$ and PhMe$_2$P·BH$_3$, and the structures of these adducts are shown in **2.6** and **2.7**.

(a) The $^{11}$B nucleus couples to 3 equivalent $^{1}$H nuclei to give a binomial quartet. This corresponds to the observed spectrum on Figure 2.6a, and a value of $J_{BH}$ can be measured between any pair of adjacent lines, e.g. as shown in the figure.

(b) The $^{11}$B nucleus couples to 3 equivalent $^{1}$H nuclei to give a binomial quartet, *and* couples to the $^{31}$P nucleus to give a doublet. The overall signal is a doublet of quartets, which can be represented in 2 stages as follows. Comparison of the 'stick' spectrum with Figure 2.6b allows you to see where to measure values of $J_{BH}$ and $J_{BP}$. (*Exercise*: The coupling constants can also be measured between other pairs of lines in the spectrum – where are these pairs of lines?)

2.24    (a) SF$_4$

Central atom is S (group 16), so number of valence electrons = 6
Number of S–F single bonds = 4
Number of lone pairs = 1
Total number of electron pairs = 5 = 1 lone and 4 bonding pairs
'Parent' shape = trigonal bipyramidal, see structure **2.8**
*Molecular* shape = disphenoidal

(b) Temperature dependent $^{19}$F NMR spectrum indicates that SF$_4$ is stereochemically non-rigid (fluxional). There is more thermal energy at 298 K than at 175 K.  At

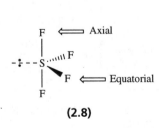

F ⟸ Axial

F ⟸ Equatorial

**(2.8)**

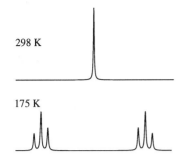

298 K

175 K

**Figure 2.7** Simulations of $^{19}$F NMR spectra of SF$_4$ at 298 K and 175 K.

298 K, a singlet is observed (Figure 2.7, top), therefore all F atoms must be equivalent, and the axial and equatorial F atoms are exchanging positions. (A mechanism is shown in Figure 2.13 in H&S).

At 175 K, two equal intensity triplets are observed (Figure 2.7, bottom). This arises as follows. There are two F environments (axial and equatorial) as shown in diagram **2.8**. Each axial F couples to 2 equatorial F's to give a binomial triplet; each equatorial F couples to 2 axial F's to give a binomial triplet. The two triplets occur at different chemical shifts.

Because the two site exchange involves *equally* populated sites (i.e. 2 F's in each environment), the chemical shift of the singlet at 298 K is the average value of the chemical shifts of the two triplets. The two spectra in Figure 2.7 are aligned to show the relative positions of the signals.

2.25    (a) SiF$_4$

Si (group 14) forms 4 Si–F single bonds; no lone pairs
'Parent' shape = *molecular* shape = tetrahedral; see structure **2.9**

All F atoms equivalent, so a singlet in the $^{19}$F NMR spectrum is consistent with a static molecular structure.

(b) PF$_5$

P (group 15) forms 5 P–F single bonds; no lone pairs
'Parent' shape = *molecular* shape = trigonal bipyramidal; see structure **2.10**

Two F environments (axial and equatorial), so a singlet in the $^{19}$F NMR spectrum is not consistent with a static molecular structure.

(c) SF$_6$

S (group 16) forms 6 S–F single bonds; no lone pairs
'Parent' shape = *molecular* shape = octahedral; see structure **2.11**

All F atoms equivalent, so a singlet in the $^{19}$F NMR spectrum is consistent with a static molecular structure.

(d) SOF$_2$

S (group 16) forms 2 S–F single bonds and 1 S=O double bond; 1 lone pair
'Parent' shape = tetrahedral
*Molecular* shape = trigonal pyramidal; see structure **2.12**

The F atoms are equivalent, so a singlet in the $^{19}$F NMR spectrum is consistent with a static molecular structure.

(e) CF$_4$

C (group 14) forms 4 C–F single bonds; no lone pairs
'Parent' shape = *molecular* shape = tetrahedral; see structure **2.13**

All F atoms equivalent, so a singlet in the $^{19}$F NMR spectrum is consistent with a static molecular structure. Notice that because of the periodic relationship between C and Si, answer (e) is effectively the same as part (a).

**(2.9)**    **(2.10)**

**(2.11)**    **(2.12)**

**(2.13)**

2.26    Berry pseudo-rotation is the dynamic process usually invoked to explain the exchange of axial and equatorial atoms in a trigonal bipyramidal species (or axial and basal atoms in a square-based pyramidal species). Figure 2.8 illustrates the process. *Only a small amount of energy is needed to interconvert trigonal bipyramidal and square-based pyramidal structures.* No bonds are broken and the interconversion involves small changes in the bond angles subtended at the central atom. Sequential bond angle changes in the directions shown in Figure 2.8 result in atoms 4 and 5 moving from equatorial positions to axial sites, and atoms 2 and 3 moving from axial to equatorial positions. Note that the intermediate structure in Figure 2.8 is a square-based pyramid, atom 1 being the axial atom. Continued

**Figure 2.8** A schematic representation of the Berry pseudo-rotational process for a 5-coordinate species, starting on the left-hand side with a trigonal bipyramidal structure.

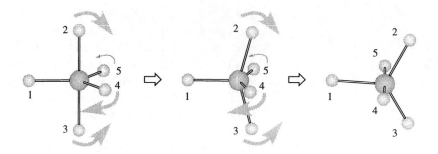

repetition of these exchanges means that every substituent 'visits' all the equatorial and axial sites in the trigonal bipyramidal structure. Five-coordinate species which undergo Berry pseudo-rotation include $PF_5$ and $Fe(CO)_5$.

2.27    'Static solution structure' is somewhat misleading, because low energy dynamic processes, in particular free rotation about single bonds, still occur on the NMR spectroscopic timescale. Take the examples given in the question to illustrate the answer.

$PMe_3$: see structure **2.14** which shows one $CH_3$ group in full. In solution, rotation about the single C–H and P–C bonds occurs, and the P–C bond rotation makes all 9 H atoms equivalent in solution, although they may not necessarily be equivalent in the solid state.

In $OPMe_3$ (**2.15**), rotations about the C–P and C–H bonds again occur. Rotations about P–C bonds mean that the 9 H atoms are equivalent. The discussion above for $PMe_3$ applies to $OPMe_3$.

In $PPh_3$, the 3 phenyl rings adopt a paddle-wheel arrangement (Figure 2.9). Any rotation about the P–C bond is hindered – compared to $PMe_3$, the barrier to rotation is higher.

$SiMe_4$ contains a tetrahedral Si centre. There is free rotation about the Si–C and C–H bonds. In interpreting $^1H$ NMR spectra, we assume that the H atoms are equivalent. This arises because the rotation about the Si–C bond occurs on the NMR spectroscopic timescale.

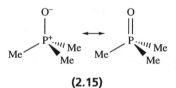

**(2.14)**

**(2.15)**

**Figure 2.9** Two views of space-filling models of $PPh_3$ showing the paddle-wheel arrangement of the Ph groups.

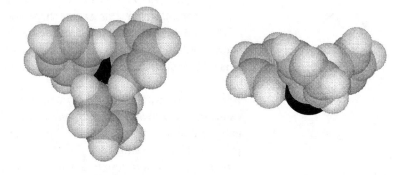

# 3 An introduction to molecular symmetry

3.1

➢

Pauling electronegativity values are listed in Table 1.7 of H&S

(3.1)    (3.2)

(3.3)    (3.4)

S $=$ C $=$ S

(3.5)

(3.6)    (3.7)

Use VSEPR theory to obtain the molecular shapes. The method of working is as in answer 1.29. After you have the shape of the molecule, if necessary, use Pauling electronegativity values to obtain the bond dipole moments and then consider the molecular dipole moment. A detailed method of working was given in answer 1.31.

(a) $BCl_3$
B (group 13), 3 valence electrons; 3 bonding pairs and no lone pairs; trigonal planar molecule (3.1). Each B–Cl bond is polar: $\chi^P(B) = 2.0$, $\chi^P(Cl) = 3.2$, but the molecule is non-polar.
See answer 1.31d for resolution of vectors in the related $BF_3$.

(b) $SO_2$
Bent molecule (3.2); see answer 1.31c for detailed answer.

(c) $PBr_3$
P (group 15), 5 valence electrons; 3 bonding pairs and 1 lone pair; trigonal pyramidal molecule (3.3). From electronegativity values, each P–Br bond is polar: $\chi^P(P) = 2.2$, $\chi^P(Br) = 3.0$. The net size and direction of the molecular dipole moment depends on the extent to which the lone pair offsets (or not) the resultant moment due to the bond dipoles. The result is shown in structure 3.4.

(d) $CS_2$
C (group 16), 4 valence electrons; forms two C=S double bonds; there are no lone pairs on the C atom, so the molecule is linear (3.5). Each bond is ≈ non-polar: $\chi^P(C) = \chi^P(S) = 2.6$. The molecule is non-polar.
See also answer 1.31b for a detailed answer to $CO_2$, which is related to $CS_2$.

(e) $CHF_3$
The molecule is tetrahedral (like $CH_4$ or $CF_4$). Two types of bond must be considered:
C–H: $\chi^P(C) = 2.6$, $\chi^P(H) = 2.2$, so slightly polar bond (see 3.6).
C–F: $\chi^P(C) = 2.6$, $\chi^P(F) = 4.0$, so polar bond (see 3.6).
The resultant molecular dipole moment is as shown in diagram 3.7.

**Questions 3.2-3.13: some general notes**

A *symmetry operation* is an operation performed on an object that leaves it in a configuration indistinguishable from, and superimposable on, the original configuration.

A symmetry operation is carried out with respect to points, lines or planes, the latter being the *symmetry elements*.

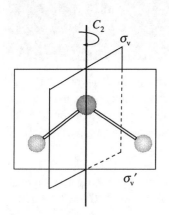

**Figure 3.1** The principal axis of rotation, and the two mirror planes in $H_2O$.

3.2    (a) $E$ is the identity operator. It effectively identifies the molecular configuration. The operator $E$ leaves the molecule unchanged. *All* objects can be operated upon by the identity operator $E$.
(b) A plane of symmetry (mirror plane) is denoted by $\sigma$.
(c) The symmetry operation of rotation about an $n$-fold axis (the symmetry element) is denoted by $C_n$, in which the angle of rotation is $360°/n$, where $n = 1, 2, 3, 4 \ldots$
(d) An $S_n$ axis is an $n$-fold improper rotation axis: rotation through $360°/n$ about the axis is followed by reflection through a plane perpendicular to the same axis.

If the mirror plane lies perpendicular to the principal axis, it is denoted $\sigma_h$.
If the plane contains the principal axis, it is denoted $\sigma_v$.
To see the difference between $\sigma_v$ and $\sigma_v'$, it is best to use an example: $H_2O$ is a simple example. Figure 3.1 shows the principal ($C_2$) axis in a molecule of $H_2O$, and the two mirror planes, *both* of which contain the principal rotation axis. The plane which *bisects* the molecule is labelled $\sigma_v$, and the plane in which the molecule lies is labelled $\sigma_v'$.
A $\sigma_d$ plane contains the principal rotation axis, and also bisects the angle between two adjacent 2-fold axes.

3.3    (a) An 8-vertex star. The symmetry of the star is such that rotation through $(360/8)°$ $= 45°$ gives another star superimposable on the first one; the * is used in the diagram to clarify the rotation that has occurred. This operation is repeated 7 more times to get back to the first orientation. The highest-order axis is an 8-fold axis ($C_8$) running through the centre of the star, perpendicular to the plane of the paper.
(b) An ellipse. The symmetry is such that rotation through $(360/2)° = 180°$ as shown in the diagram gives another ellipse superimposable on the first one. This operation is repeated once more to get back to the first orientation. The highest-order rotation axis of the ellipse is a 2-fold axis ($C_2$). There are 2 other $C_2$ axes: both lie in the plane of the paper, one horizontal, and one vertical with respect to the ellipse drawn.
(c) A pentagon. The symmetry is such that rotation through $(360/5)° = 72°$ gives another pentagon superimposable on the first one. This operation is repeated 4 more times to get back to the first orientation. The principal axis is a 5-fold axis ($C_5$).
(d) The symmetry of this shape is such that rotation through $(360/3)° = 120°$ gives a shape superimposable on the first one. This operation is repeated twice more to get back to the first orientation. The principal axis is a 3-fold axis ($C_3$).

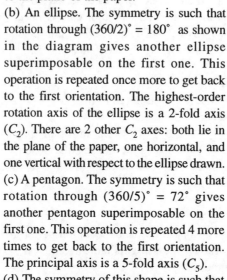

3.4 $SO_2$ is a bent molecule (see answer 1.31c).
It must possess an $E$ operator – all molecules do.
The other symmetry operators are:

A $C_2$ axis:                                    A $\sigma_v$ plane:

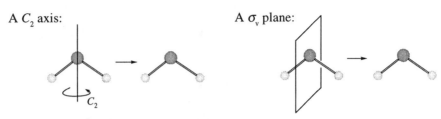

A $\sigma_v'$ plane:                    The symmetry operators are $E$, $C_2$, $\sigma_v$ and $\sigma_v'$.

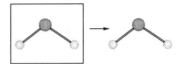

3.5 Figure 1.16 in H&S shows one view of the $H_2O_2$ molecule. Another 2 views are shown here in structures **3.8** and **3.9**; in **3.9**, the molecule is viewed along the O–O bond. The symmetry operator that $H_2O_2$ possesses is a $C_2$ axis running through the midpoint of the O–O bond in the direction shown in the left-hand diagram below. Its operation is shown in the right-hand diagram below – look carefully at the perspective in the diagrams.

**(3.8)**    **(3.9)**

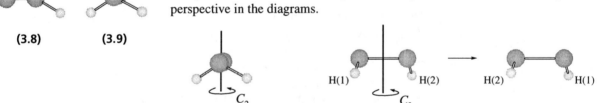

3.6 The diagrams below summarize the answer.

F
|
B
F     F

**(3.10)**

Cl
|
B
F     F

**(3.11)**

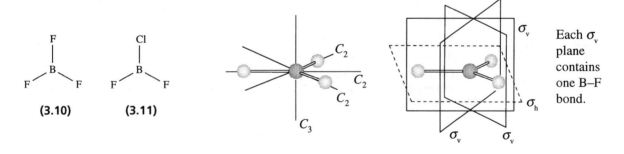

Each $\sigma_v$ plane contains one B–F bond.

3.7 (a) On going from $BF_3$ to $BClF_2$ (**3.10** to **3.11**): $C_3$ axis is lost, two $C_2$ axes are lost, and two $\sigma_v$ planes are lost. (Re-colour the left hand F atom in the diagrams in answer 3.6 making it a Cl, and confirm which of the symmetry operators are no longer valid).

Cl
|
B
F     Br

**(3.12)**

(b) On going from $BClF_2$ to $BBrClF$ (**3.11** to **3.12**): $C_2$ axis is lost, and the $\sigma_v$ plane is lost. (Use the diagrams above again to help you confirm these).
(c) Each molecule has a $\sigma_h$ plane – the plane containing the atoms.

**3.8**    First, draw out the structures of the molecules and ions:

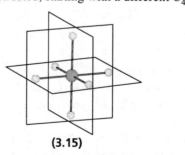

Bond orders in $[SO_4]^{2-}$ and $[NO_3]^-$ are not shown because the bonds are equivalent (delocalized bonding) and these representations emphasize the symmetry.

Note:  (i) that they *all* possess a $C_3$ axis;
         (ii) only the planar molecules possess a $\sigma_h$ plane.

The answers are therefore:

(a) $C_3$ axis but no $\sigma_h$ plane: $NH_3$, $PBr_3$, $[SO_4]^{2-}$.

(b) $C_3$ axis and a $\sigma_h$ plane: $SO_3$, $AlCl_3$, $[NO_3]^-$.

**3.9**    First, draw out the structures of the molecules and ions; the VSEPR model shows why $[ICl_4]^-$ and $XeF_4$ are square planar (2 lone pairs and 4 bonding pairs):

The only two species with a $C_4$ axis and a $\sigma_h$ plane are the square planar ones: $[ICl_4]^-$ and $XeF_4$.

**3.10**    For each, first draw out the structure (using VSEPR to help you) and then work out the number of mirror planes.

(a) $SF_4$: the structure is disphenoidal (**3.13**). $SF_4$ contains 2 mirror planes as shown in **3.14**: one plane contains the S and 2 F(axial) atoms, and the other contains the S and 2 F(equatorial) atoms.

(b) $H_2S$: the bent structure is analogous to that of $H_2O$ (although the bond angle is different) and so the 2 mirror planes are as shown in Figure 3.1.

(c) $SF_6$: octahedral structure (6 bonding pairs, 0 lone pairs). There are 9 mirror planes. The 3 $\sigma_h$ planes are shown in **3.15**, and each contains the S and 4 F atoms. The 6 $\sigma_d$ planes can be considered in 3 sets of 2. One set is shown in **3.16**; each plane contains the S and two opposite F atoms (i.e. it contains a $C_4$ axis), and bisects the other two F–S–F axes. The other 2 sets of $\sigma_d$ planes are similarly constructed, starting with a different $C_4$ axis (i.e. F–S–F axis).

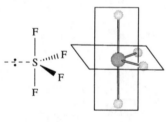

**(3.13)**      **(3.14)**

**(3.15)**          **(3.16)**

(d) $SOF_4$: the structure is disphenoidal (Figure 1.10, p.13). $SOF_4$ contains 2 mirror planes, analogous to those of $SF_4$ (**3.14**).

(e) $SO_2$: this has a bent structure and so has the same symmetry as $H_2S$ in part (b) – 2 mirror planes.

(f) $SO_3$: the structure is trigonal planar (see answer 1.29i, p. 12). There are 4 mirror planes: one $\sigma_h$ and 3 $\sigma_v$ planes. Their positions are analogous to those in $BF_3$, answer 3.6.

---

**Questions 3.11-3.12: some general notes**

*A centre of inversion:*

If reflection of all parts of a molecule through the centre of the molecule produces an indistinguishable configuration, the centre is a centre of inversion (centre of symmetry), designated by the symbol *i*.

---

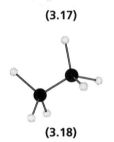

**(3.17)**

**(3.18)**

3.11　(a) Si is in the same group as C, so $Si_2H_6$ is expected to have the same structure as ethane.

(b) A staggered conformer (**3.17**) will be the most favoured in terms of steric energy.

(c) The midpoint of the Si–Si bond is an inversion centre. This point is shown in structure **3.17**. To confirm that this is a centre of inversion, reflect each point of the molecule through centre *i* and show that an identical point is generated. For example, take H atom **a** ... reflect through *i* ... you end up at H atom **a′** which is indistinguishable from atom **a**.

(d) The eclipsed conformer **3.18** is the least favoured in terms of steric energy.

(e) This conformer does not contain a centre of inversion. To confirm this, try taking the midpoint of the Si–Si bond again as in part (c); reflection of all parts of the molecule through this point does *not* lead to an indistinguishable configuration.

3.12　In each part of this answer, if you are unsure of the reason why a molecule does or does not have an inversion centre, take the central atom as a trial inversion centre, and try reflecting parts of the molecule through this point.

(a) $BF_3$: B, group 13; 3 valence electrons; trigonal planar; structure **3.19**. No inversion centre.

(b) $SiF_4$: Si, group 14; 4 valence electrons; tetrahedral; structure **3.20**. No inversion centre.

(c) $XeF_4$: Xe, group 18; 8 valence electrons; square planar (4 bonding and 2 lone pairs); structure **3.21**. The Xe centre corresponds to an inversion centre.

(d) $PF_5$: P, group 15; 5 valence electrons; trigonal bipyramidal; structure **3.22**. No inversion centre.

(e) $[XeF_5]^-$: Xe, group 18; 8 valence electrons; pentagonal planar (5 bonding and 2 lone pairs); structure **3.23**. No inversion centre.

(f) $SF_6$: S, group 16; 6 valence electrons; octahedral; structure **3.24**. The S centre corresponds to an inversion centre.

(g) $C_2F_4$: planar structure **3.25**; this is analogous to ethene. The midpoint of the C=C bond is an inversion centre.

**(3.19)**

**(3.20)**

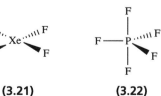

**(3.21)**

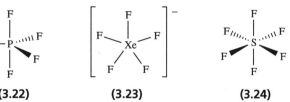

**(3.22)**　　**(3.23)**　　**(3.24)**　　**(3.25)**

(3.26)

(h) The C=C=C unit is linear, but the presence of two adjacent $\pi$-bonds places a restriction on the orientation of the two $CH_2$ groups: they are orthogonal. Therefore, there is no inversion centre.

3.13    A linear molecule possesses an $\infty$-fold axis of rotation. This applies to both symmetrical (e.g. $F_2$, **3.27**) and asymmetrical (e.g. HCN, **3.28**) molecules.

(3.27)                              (3.28)

**Questions 3.14-3.21: some general notes**

Before attempting questions 3.14-3.21, make sure that you have studied worked examples 3.4-3.7 in H&S, in which point group assignments are made with accompanying explanations. When reading through answers 3.14-3.21, make sure that you have Figure 3.9 from H&S available for reference. This gives a flowchart for assigning point groups.

3.14    $NF_3$: trigonal pyramidal structure **3.29**.
To determine the point group, apply the strategy shown in Figure 3.9 in H&S:

(3.29)

| START $\Longrightarrow$ Is the molecule linear? | No |
| Does it have $T_d$, $O_h$ or $I_h$ symmetry? | No |
| Is there a $C_n$ axis? | Yes: $C_3$ axis |
| Are there 3 $C_2$ axes perpendicular to the principal axis? | No |
| Is there a $\sigma_h$ plane? | No |
| Are there $n$ (i.e. 3) $\sigma_v$ planes containing the $C_n$ axis? | Yes $\Longrightarrow$ STOP |

Conclusion: the point group is $C_{3v}$.

3.15    A member of the $D_{\infty h}$ point group must contain a $C_\infty$ axis and, therefore, the species is linear. See answer 3.13.

3.16    $SF_5Cl$: structure **3.30**. This is an example of a molecule that we loosely call 'octahedral' but which actually does not possess octahedral symmetry, i.e. it does not belong to the $O_h$ point group.
To determine the point group, apply the strategy in Figure 3.9 in H&S:

(3.30)

(3.31)

| START $\Longrightarrow$ Is the molecule linear? | No |
| Does it have $T_d$, $O_h$ or $I_h$ symmetry? | No |
| Is there a $C_n$ axis? | Yes: $C_4$ axis (see **3.31**) |
| Are there 4 $C_2$ axes perpendicular to the principal axis? | No |
| Is there a $\sigma_h$ plane? | No |
| Are there $n$ (i.e. 4) $\sigma_v$ planes containing the $C_n$ axis? | Yes $\Longrightarrow$ STOP |

Conclusion: the point group is $C_{4v}$.

3.17    If $BrF_3$ were trigonal planar or trigonal pyramidal, the molecule would have a $C_3$ axis. The fact that it belongs to the $C_{2v}$ point group means that the principal axis is a $C_2$ axis. The other 3-coordinate structure possibility is T-shaped, so the next step in the answer is to work out the symmetry properties of a T-shaped $BrF_3$ molecule. Apart from the $E$ operator, T-shaped $BrF_3$ contains a $C_2$ axis, a $\sigma_v$ plane, and a $\sigma_v'$ plane as shown in Figure 3.2. These are consistent with the $C_{2v}$ point group.

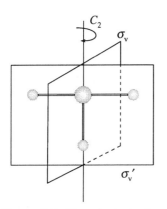

Now see if this agrees with VSEPR theory:
    Central atom is Br
    Br is in group 17; number of valence electrons = 7
    Number of bonding pairs (3 Br–F bonds) = 3
    Number of lone pairs = 2
    Total number of electron pairs = 5
    'Parent' shape = trigonal bipyramidal
    *Molecular* shape = T-shaped,  see structure **3.32**

**(3.32)**

**Figure 3.2** The principal axis of rotation, and the two mirror planes in $BrF_3$.

3.18    The structure of $[XeF_5]^-$ is shown in Figure 3.3.
To determine the point group, apply the strategy shown in Figure 3.9 in H&S:

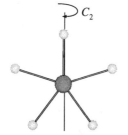

| START $\Longrightarrow$ Is the molecule linear? | No |
|---|---|
| Does it have $T_d$, $O_h$ or $I_h$ symmetry? | No |
| Is there a $C_n$ axis? | Yes: $C_5$ axis (runs perpendicular to plane of paper) |
| Are there 5 $C_2$ axes perpendicular to the principal axis? | Yes (see Figure 3.3) |
| Is there a $\sigma_h$ plane perpendicular to the principal axis? | Yes $\Longrightarrow$ STOP |

Conclusion: the point group is $D_{5h}$.

**Figure 3.3** The pentagonal planar structure of $[XeF_5]^-$. One of the 5 $C_2$ axes is shown; each of the other 4 $C_2$ axes coincides with a different Xe–F bond vector.

3.19    This question is best done by first grouping compounds as follows, taking into account the number of different substituents:
(a) $CCl_4$ and (e) $CF_4$
(b) $CCl_3F$ and (d) $CClF_3$
(c) $CCl_2F_2$
Take (a) and (e) together: each is *regular* tetrahedral and belongs to the $T_d$ point group.
Take (b) and (d) together: molecules **3.33** and **3.34** belong to the same point group, which is assigned as follows:

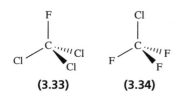

**(3.33)**    **(3.34)**

| START $\Longrightarrow$ Is the molecule linear? | No |
|---|---|
| Does it have $T_d$, $O_h$ or $I_h$ symmetry? | No |
| Is there a $C_n$ axis? | Yes: $C_3$ axis |
| Are there 3 $C_2$ axes perpendicular to the principal axis? | No |
| Is there a $\sigma_h$ plane? | No |
| Are there $n$ (i.e. 3) $\sigma_v$ planes containing the $C_n$ axis? | Yes $\Longrightarrow$ STOP |

Conclusion: the point group is $C_{3v}$.

**(3.35)**

(c) $CCl_2F_2$: see structure **3.35** in which the orientation has been chosen so as to help in finding the symmetry elements. To assign the point group:

| START $\Longrightarrow$ | Is the molecule linear? | No |
|---|---|---|
| | Does it have $T_d$, $O_h$ or $I_h$ symmetry? | No |
| | Is there a $C_n$ axis? | Yes: $C_2$ axis (horizontal through C atom in **3.35**) |
| | Are there 2 $C_2$ axes perpendicular to the principal axis? | No |
| | Is there a $\sigma_h$ plane? | No |
| | Are there $n$ (i.e. 2) $\sigma_v$ planes containing the $C_2$ axis? | Yes $\Longrightarrow$ STOP |

Conclusion: the point group is $C_{2v}$.

3.20    (a) To assign a point group to $SF_4$, first draw the structure (see answer 2.24, p. 22) which is shown in **3.36**.

**(3.36)        (3.37)**

The $\sigma_v$ planes are shown in structure **3.14**, p. 28

| START $\Longrightarrow$ | Is the molecule linear? | No |
|---|---|---|
| | Does it have $T_d$, $O_h$ or $I_h$ symmetry? | No |
| | Is there a $C_n$ axis? | Yes: $C_2$ axis (horizontal through S atom in **3.36**) |
| | Are there 2 $C_2$ axes perpendicular to the principal axis? | No |
| | Is there a $\sigma_h$ plane? | No |
| | Are there $n$ (i.e. 2) $\sigma_v$ planes containing the $C_2$ axis? | Yes $\Longrightarrow$ STOP |

Conclusion: the point group is $C_{2v}$.

(b) The structures of $SOF_4$ and $SF_4$ are related – compare **3.37** with **3.36**. The presence of the O atom does not change the symmetry – the O atom lies on the $C_2$ axis. The point group is still $C_{2v}$.

3.21    The tetrahedron, octahedron and icosahedron are drawn below:

Tetrahedron         Octahedron         Icosahedron

The highest symmetry is possessed by the icosahedron – the $I_h$ point group possesses the highest number of symmetry elements.

---

**Questions 3.22-3.23: some general notes**

For a molecule containing $n$ atoms, the number of degrees of vibrational freedom can be determined by using the equations:

*Linear molecule*:        Number of degrees of vibrational freedom = $3n - 5$
*Non-linear molecule*:        Number of degrees of vibrational freedom = $3n - 6$

For a vibrational mode to be *IR active*, the vibration must give rise to a *change* in the molecular dipole moment.

---

H——C≡N
**(3.38)**

3.22    (a) $SO_2$ is non-linear (see answer 1.31c, p. 13).
Number of degrees of vibrational freedom $= 3n - 6 = (3 \times 3) - 6 = 3$

(b) $SiH_4$ is non-linear (tetrahedral, like $CH_4$).
Number of degrees of vibrational freedom $= 3n - 6 = (3 \times 5) - 6 = 9$

(c) HCN is linear (see **3.38**).
Number of degrees of vibrational freedom $= 3n - 5 = (3 \times 3) - 5 = 4$

(d) $H_2O$ is non-linear.
Number of degrees of vibrational freedom $= 3n - 6 = (3 \times 3) - 6 = 3$

(e) $BF_3$ is non-linear (trigonal planar, see **1.27**, p. 13)
Number of degrees of vibrational freedom $= 3n - 6 = (3 \times 4) - 6 = 6$

3.23    First draw out the structure of the molecule and determine if there is a molecular dipole moment ($\mu$). A vibrational mode is only IR active if it gives rise to a change in molecular dipole moment ($\Delta\mu > 0$). For a non-linear molecule, total number of modes of vibrational freedom $= 3n - 6$.
(a) $H_2O$ is polar (**3.39**). Now consider the 3 modes of vibrational freedom:

      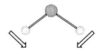

| Symmetric stretch: | Asymmetric stretch: | Deformation: |
|---|---|---|
| $\Delta\mu > 0$ | $\Delta\mu > 0$ | $\Delta\mu > 0$ |
| IR active | IR active | IR active |

(b) $SiF_4$ is tetrahedral and non-polar (like $CH_4$); $T_d$ symmetry. Vibrational modes are shown in Figure 3.13 in H&S; 4 modes of vibrational freedom (2 degenerate pairs) are IR active and give rise to 2 bands in the IR spectrum.
(c) $PCl_3$ has a trigonal pyramidal structure and is polar (**3.40** shows two views of $PCl_3$). The vibrational modes are:

Symmetric stretch: $\Delta\mu > 0$                IR active

Symmetric deformation: $\Delta\mu > 0$                IR active

Doubly degenerate (asymmetric) stretch: $\Delta\mu > 0$                IR active
Doubly degenerate deformation: $\Delta\mu > 0$                IR active
∴ There are 6 modes of vibrational freedom giving rise to 4 bands in the IR spectrum.
(d) $AlCl_3$ has a trigonal planar structure and is non-polar (**3.41** shows two views of $AlCl_3$). The modes of vibrational freedom can be represented in a simialr way to those for $PCl_3$, *but* the molecular framework of $AlCl_3$ is planar. As a consequence of this, the symmetric stretch in $AlCl_3$ is IR inactive because $\Delta\mu = 0$. The IR active

**(3.39)**

**(3.40)**

**(3.41)**

modes for $AlCl_3$ are the symmetric deformation, doubly degenerate stretch, and doubly degenerate deformation – 5 modes of vibrational freedom giving rise to 3 bands in the IR spectrum.

(d) $CS_2$ is linear and non-polar. Now consider the vibrational modes:

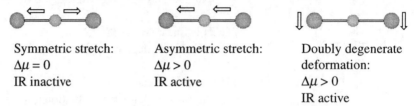

Symmetric stretch:
$\Delta\mu = 0$
IR inactive

Asymmetric stretch:
$\Delta\mu > 0$
IR active

Doubly degenerate deformation:
$\Delta\mu > 0$
IR active

The deformation is a bending mode and is doubly degenerate (see Figure 3.10 in H&S for more detail). Two fundamental bands are therefore seen in the IR spectrum due to the asymmetric stretch and the deformation.

(e) HCN is linear and polar. Now consider the vibrational modes: like $CS_2$, there are 3 vibrational modes, but an important point about a linear XYZ molecule in which X and Z are of significantly different mass is that the modes can essentially be assigned to an X–Y stretch, a Y–Z stretch, and an XYZ deformation (doubly degenerate). This is because each of the symmetric and asymmetric stretches is dominated by the stretch of one or other of the bonds. Thus, in HCN the vibrational modes are:

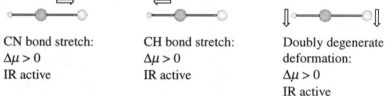

CN bond stretch:
$\Delta\mu > 0$
IR active

CH bond stretch:
$\Delta\mu > 0$
IR active

Doubly degenerate deformation:
$\Delta\mu > 0$
IR active

Three fundamental absorptions are seen in the IR spectrum.

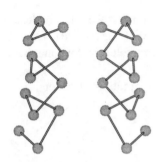

**Figure 3.4** Two helical $S_\infty$ chains which are enantiomers because one has a left-handed twist and the other, a right-handed twist.

3.24    For (a) and (b), read the first part of Section 3.8 in H&S.
(c) A helical chain is one which has a sense of handedness: a right-hand or left-handed helix. The handedness renders the species chiral. Figure 3.4 shows parts of two *catena*-sulfur ($S_\infty$) chains; the caption explains how the chirality arises.

3.25    To answer this question, choose suitable compounds on which to test the operations.
(a) Choose a molecule with a centre of inversion ($i$), e.g. *trans*-$N_2F_2$. In **3.42**, inversion through point $i$ takes one F atom to the position of the other F atom. An $S_2$ improper rotation is a $C_2$ rotation (i.e. through 180°) followed by a reflection through a plane perpendicular to this axis. Comparing the diagrams below with **3.42** shows that inversion is equivalent to an $S_2$ improper rotation.

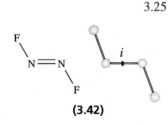

(3.42)

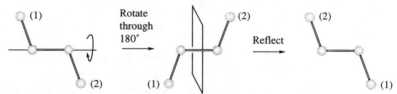

(b) An $S_1$ improper rotation is a $C_1$ rotation (i.e. through 360°) followed by a reflection through a plane perpendicular to this axis. But a $C_1$ rotation just brings the molecule back to where is started. Hence, only the reflection is significant, and an $S_1$ improper rotation is equivalent to a reflection.

# 4 Bonding in polyatomic molecules

4.1 (a), (b) Hybrid orbitals are generated by mixing atomic orbitals that are close in energy. A set of spatially-directed orbitals, for application within VB theory, is derived. The character of a hybrid orbital depends on the atomic orbitals involved and their percentage contributions, e.g. an $sp^2$ hybrid comprises $^1/_3 s$ and $^2/_3 p$ orbital character. Each hybrid orbital points along an internuclear vector or towards a lone pair within a molecule, e.g. in $CH_4$, $sp^3$ hybridization is used to obtain 4 equivalent hybrid orbitals, each pointing along a different C–H internuclear vector. Hybridization provides a convenient way to develop a bonding picture using localized $\sigma$-bonds; unused orbitals such as $p$ atomic orbitals can be used to form $\pi$-bonds, e.g. in $CO_2$ (see answer 4.9).

(c) Equations 4.1 and 4.2 in H&S are:

➢
**Normalization:**
see Box 1.3 and discussion with equation 1.24 in H&S

$$\psi_{sp \text{ hybrid}} = \frac{1}{\sqrt{2}}\left(\psi_{2s} + \psi_{2p_x}\right)$$

$$\psi_{sp \text{ hybrid}} = \frac{1}{\sqrt{2}}\left(\psi_{2s} - \psi_{2p_x}\right)$$

The equations refer to two $sp$ hybrid orbitals. For normalized wavefunctions, the sum of the squares of the normalization factors must equal unity. Check this is true. For the first wavefunction:

$$\psi_{sp \text{ hybrid}} = \frac{1}{\sqrt{2}}\psi_{2s} + \frac{1}{\sqrt{2}}\psi_{2p_x}$$

$$\left(\frac{1}{\sqrt{2}}\right)^2 + \left(\frac{1}{\sqrt{2}}\right)^2 = \frac{1}{2} + \frac{1}{2} = 1 \qquad \therefore \text{ Normalized.}$$

Similarly for the second wavefunction.

---

**Questions 4.2-4.4: some general notes**

In hybridization schemes:
• mixing orbitals of the same phase corresponds to constructive interference of waves;
• mixing orbitals of opposite phases corresponds to destructive interference of waves;
• changing the sign of a wavefunction changes its phase.

---

4.2 (a) The information in Figure 4.4 in H&S only allows a qualitative answer.
It is easily seen that the first (additive) combination yields the hybrid orbital shown:

For the second hybrid, take the combination of atomic orbitals in 2 stages. Step 1:

➢
Watch signs!
Think about phases!

Now add in the $2p_y$ character:

For the last hybrid, again take the orbital combinations in 2 stages. Step 1:

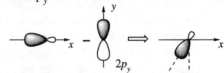

$$2s \qquad 2p_x$$

**Watch signs!**
**Think about phases!**

Now add in the $2p_y$ character:

(b) The method is as in answer 4.1c. Take each wavefunction in turn:

$$\psi_{sp^2 \text{ hybrid}} = \frac{1}{\sqrt{3}} \psi_{2s} + \sqrt{\frac{2}{3}} \psi_{2p_x}$$

Sum of the squares of the normalization factors is: $\frac{1}{3} + \frac{2}{3} = 1 \qquad \therefore$ Normalized.

$$\psi_{sp^2 \text{ hybrid}} = \frac{1}{\sqrt{3}} \psi_{2s} - \frac{1}{\sqrt{6}} \psi_{2p_x} + \frac{1}{\sqrt{2}} \psi_{2p_y}$$

Sum of the squares of the normalization factors is: $\frac{1}{3} + \frac{1}{6} + \frac{1}{2} = 1 \quad \therefore$ Normalized.

$$\psi_{sp^2 \text{ hybrid}} = \frac{1}{\sqrt{3}} \psi_{2s} - \frac{1}{\sqrt{6}} \psi_{2p_x} - \frac{1}{\sqrt{2}} \psi_{2p_y}$$

Sum of the squares of the normalization factors is: $\frac{1}{3} + \frac{1}{6} + \frac{1}{2} = 1 \quad \therefore$ Normalized.

4.3    The method of working is as answer 4.2a, but now you must work in 3-dimensions – use Figure 4.1 to help you. Equation 4.6 in H&S is:

$$\psi_{sp^3 \text{ hybrid}} = \frac{1}{2} \left( \psi_{2s} + \psi_{2p_x} + \psi_{2p_y} + \psi_{2p_z} \right)$$

Take the combinations below in a stepwise manner as in answer 4.2a. Because of the vector properties of the $2p$ orbitals, as each $2p$ contribution is added in, the resultant hybrid orbital changes direction:

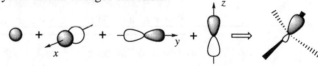

Equation 4.7 in H&S is:

$$\psi_{sp^3 \text{ hybrid}} = \frac{1}{2} \left( \psi_{2s} + \psi_{2p_x} - \psi_{2p_y} - \psi_{2p_z} \right)$$

Again, consider the contributions from the atomic orbitals in a stepwise manner and *watch the signs*:

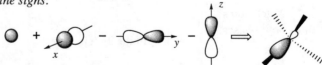

Equations 4.8 and 4.9 in H&S can be correlated to the lower two diagrams in Figure 4.6a in H&S in a similar way to the worked answers above.

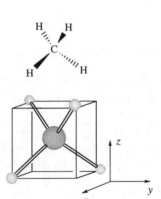

**Figure 4.1** The relationship between a cube and a tetrahedron. The edges of the cube coincide with a set of Cartesian axes.

4.4    (a) Take the shaded lobes of the $p_x$ and $d_{x^2-y^2}$ orbital to point along the $+x$ axis, and the shaded lobe of the $p_y$ orbital to point along the $+y$ axis. In the $xy$ plane, the orbital combinations to give 4 $sp^2d$ hybrid orbitals are:

Watch signs!
Think about phases!

(b) Available for hybridization are one $s$, two $p$, and one $d$ orbital. Each hybrid orbital must contain the same amount of $s$ character; since there are 4 hybrid orbitals, each contains 25% $s$ character. Each hybrid orbital must contain the same amount of $p$ character, i.e. 50% $p$ character. Each hybrid contains 25% $d$ character.

4.5    (a) $SiF_4$: Si, group 14, 4 valence electrons, see structure **4.1**. Molecular structure is tetrahedral; 4 substituents and no lone pairs, therefore $sp^3$ hybridization.

Details of how to apply VSEPR theory: see answer 1.29, p. 11

(b) $[ICl_4]^-$: I, group 17, 7 valence electrons, see structure **4.2**. Molecular structure is square planar; 4 substituents and 2 lone pairs. All bonding and lone pairs must be accommodated in the hybridization scheme, so $sp^3d^2$ is appropriate.

(c) $NF_3$: N, group 15, 5 valence electrons, see structure **4.3**. Molecular structure is trigonal pyramidal; 3 substituents and 1 lone pair, therefore $sp^3$ hybridization.

(d) $F_2O$: O, group 16, 6 valence electrons, see structure **4.4**. Molecular structure is bent; 2 substituents and 2 lone pairs, therefore $sp^3$ hybridization.

(e) $SbF_5$: Sb, group 15, 5 valence electrons, see structure **4.5**. Molecular structure is trigonal bipyramidal; 5 substituents and no lone pairs, hence $sp^3d$ hybridization.

(f) $[AsF_6]^-$: As, group 15, 5 valence electrons, see structure **4.6**. Molecular structure is octahedral; 6 substituents and no lone pairs, therefore $sp^3d^2$ hybridization.

(g) $ClF_3$: Cl, group 17, 7 valence electrons, see structure **4.7**. Molecular structure is T-shaped; 3 substituents and 2 lone pairs, therefore $sp^3d$ hybridization.

(h) $BF_3$: B, group 13, 3 valence electrons, see structure **4.8**. Molecular structure is trigonal planar; 3 substituents and no lone pairs, therefore $sp^2$ hybridization.

(4.1)

(4.2)

(4.3)    (4.4)    (4.5)    (4.6)    (4.7)    (4.8)

4.6    (a) See **1.29** and **1.31** (p. 14) for structures of *cis-* and *trans*-$N_2F_2$. Both contain N in a non-linear environment with 1 lone pair, so $sp^2$ hybridization is appropriate. (b) Figure 1.16 in H&S shows that each O atom in $H_2O_2$ is in a non-linear environment with 2 lone pairs. Hence, $sp^3$ hybridization is appropriate.

4.7    (a) Structure **4.9**; for explanation, see answer 3.20, p. 32.
(b) For the $\sigma$-bonding framework, $sp^3d$ hybridization is appropriate (trigonal bipyramidal). The $\pi$-contribution to the S=O double bond can be described in terms of the overlap of a $2p$ orbital on the O atom and an appropriate $3d$ orbital on S (diagram **4.10**).

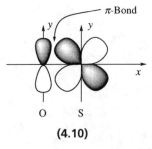

**(4.9)**

**(4.10)**

*Self-study exercise*:
In Section 4.7 in H&S, we discuss the bonding in $SF_6$ and the issue of whether or not *d*-orbitals play a role in bonding. This topic arises several times in Chapters 13, 14 and 15 when we describe the chemistry of compounds containing group 14, 15 and 16 elements. How does the valence bond treatment of $SOF_4$ in part (b) above, and also the VB treatment of molecules such as $PF_5$, $SF_6$ and $POCl_3$, fit in with arguments concerning the role of *d*-orbitals in the bonding in compounds containing S and P?

4.8    (a) Trigonal planar (isoelectronic with $[NO_3]^-$, see worked example 3.2 in H&S).
(b) The resonance structures which contribute the most are:

(c) The bonding description is like that for $[NO_3]^-$ in worked example 3.2 in H&S and can be summarized as follows:

> Check that the number of electrons used in the scheme = number of valence electrons available (24)

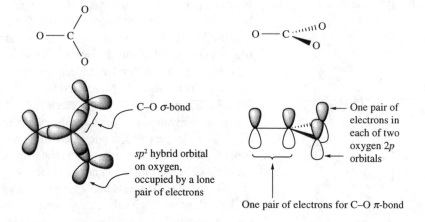

This scheme corresponds to one resonance structure (1 C=O double and 2 C–O single bonds). Each resonance structure may be similarly described.

4.9    (a) $CO_2$ is linear (see answer 1.31b, p.13).
(b) For a linear triatomic molecule, the central atom can be considered to be *sp* hybridized.

(c) An appropriate bonding scheme is summarized below:

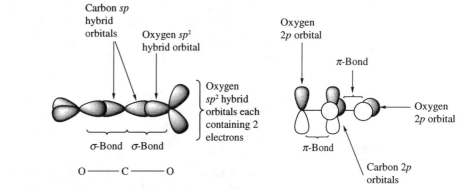

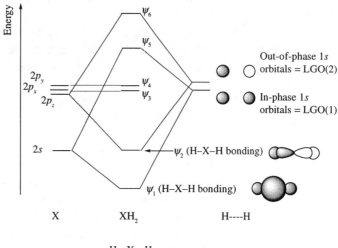

(4.11)

(d) The scheme shows the formation of two C=O double bonds.

(e) Lewis structures are shown in **4.11**. These are consistent with the bonding schemes developed using hybridized atomic orbitals in the VB model.

4.10    Read through the subsections *'Molecular orbital diagrams: moving from a diatomic to polyatomic species'* and *'Ligand group orbitals: MO approach to the bonding in linear XH$_2$'* in order to answer this question. Your answer should be constructed from these subsections and include a simple example such as an MO diagram for linear XH$_2$.

4.11    In VB theory, localized bonds arise because a wavefunction is set up to describe each X–H interaction. Using a hybrid orbital approach for linear XH$_2$, the X atom can be considered to be *sp* hybridized and each X–H interaction described as in Figure 4.2. In MO theory, molecular orbitals are constructed using contributions from atomic orbitals of all the atoms (where this is allowed by symmetry). For linear XH$_2$, the interactions between the atomic orbitals of X and the ligand group orbitals (LGOs) of the H----H fragment are considered (Figure 4.3). Each of the bonding MOs possesses bonding character spread out (delocalized) over all 3 atoms. Assuming X has at least 2 valence electrons (e.g. Be), then the bonding MOs are filled. MOs $\psi_3$ and $\psi_4$ are non-bonding, and $\psi_5$ and $\psi_6$ are antibonding.

**Figure 4.2** In linear XH$_2$, each X–H bond can be described by the overlap of an *sp* hybrid orbital and H 1*s* orbital. Each bond comprises a localized 2c-2e interaction.

**Figure 4.3** An MO diagram for the formation of linear XH$_2$; the atoms are defined to lie along the *z* axis.

**(4.12)**

4.12  (a) First draw the $H_2O$ molecular framework with respect to the axis set specified in the question: see **4.12**. The left-hand column in Table 4.2 in H&S lists the atomic or ligand group orbitals, and reading down each of the next 6 columns gives the % composition of each MO along with the eigenvector (i.e. sign of each contributing wavefunction). MOs can be constructed as follows, with the relative sizes of the lobes reflecting the % contributions:

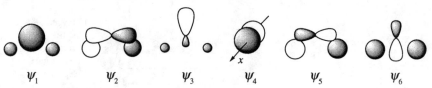

(b) The number of valence electrons available = 6 (from O) + 2 (from 2H) = 8. Hence MOs $\psi_1$, $\psi_2$, $\psi_3$ and $\psi_4$ are occupied; $\psi_1$ and $\psi_2$ provide the O–H bonding character, and $\psi_3$ and $\psi_4$ correspond to the lone pairs.

4.13  (a) The $BH_3$ molecule is defined as lying in the $xy$ plane. The H atoms lie in the nodal plane of the B $2p_z$ orbital; there is no net overlap between the H $1s$ and B $2p_z$ orbitals, so the B $2p_z$ orbital becomes a non-bonding MO in $BH_3$.
(b) Schematic representations are:

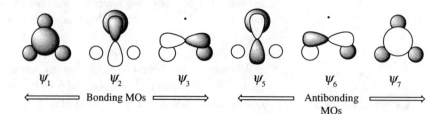

$\Longleftarrow$ Bonding MOs $\Longrightarrow$ $\qquad$ $\Longleftarrow$ Antibonding MOs $\Longrightarrow$

4.14  (a) Schematic representations are:

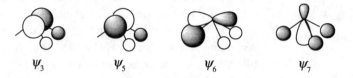

(b) A small degree of N–H bonding character.

4.15  $[NH_4]^+$ is isoelectronic with $CH_4$; the description of the bonding in $[NH_4]^+$ is essentially the same as that for $CH_4$. Refer to the discussion in Section 4.5 in H&S and to Figures 4.17 and 4.18 in H&S. In each, replace C by $N^+$ (isoelectronic species).

4.16  The Lewis structures give 2c-2e localized I–I bonds in each case. This does not explain the variation in bond lengths.

(b) The bonding in $I_2$ can be described as in $F_2$ (Figure 1.21 in H&S, replacing $2s$ and $2p$ orbitals by $5s$ and $5p$). This gives a bond order of 1. For bent $[I_3]^+$, an MO diagram can be constructed using that for $H_2O$ (Figure 4.14 in H&S) as a basis, because $I^+$ is isoelectronic (in terms of valence electrons) with O. Each I–I bond order is therefore 1. For linear $[I_3]^-$, an MO diagram can be constructed using that for $XeF_2$ (Figure 4.22 in H&S) as a basis, because $I^-$ is isoelectronic with Xe. This leads to one occupied bonding MO which is delocalized over all three atoms, and therefore an I–I bond order of $^1/_2$. The conclusion from MO theory is:

I–I bond orders:    $I_2 \approx [I_3]^+ > [I_3]^-$

Expected trend in I–I bond lengths:    $I_2 \approx [I_3]^+ < [I_3]^-$

This agrees with the experimental values of 267 pm in $I_2$, 268 pm in $[I_3]^+$, and 290 pm in $[I_3]^-$.

4.17    (a) Xe: group 18, 8 valence electrons; F: group 17, 7 valence electrons. Total = 22.
(b) First, read the appropriate text in H&S to which the question refers: '$XeF_2$' in Section 4.7. In constructing Figure 4.22 in H&S, all $\pi$-type interactions were ignored. If these are added in, a more representative (but still qualitative) MO diagram can be constructed. This is shown in Figure 4.4 below. The ordering of the MOs is approximate.
(c)-(d) There are 22 valence electrons and these occupy all but the highest lying $\sigma^*$ MO in Figure 4.4. Therefore, there is no net $\pi$-bonding interaction, and the same conclusion about the bonding in $XeF_2$ is reached using either the simplified Figure 4.22 in H&S, or the more complete Figure 4.4 shown here.

4.18    Have available for reference Figures 4.19, 4.20 and 4.23 in H&S.
$BF_3$ and $[NO_3]^-$ are isoelectronic, and both are trigonal planar ($D_{3h}$). Both have available the same basis sets of atomic orbitals ($2s$ and $2p$ on B, F, N and O). Bonding descriptions should therefore be analogous and your description of the bonding in $[NO_3]^-$ can follow that of $BF_3$. However, the energies of the atomic orbitals and of the MOs will not be the same: effective nuclear charges are different. MO diagrams need modifying to take this into account. The extent to which orbital mixing occurs is not accounted for in a simple treatment.

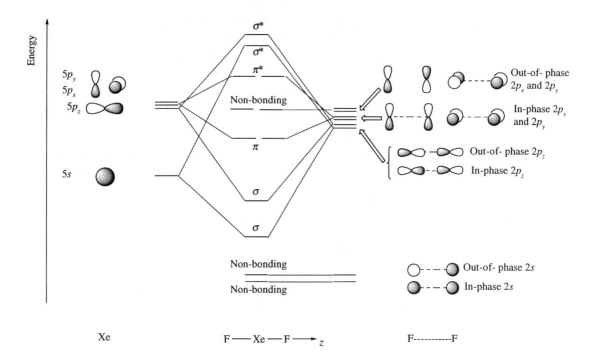

**Figure 4.4** An approximate, qualitative MO diagram for the formation of $XeF_2$; this includes $\pi$-type interactions which are omitted from the simplified Figure 4.22 in H&S. The $XeF_2$ molecule is defined to lie on the $z$-axis.

4.19    (a) Structure of $BH_3$, see **4.13**. To determine the point group, use the scheme in Figure 3.9 in H&S:

**(4.13)**

| | |
|---|---|
| START $\Longrightarrow$ Is the molecule linear? | No |
| Does it have $T_d$, $O_h$ or $I_h$ symmetry? | No |
| Is there a $C_n$ axis? | Yes: $C_3$ axis (runs perpendicular to plane of paper) |
| Are there 3 $C_2$ axes perpendicular to the principal axis? | Yes (see diagrams on p. 27) |
| Is there a $\sigma_h$ plane perpendicular to the principal axis? | Yes $\Longrightarrow$ STOP |

Conclusion: the point group is $D_{3h}$.

(b) Working out the point group is the same as for $NF_3$ – see answer 3.14 (p. 30).

(c) Structure of $B_2H_6$, see **4.14**.

| | |
|---|---|
| START $\Longrightarrow$ Is the molecule linear? | No |
| Does it have $T_d$, $O_h$ or $I_h$ symmetry? | No |
| Is there a $C_n$ axis? | Yes: $C_2$ axis (one runs horizontally through B atoms) |
| Are there 2 $C_2$ axes perpendicular to the principal axis? | Yes (perpendicular to plane of paper, vertically through bridging H) |
| Is there a $\sigma_h$ plane perpendicular to the principal axis? | Yes $\Longrightarrow$ STOP |

**(4.14)**

**(4.15)    (4.16)**

Conclusion: the point group is $D_{2h}$.

4.20    Refer to Figure 4.28 in H&S.
(a) The 2 bonding MOs are represented in **4.15** and **4.16**. In addition to B–H–B bonding character, **4.16** possesses B–H terminal and B–B bonding character.
(b) No; difficult in this qualitative picture to assess extent of B–B bonding character in **4.16**.

4.21    The structure of $[B_2H_7]^-$ is shown in **4.17**.
(a) 2B = 6 valence electrons; 7H = 7 electrons; negative charge. Total = 14 electrons.
(b) Each B has left one $sp^3$ hybrid orbital.
(c) See MO diagram below; $H^-$ provides 2 electrons, but all electrons in each $BH_3$ unit are used for terminal B–H bonding (MOs for this are not shown). The B–H–B bridge is therefore a 3c-2e interaction – the $\sigma$-MO is delocalized over three atoms.

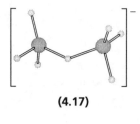

**(4.17)**

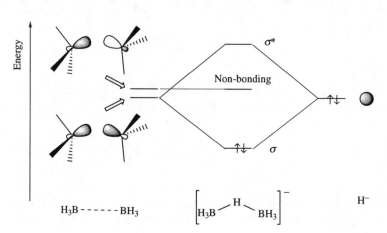

# 5  Structures and energetics of metallic and ionic solids

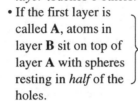
(a)

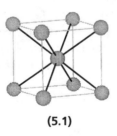
(b)

**Figure 5.1** Close-packing of spheres: (a) cubic close-packed (ccp), and (b) hexagonal close-packed (hcp) arrangements.

✓ 5.1  Your answer to this question should include the following points, and give diagrams similar to those shown here.

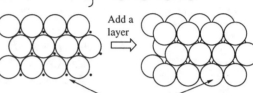

- Both cubic and hexagonal close-packed lattices are made up of stacked layers of close-packed spheres; each sphere *in one layer* touches 6 others.
- If the first layer is called **A**, atoms in layer **B** sit on top of layer **A** with spheres resting in *half* of the holes.

Add a layer

Hole between 3 spheres in layer **A** above which an atom in layer **B** sits.

- Atoms in the third layer may sit above atoms in layer **A** (this gives an **ABAB**... arrangement = hexagonal close-packing), or may sit above the unoccupied holes in layer **A** to give a new layer **C** (this gives an **ABCABC**... arrangement = cubic close-packing = face-centred cubic structure).
- In both ccp and hcp arrays, the spheres occupy 74% of the available space.
- Coordination number of each atom in a ccp or hcp array is 12.
- Close-packed structures contain octahedral and tetrahedral interstitial holes; there are twice as many tetrahedral as octahedral holes in a close-packed array; octahedral holes are larger than the tetrahedral sites.
- A unit cell is the smallest repeating unit in the 3D-array; Figure 5.1 shows unit cells of ccp and hcp arrangements.

✓ 5.2  (a), (b) and (d) are all close-packed arrays with 12-coordinate atoms; recall that face-centred cubic is an alternative name for cubic close-packing. See Figure 5.1.
(c) 8-Coordination in bcc; central atom in unit cell shows this (diagram **5.1**).
(e) Simple cubic, 6-coordinate atoms; consider next unit cells as in diagram **5.2**.

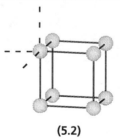

**(5.1)**                **(5.2)**

5.3  (a) The $\alpha \rightarrow \beta$ phase transition is caused by a temperature increase and would be expected to give rise to a *less dense* phase. Since bcc is less dense than ccp, suggest that $\alpha$-Li is ccp and $\beta$-Li is bcc. The change in structure illustrates *polymorphism*.
(b) $\beta$-Sn is the thermodynamically stable polymorph but on lowering the temperature to 286 K, a slow transition to $\alpha$-Sn occurs; density of Sn *decreases* on going from higher ($\beta$) to lower ($\alpha$) temperature form, causing the buttons to crumble.

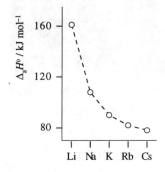

**Figure 5.2** Trend in values of $\Delta_a H^\circ$ for the group 1 metals.

**Figure 5.3** Trend in values of $\Delta_a H^\circ$ across the period from Cs to Bi. The lanthanoids (see Chapter 24) between La and Hf are omitted.

5.4    (a) Standard enthalpy of atomization ($\Delta_a H^\circ$) is defined per mole of cobalt atoms:

$$Co(s) \rightarrow Co(g) \qquad or \qquad {}^1/_n Co_n(s) \rightarrow nCo(g)$$

(b) Trend is shown in Figure 5.2 (values from Table 5.2 in H&S). All the metals crystallize with a bcc structure (consistency in structure type crucial for a meaningful comparison), so the decrease in $\Delta_a H^\circ$ indicates that the M–M bonding becomes weaker down the group. Atom size increases down group, valence atomic orbitals become more diffuse (principal quantum number goes from 2 for Li to 6 for Cs), and orbital overlap becomes less efficient.

(c) Trend is shown in Figure 5.3. First note that the structure types are *not* constant across the period. In the *s*-block, Cs and Ba are both bcc, so can be compared directly; addition of an extra valence electron into the 6*s* orbital in going from Cs to Ba leads to stronger M–M bonding and an increase in $\Delta_a H^\circ$. Going from La to Hg across the *d*-block, all metals have close-packed lattices (hcp or ccp) except for Ta and W (bcc) and Hg (unique cubic structure, but a liquid at 298 K, see Section 5.3 in H&S); there is a rough correlation between the number of unpaired electrons and the value of $\Delta_a H^\circ$, with the maximum values being for middle *d*-block metals.

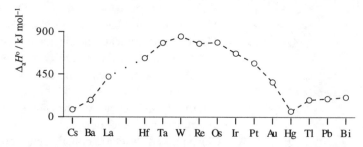

5.5    Ti metal adopts an hcp lattice, and forms a *solid solution* of stoichiometry $TiN_{0.2}$. A solid solution can be of (i) the substitutional type in which sites in the host lattice are occupied by solute atoms, or (ii) the interstitial type in which solute atoms enter the interstitial holes in the host lattice. Except when very small amounts of solute are present, the host lattice type usually changes in going from the pure host (pure metal) to the solid solution. In $TiN_{0.2}$, it is likely that the N atoms occupy octahedral (rather than the smaller tetrahedral) holes in the Ti hcp lattice.

5.6    Answer should be based on Section 5.8 in H&S, subsection '*Band theory of metals and insulators*', and discuss the following points:
- application of molecular orbital theory;
- for *n* metal (M) atoms, overlap of atomic orbitals gives rise to *bands* composed of MOs which are extremely close in energy;
- population of bands;
- a partially occupied band (or overlap of a filled and vacant band) is characteristic of metal;
- delocalized and non-directional bonding;
- band gap;
- electrical conductivity is a property of partially occupied bands.

5.7    (a) See Figure 5.4. Each C atom is tetrahedral and bonds are localized 2c-2e interactions, using all 4 of the valence electrons (C, group 14, $[He]2s^2 2p^2$). Therefore, diamond has a covalently bonded, extended lattice.

**Figure 5.4** Part of the 3D-lattice of diamond and silicon.

(b) The bonding in Si (Figure 5.4) can be described in a similar manner to diamond; Si is also in group 14. However, unlike diamond which is an insulator, Si is an intrinsic semiconductor – band theory can be used to rationalize this difference in properties. In both bulk diamond and Si, the use of all the valence orbitals and electrons leads to the formation of a fully occupied band and a vacant band, between which is a band gap of 520 kJ mol⁻¹ for diamond and 106 kJ mol⁻¹ for Si. In Si, the band gap is small enough to allow the upper band to be *thermally populated* as the temperature increases; this generates a conduction band (partially occupied band) with electrons as charge carriers. Positive holes left in the originally fully occupied band also act as charge carriers.

**5.8**

> More information at the beginning of Section 5.8 in H&S

(a) Electrical resistivity ($\rho$) of a substance measures its resistance to an electrical current; for a wire of uniform cross section ($a$ m²), units of resistivity are $\Omega$ m. Resistance ($R$) is related to $\rho$ by the equation:

$$R = \frac{\rho l}{a}$$

where $l$ = length of wire in m

Electrical conductivity is the inverse of resistance.

(b) Trend in resistivities corresponds to decrease in the band gap (see answer 5.7); diamond is an insulator, Si and Ge are intrinsic semiconductors, and Sn is metallic.

(c) For a pure metal, $\rho$ increases with temperature; for a semiconductor, it decreases as temperature increases.

**5.9**

Intrinsic semiconductor (e.g. Si, Ge): charge carriers generated by thermal population of vacant band which becomes the conduction band (see answer 5.7b). Extrinsic semiconductor, illustrated by Ga-doped or As-doped Si. Doping Si (group 14) with Ga (group 13) introduces an unoccupied band (*acceptor band*) into the band structure producing a p-type semiconductor. Acceptor band lies just above the fully occupied band and band gap is small enough to allow its ready thermal population. Charge carriers arising are (i) electrons in acceptor band and (ii) positive holes left behind in lower band. Doping Si with As (group 15) introduces a new filled band (*donor band*) just under the conduction band and produces an n-type semiconductor. Small band gap allows easy thermal population of conduction band.

**5.10**

(Me₃Si)₂HC          CH(SiMe₃)₂
       \\        /
        Al — Al
       /        \\
(Me₃Si)₂HC          CH(SiMe₃)₂

**(5.3)**

(a) Metallic radius ($r_{metal}$) defined as half Al–Al internuclear separation in the close-packed (bulk) metal. Covalent radius ($r_{cov}$) defined as half Al–Al internuclear separation in an Al–Al single bond in a covalent compound. Few suitable compounds are available: in **5.3**, the Al–Al bond length is 266 pm. In an ionic lattice, the internuclear separation of adjacent, oppositely charged ions is the sum of the radii of cation and anion. Partitioning the internuclear distance into components of $r_{cation}$ and $r_{anion}$ is not straightforward. To find $r_{ion}$ for Al³⁺ is even more difficult since a suitable 'ionic' lattice is not available. One possible candidate for an essentially 'ionic' 6-coordinate Al centre could be crystalline AlF₃, but this probably has significant covalent character.

(b) Trend is: $r_{metal} > r_{cov} > r_{ion}$. Expect $r_{ion}$ to be the smallest – 3 electrons removed from atom results in increased effective nuclear charge. In the Al–Al single bond, pair of electrons are shared between atoms, so effective radius of each atom in the compound is reduced compared to atom in the bulk metal.

**Figure 5.5** Unit cells of (a) NaCl (rock salt), (b) CsCl, and (c) TiO$_2$ (rutile).

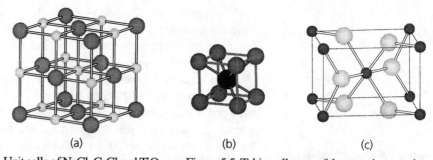

(a)                    (b)                    (c)

✓ 5.11

Unit cells of NaCl, CsCl and TiO$_2$: see Figure 5.5. Taking all parts of the question together: The coordination number of an ion in a lattice is its number of nearest neighbours of opposite charge. In some cases, you have to extend the lattice from the diagram given (the unit cell) to work out the coordination number (see below). A unit cell is the smallest repeating unit in a 3D-lattice; in any ionic compound with 1:1 stoichiometry (e.g. NaCl, CsCl), positions of cations and anions can be interchanged. Unit cell of CsCl consists of a Cs$^+$ (or Cl$^-$) surrounded by cubic array of Cl$^-$ (or Cs$^+$); coordination number of each ion is 8. Unit cell of NaCl consists of Na$^+$ (or Cl$^-$) surrounded by octahedral arrangement of Cl$^-$ (or Na$^+$); coordination number of each ion is 6. Unit cell of TiO$_2$ is tetragonal ('square-ended box') with central Ti(IV) octahedrally sited with respect to 6 O$^{2-}$; each O$^{2-}$ is 3-coordinate. Note relationship between stoichiometry of compound X$_m$Y$_n$, and ratio of coordination numbers:

*Self-study exercise*
Why is the unit cell of NaCl not just a simple cube as shown here?

| Stoichiometry X : Y | Ratio coordination numbers X : Y |
|---|---|
| 1 : 1 | 1 : 1 |
| 1 : 2 | 2 : 1 |
| 2 : 1 | 1 : 2 |

Some ions in a lattice lie completely within a unit cell, but others are shared between unit cells: types of shared sites are (i) face (shared between 2 unit cells, see **5.4**), (ii) edge (shared between 4), (iii) corner (shared between 8). NaCl unit cell contains all these sites; in CsCl, the only shared sites are corner ones. In TiO$_2$, shared sites are on faces and at corners. Determining the stoichiometry takes into account the sharing of ions. e.g. CsCl (with Cs$^+$ in centre site):

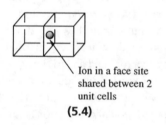

Ion in a face site shared between 2 unit cells

**(5.4)**

| Site | Number of Cs$^+$ | Number of Cl$^-$ | |
|---|---|---|---|
| Within unit cell | 1 | 0 | |
| Corner site | 0 | $8 \times 1/8 = 1$ | |
| Total | 1 | 1 | i.e. stoichiometry = 1:1 |

➤ For ion sharing in NaCl: see worked example 5.2 in H&S

e.g. TiO$_2$:

| Site | Number of Ti$^{4+}$ | Number of O$^{2-}$ | |
|---|---|---|---|
| Within unit cell | 1 | 2 | |
| Corner site | $8 \times 1/8 = 1$ | 0 | |
| Face site | 0 | $4 \times 1/2 = 2$ | |
| Total | 2 | 4 | i.e. stoichiometry = 1:2 |

(a) CaF$_2$: see Figure 5.6.

✓ 5.12

| Site | Number of Ca$^{2+}$ | Number of F$^-$ |
|---|---|---|
| Within unit cell | 0 | 8 |
| Corner site | $8 \times 1/8 = 1$ | 0 |
| Face site | $6 \times 1/2 = 3$ | 0 |
| Total | 4 | 8 |

∴ There are 4 formula units per unit cell.

(b) TiO$_2$: see the end of answer 5.11 – there are 2 formula units per unit cell.

**Figure 5.6** Unit cell of CaF$_2$ (fluorite); Ca$^{2+}$ ions are shown in dark grey.

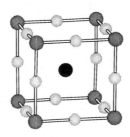

**Figure 5.7** Unit cell of perovskite; colour code: Ca$^{2+}$, black; Ti$^{4+}$, dark grey; O$^{2-}$, pale grey. Ca–O near neighbour contacts are omitted for clarity.

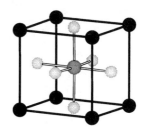

**Figure 5.8** Alternative unit cell for perovskite; colour code: Ca$^{2+}$, black; Ti$^{4+}$, dark grey; O$^{2-}$, pale grey. Ca–O near neighbour contacts are omitted for clarity.

**5.13** (a) The unit cell of perovskite is drawn in Figure 5.7. Consider sharing of ions:

| Site | Number of Ca$^{2+}$ | Number of Ti$^{4+}$ | Number of O$^{2-}$ |
|------|------|------|------|
| Central | 1 | 0 | 0 |
| Corner | 0 | $8 \times 1/8 = 1$ | 0 |
| Edge | 0 | 0 | $12 \times 1/4 = 3$ |
| Total | 1 | 1 | 3 |

i.e. stoichiometry = 1:1:3

(b) If Ti$^{4+}$ is in the central site and is octahedrally surrounded by O$^{2-}$, each O$^{2-}$ will be in a face-sharing site. This gives $(6 \times 1/2) = 3$ O$^{2-}$. The Ti$^{4+}$:O$^{2-}$ ratio is 1:3 and the Ca$^{2+}$ ions must be in corner sites so as to give $(8 \times 1/8) = 1$ Ca$^{2+}$ per unit cell, and a stoichiometry of Ca$^{2+}$:Ti$^{4+}$:O$^{2-}$ = 1:1:3. Figure 5.8 shows the diagram for this unit cell.

**5.14** (a) The lattice energy, $\Delta_{\text{lattice}}U(0\ \text{K})$, of an ionic compound is the change in internal energy that accompanies the formation of one mole of the solid from its constituent gas-phase ions at 0 K. This definition is consistent with an associated exothermic reaction, i.e. negative $\Delta H$.

See Box 1.6 in H&S:
*'The relationship between $\Delta U$ and $\Delta H$'*

$$\Delta_{\text{lattice}}H(298\ \text{K}) \approx \Delta_{\text{lattice}}U(0\ \text{K})$$

(b) The Born-Landé equation is:

Physical constants: see back inside cover of this book; Born exponents: see Table 5.3 in H&S; Madelung constants: see Table 5.4 in H&S

$$\Delta U(0\ \text{K}) = -\frac{LA|z_+||z_-|e^2}{4\pi\varepsilon_0 r_0}\left(1 - \frac{1}{n}\right)$$

Since KBr adopts an NaCl lattice, use the Madelung constant, $A$, for NaCl. The Born exponent, $n$, for KBr = $1/2(9 + 10) = 9.5$. Convert 328 pm to m.

$$\Delta U(0\ \text{K}) = -\frac{(6.022 \times 10^{23})(1.7476)(1)(1)(1.602 \times 10^{-19})^2}{4\pi(8.854 \times 10^{-12})(328 \times 10^{-12})}\left(1 - \frac{1}{9.5}\right)$$

$$= -662\,000\ \text{J mol}^{-1} \quad \text{(to 3 sig. fig.)}$$

$$= -662\ \text{kJ mol}^{-1}$$

5.15    Scheme needed for this question is:

$$
\begin{array}{l}
Ba(s) \xrightarrow{\Delta_a H^\circ(Ba,s)} Ba(g) \xrightarrow{IE_1 + IE_2} Ba^{2+}(g) \\
Cl_2(g) \xrightarrow{D(Cl_2,g)} 2Cl(g) \xrightarrow{2\Delta_{EA}H(Cl,g)} 2Cl^-(g)
\end{array}
\left.\right\}
\xrightarrow{\Delta_{lattice}H(BaCl_2,s)} BaCl_2(s)
$$

$\Delta_f H^\circ(BaCl_2,s)$

*Remember!*
For a diatomic $X_2$:
$D(X_2,g) = 2\Delta_a H(X,g)$

$$\Delta_{lattice}H = \Delta_f H^\circ(BaCl_2,s) - \Delta_a H^\circ(Ba,s) - IE_1 - IE_2 - D(Cl_2,g)$$
$$-2\Delta_{EA}H(Cl,g)$$

$$= -859 - 178 - 502.8 - 965.2 - 242 - 2(-349)$$

$$= -2050 \text{ kJ mol}^{-1} \quad \text{(to 3 sig. fig.)}$$

This method gives an enthalpy value, but the internal energy change $\Delta U(0 \text{ K}) \approx \Delta H(298 \text{ K})$. A similar approximation is made for electron affinity and ionization energy values where the thermochemical cycle uses an enthalpy rather than energy change.

5.16    (a) Scheme needed for this question is:

$$
\begin{array}{l}
Mg(s) \xrightarrow{\Delta_a H^\circ(Mg,s)} Mg(g) \xrightarrow{IE_1 + IE_2} Mg^{2+}(g) \\
{}^1\!/_2 O_2(g) \xrightarrow{\Delta_a H^\circ(O,g)} O(g) \xrightarrow{\Delta_{EA}H^\circ} O^{2-}(g)
\end{array}
\left.\right\}
\xrightarrow{\Delta_{lattice}H(MgO,s)} MgO(s)
$$

$\Delta_f H^\circ(MgO,s)$

where $\Delta_{EA}H^\circ$ is the standard enthalpy change at 298 K for the reaction:

$$O(g) + 2e^- \rightarrow O^{2-}(g)$$

From the thermochemical cycle:

$$\Delta_{lattice}H = \Delta_f H^\circ(MgO,s) - \Delta_a H^\circ(Mg,s) - IE_1 - IE_2 - \Delta_a H^\circ(O,g) - \Delta_{EA}H^\circ$$

$$\Delta_{EA}H^\circ = \Delta_f H^\circ(MgO,s) - \Delta_a H^\circ(Mg,s) - IE_1 - IE_2 - \Delta_a H^\circ(O,g) - \Delta_{lattice}H$$

$$= -602 - 146 - 737.7 - 1451 - 249 - (-3795)$$

$$= 609 \text{ kJ mol}^{-1} \quad \text{(to 3 sig. fig.)}$$

(b) From $\Delta_{EA}H$ values in Appendix 9 in H&S, $\Delta_{EA}H = 657$ kJ mol$^{-1}$. Approximation in the calculation was to take $\Delta_{lattice}U(0 \text{ K}) \approx \Delta_{lattice}H(298 \text{ K})$, but the real uncertainty is in the value of $\Delta_{lattice}H^\circ(MgO,s)$ – a calculated value from an electrostatic model.

5.17    (a) Look at the contributing factors in the relevant Born-Haber cycles:

$$\Delta_f H^\circ(MX,s) = \Delta_{lattice}H(MX,s) + \Delta_a H^\circ(M,s) + IE_1 + \Delta_a H^\circ(X,g) + \Delta_{EA}H(X,g)$$

For the fluorides, MF, $\Delta_a H^\circ(F,g)$ and $\Delta_{EA}H(F,g)$ are constant factors. Similarly for the iodides, MI, $\Delta_a H^\circ(I,g)$ and $\Delta_{EA}H(I,g)$ are constant factors. The factors

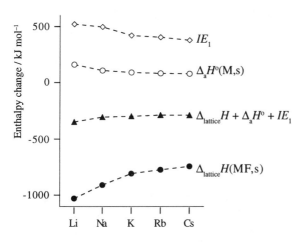

**Figure 5.9**  Trends in values of $\Delta_a H^\circ$ and $IE_1$ for the group 1 metals (M), and in $\Delta_{lattice} H$ for the fluorides MF. The overall trend in the sum of these values is also shown.

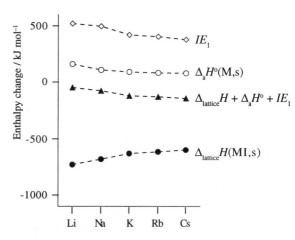

**Figure 5.10**  Trends in values of $\Delta_a H$ and $IE_1$ for the group 1 metals (M), and in $\Delta_{lattice} H$ for the iodides MI. The overall trend in the sum of these values is also shown.

➤ Values of $\Delta_a H^\circ$ and $IE_1$ come from Appendices in H&S. Values of lattice energies or enthalpies can be found in standard reference tables, or estimated using the Born-Landé equation

that contribute towards the trend in $\Delta_f H^\circ (MF,s)$ are $\Delta_a H^\circ$ and $IE_1$ for the alkali metals, and $\Delta_{lattice} H(MF,s)$. For the iodides, the trend in $\Delta_f H^\circ (MI,s)$ is determined by trends in $\Delta_a H^\circ$ and $IE_1$ for the alkali metals, and $\Delta_{lattice} H(MI,s)$. On descending group 1, values of $\Delta_a H^\circ$ and $IE_1$ for the metals become less positive: these trends are shown in both Figures 5.9 and 5.10. Figure 5.9 also shows the trend in values of $\Delta_{lattice} H(MF,s)$, and in Figure 5.10, the trend in values of $\Delta_{lattice} H(MI,s)$ is shown. In both cases, $\Delta_{lattice} H$ becomes less negative on going from Li to Cs. Difference between the trends in $\Delta_f H^\circ (MF,s)$ and $\Delta_f H^\circ (MI,s)$ can be seen in the lines labelled '$\Delta_{lattice} H + \Delta_a H^\circ + IE_1$' in Figures 5.9 and 5.10. This sum of enthalpy values appears in the Born-Haber cycle. Figure 5.9 shows that for the *fluorides*, this sum becomes *less negative* on descending group 1, whereas for the *iodides* it becomes *more negative*. This is the origin of the observed trends in enthalpies of formation of MF and MI.

(b) Reaction to consider:

$$MSO_4(s) \rightarrow MO(s) + SO_3(g)$$

The trend in stabilities of the sulfates with respect to conversion to oxides depends on the values of $\Delta_r G$ on going from M = Ca to Sr to Ba. Assuming entropy effects are similar, you can look at values of $\Delta_r H$ in place of $\Delta_r G$:

$$\Delta_r H = \Delta_f H(MO,s) + \Delta_f H(SO_3,g) - \Delta_f H(MSO_4,s)$$

$\Delta_f H(SO_3,g)$ is a constant factor, and so trend in $\Delta_r H$ values depends on difference between $\Delta_f H(MO,s)$ and $\Delta_f H(MSO_4,s)$ which in turn depends on difference between $\Delta_{lattice} H(MO,s)$ and $\Delta_{lattice} H(MSO_4,s)$ since other contributing quantities in the Born-Haber cycles are constant factors; see part (a) of this answer to follow the reasoning. The sulfates are isomorphous, as are the oxides; therefore, Madelung constants are the same for each series of salts. The cation radii follow the ordering:

$$Ca^{2+} < Sr^{2+} < Ba^{2+}$$

and the lattice energy depends on the internuclear separation according to:

$$\Delta U \propto \frac{1}{r}$$

However, for each series of salts, the anion radius is a constant factor, and it is the variation in cation radius that is of importance: for both oxides and sulfates, values of $\Delta U$ (or $\Delta_{lattice}H$) become less negative on going from $Ca^{2+}$ to $Sr^{2+}$ to $Ba^{2+}$. The oxide and sulfate lattice energies are:

$$\Delta U \propto \frac{1}{r_{M^{2+}} + r_{O^{2-}}} \qquad \text{and} \qquad \Delta U \propto \frac{1}{r_{M^{2+}} + r_{SO_4^{2-}}}$$

Since $r_{SO_4^{2-}} \gg r_{O^{2-}}$, we can consider that the lattice energies of the sulfates to be relatively similar along the series $Ca^{2+}$ to $Sr^{2+}$ to $Ba^{2+}$, whereas the variation in lattice energies of oxides is more significant. Since trend in $\Delta_r H$ values depends on difference between $\Delta_{lattice}H(MO,s)$ and $\Delta_{lattice}H(MSO_4,s)$, and $\Delta_{lattice}H(MO,s)$ is taken to be the crucial variable, becoming less negative from $Ca^{2+}$ to $Sr^{2+}$ to $Ba^{2+}$, then it follows that the thermal stability of the sulfates increases in the order $CaSO_4 <$ $SrSO_4 < BaSO_4$.

5.18    (a) Born-Landé equation is:

Physical constants: see back inside cover of this book; Born exponents: see Table 5.3 in H&S; Madelung constants: see Table 5.4 in H&S

$$\Delta U(0\ K) = -\frac{LA|z_+||z_-|e^2}{4\pi\varepsilon_0 r_0}\left(1 - \frac{1}{n}\right)$$

For CsCl with a CsCl lattice:

$$\Delta U(0\ K) = -\frac{(6.022 \times 10^{23})(1.7627)(1)(1)(1.602 \times 10^{-19})^2}{4\pi(8.854 \times 10^{-12})(356.6 \times 10^{-12})}\left(1 - \frac{1}{10.5}\right)$$

$$= -621200 \text{ J mol}^{-1} \text{ (to 4 sig. fig.)}$$
$$= -621.2 \text{ kJ mol}^{-1}$$

(b) For CsCl with an NaCl lattice:

$$\Delta U(0\ K) = -\frac{(6.022 \times 10^{23})(1.7476)(1)(1)(1.602 \times 10^{-19})^2}{4\pi(8.854 \times 10^{-12})(347.4 \times 10^{-12})}\left(1 - \frac{1}{10.5}\right)$$

$$= -632200 \text{ J mol}^{-1} \text{ (to 4 sig. fig.)}$$
$$= -632.2 \text{ kJ mol}^{-1}$$

(c) From lattice energy considerations (and assuming an electrostatic model which is realistic in this case), the energy difference between the CsCl and NaCl structures for CsCl is very small, indicating only a small preference for one lattice type over the other. In the main text H&S, we discussed the fact that at 298 K, $NH_4Cl$ and $NH_4Br$ have CsCl lattices, but above 457 and 411 K respectively, they crystallize with NaCl lattices.

# 6 Acids, bases and ions in aqueous solution

**6.1** (a) $pK_a = -\log K_a$ or $K_a = 10^{-pK_a}$

For the first dissociation step: $pK_a(1) = 0.74$ $\therefore K_a(1) = 10^{-0.74} = 0.18$
For the second dissociation step: $pK_a(2) = 6.49$ $\therefore K_a(2) = 10^{-6.49} = 3.24 \times 10^{-7}$
(b)

$$H_2CrO_4(aq) + H_2O(aq) \rightleftharpoons [H_3O]^+(aq) + [HCrO_4]^-(aq)$$
$$[HCrO_4]^-(aq) + H_2O(aq) \rightleftharpoons [H_3O]^+(aq) + [CrO_4]^{2-}(aq)$$

**6.2** Structure of $H_4P_2O_7$ is shown in **6.1**. The 4 dissociation steps with associated $pK_a$ values are:

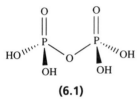

**(6.1)**

$$H_4P_2O_7(aq) + H_2O(l) \rightleftharpoons [H_3O]^+(aq) + [H_3P_2O_7]^-(aq) \qquad pK_a(1)$$
$$[H_3P_2O_7]^-(aq) + H_2O(l) \rightleftharpoons [H_3O]^+(aq) + [H_2P_2O_7]^{2-}(aq) \qquad pK_a(2)$$
$$[H_2P_2O_7]^{2-}(aq) + H_2O(l) \rightleftharpoons [H_3O]^+(aq) + [HP_2O_7]^{3-}(aq) \qquad pK_a(3)$$
$$[HP_2O_7]^{3-}(aq) + H_2O(l) \rightleftharpoons [H_3O]^+(aq) + [P_2O_7]^{4-}(aq) \qquad pK_a(4)$$

$$pK_a(1) = 1.0 \qquad pK_a(2) = 2.0 \qquad pK_a(3) = 7.0 \qquad pK_a(4) = 9.0$$

These can be assigned on the basis that generally:
$$K_a(1) > K_a(2) > K_a(3) > K_a(4)$$
removal of $H^+$ being increasingly more difficult as the negative charge on the anion increases. The larger $pK_a$, the smaller $K_a$: e.g. $pK_a(4) = 9.0$ corresponds to $K_a = 1 \times 10^{-9}$, while $pK_a(1) = 1.0$ corresponds to $K_a = 0.1$.

**6.3** For $CH_3CO_2H$, $pK_a = 4.75$, and for $CF_3CO_2H$, $pK_a = 0.23$. Smaller $pK_a$ corresponds to larger $K_a$, since $K_a = 10^{-pK_a}$. Greater acid strength of $CF_3CO_2H$ can be explained in terms of the inductive effect (**6.2**). A physicochemical interpretation follows from studies of the temperature dependence of $pK_a$ which show that for dissociation of $HCO_2H$, $CH_3CO_2H$ and $CCl_3CO_2H$, $\Delta H° \approx 0$; the variation in $pK_a$ arises from variation in entropy of dissociation which becomes less negative along the series $CH_3CO_2H$, $HCO_2H$, $CCl_3CO_2H$, $CF_3CO_2H$ – withdrawal of electrons away from carboxylate end of the anion results in less orientation of surrounding solvent molecules.

**(6.2)**

**6.4** (a) $H_2NCH_2CH_2NH_2$ is a Brønsted base. The question asks about $pK_a$ values of the conjugate acid of $H_2NCH_2CH_2NH_2$; these values correspond to equilibria involving $H^+$ loss from the protonated base:

$$[H_3NCH_2CH_2NH_3]^{2+}(aq) + H_2O(l) \rightleftharpoons [H_2NCH_2CH_2NH_3]^+(aq) + [H_3O]^+(aq)$$
$$pK_a(1) = 10.71$$

$$[H_2NCH_2CH_2NH_3]^+(aq) + H_2O(l) \rightleftharpoons H_2NCH_2CH_2NH_2(aq) + [H_3O]^+(aq)$$
$$pK_a(2) = 7.56$$

(b) The relationship between $pK_b$ and $pK_a$, or $K_b$ and $K_a$ is:
$$pK_a + pK_b = 14.00 \qquad \text{or} \qquad K_a + K_b = 10^{-14.00}$$

∴ For the $pK_b$ value corresponding to a $pK_a(1)$ of 10.71:

$pK_b = (14.00 - 10.71) = 3.29$

∴ For the $pK_b$ value corresponding to a $pK_a(2)$ of 7.56:

$pK_b = (14.00 - 7.56) = 6.44$

The equilibria to which these values refer are the reverse of those shown on p. 51. The first and second proton additions, and their corresponding $pK_b$ values are:

$$H_2NCH_2CH_2NH_2(aq) + [H_3O]^+(aq) \rightleftharpoons [H_2NCH_2CH_2NH_3]^+(aq) + H_2O(l)$$
$$pK_b(1) = 3.29$$
$$[H_2NCH_2CH_2NH_3]^+(aq) + [H_3O]^+(aq) \rightleftharpoons [H_3NCH_2CH_2NH_3]^{2+}(aq) + H_2O(l)$$
$$pK_b(2) = 6.44$$

Check that result makes sense: first proton addition should be easier than second, since in step (2), H$^+$ is being added to a *cation*. Thus, $K_b(1) > K_b(2)$, and so $pK_b(1) < pK_b(2)$ since $K_b = 10^{-pK_b}$.

6.5    The compounds for this question are redrawn here in structures **6.3-6.8**.

(a) Acid **6.3** (phosphonic acid) is *dibasic* (2 OH groups); the dissociation steps are:

$$H_3PO_3(aq) + H_2O(l) \rightleftharpoons [H_3O]^+(aq) + [H_2PO_3]^-(aq)$$
$$[H_2PO_3]^-(aq) + H_2O(l) \rightleftharpoons [H_3O]^+(aq) + [HPO_3]^{2-}(aq)$$

Acid **6.4** (phosphoric acid) is *tribasic* (3 OH groups), and the dissociation steps are:

$$H_3PO_4(aq) + H_2O(l) \rightleftharpoons [H_3O]^+(aq) + [H_2PO_4]^-(aq)$$
$$[H_2PO_4]^-(aq) + H_2O(l) \rightleftharpoons [H_3O]^+(aq) + [HPO_4]^{2-}(aq)$$
$$[HPO_4]^{2-}(aq) + H_2O(l) \rightleftharpoons [H_3O]^+(aq) + [PO_4]^{3-}(aq)$$

Acid **6.5** (thiosulfuric acid) is *dibasic* (2 OH groups), and the dissociation steps are:

$$H_2S_2O_3(aq) + H_2O(l) \rightleftharpoons [H_3O]^+(aq) + [HS_2O_3]^-(aq)$$
$$[HS_2O_3]^-(aq) + H_2O(l) \rightleftharpoons [H_3O]^+(aq) + [S_2O_3]^{2-}(aq)$$

Acid **6.6** (peroxodisulfuric acid) is *dibasic* (2 OH groups); the dissociation steps are:

$$H_2S_2O_8(aq) + H_2O(l) \rightleftharpoons [H_3O]^+(aq) + [HS_2O_8]^-(aq)$$
$$[HS_2O_8]^-(aq) + H_2O(l) \rightleftharpoons [H_3O]^+(aq) + [S_2O_8]^{2-}(aq)$$

Acid **6.7** (sulfamic acid) is *monobasic* (1 OH group):

$$HSO_3NH_2(aq) + H_2O(l) \rightleftharpoons [H_3O]^+(aq) + [SO_3NH_2]^-(aq)$$

**(6.3)**

**(6.4)**

**(6.5)**

**(6.6)**

**(6.7)**

**(6.8)**

(b) The acid (**6.8**) is *tetrabasic* and can be represented as $H_4L$. From treatment with NaOH, four salts could be isolated by the neutralization reactions:

$$H_4L + NaOH \rightarrow Na[H_3L] + H_2O$$
$$H_4L + 2NaOH \rightarrow Na_2[H_2L] + 2H_2O$$
$$H_4L + 3NaOH \rightarrow Na_3[HL] + 3H_2O$$
$$H_4L + 4NaOH \rightarrow Na_4[L] + 4H_2O$$

6.6 (a) Structures are shown in **6.9** and **6.10**; in the solid state, boric acid has a hydrogen-bonded layer structure (see Section 12.7 in H&S).

(b) Acts as a *Lewis* acid, accepting $[OH]^-$:

$$B(OH)_3(aq) + [OH]^-(aq) \rightleftharpoons [B(OH)_4]^-(aq)$$

(c) $H_3BO_3$ and $H_3PO_3$: similar formulae, but different structures and behaviour in aqueous solution. $H_3BO_3$ (see above) accepts $[OH]^-$ and is a Lewis acid. $H_3PO_3$ (see structure **6.3**) is a Brønsted acid and is dibasic:

$$H_3PO_3(aq) + H_2O(l) \rightleftharpoons [H_3O]^+(aq) + [H_2PO_3]^-(aq)$$
$$[H_2PO_3]^-(aq) + H_2O(l) \rightleftharpoons [H_3O]^+(aq) + [HPO_3]^{2-}(aq)$$

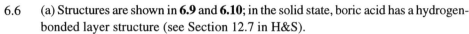

**(6.9)**

**(6.10)**

6.7 Dissolution of NaCN (ionic solid) in water:

$$NaCN(s) \xrightarrow{\text{water}} Na^+(aq) + [CN]^-(aq) \qquad (\textit{fully dissociated})$$

The $pK_a$ value of 9.31 for HCN refers to the equilibrium:

$$HCN(aq) + H_2O(l) \rightleftharpoons [H_3O]^+(aq) + [CN]^-(aq)$$

and shows that HCN is a weak acid ($K_a = 10^{-9.31} = 4.90 \times 10^{-10}$). Thus, $[CN]^-$ is the conjugate base of a weak acid, and in aqueous solution, abstracts a proton from water making the solution basic:

$$[CN]^-(aq) + H_2O(l) \rightleftharpoons HCN(aq) + [OH]^-(aq)$$

**Questions 6.8-6.9: some general notes**

Amphoteric behaviour means that a species reacts with both acids (i.e. species acts as a base) and bases (i.e. species acts as an acid). Among oxides and hydroxides:
- metal oxides and hydroxides are basic;
- non-metal oxides and hydroxides are acidic;
- oxides and hydroxides of elements near to the 'diagonal line' (see Figure 6.1) are amphoteric.

6.8 $[HCO_3]^-$ can accept a proton (acting as a base) or lose a proton (acting as an acid):

*Basic behaviour:* $\quad [HCO_3]^-(aq) + [H_3O]^+(aq) \rightleftharpoons H_2CO_3(aq) + H_2O(l)$

*Acidic behaviour:* $\quad [HCO_3]^-(aq) + H_2O(l) \rightleftharpoons [CO_3]^{2-}(aq) + [H_3O]^+(aq)$

Group number

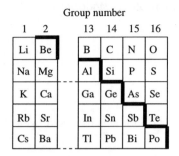

**Figure 6.1** The 'diagonal line' in the periodic table.

6.9   Classify the oxides as being those of a metal, non-metal, or an element close to the 'diagonal line' (Figure 6.1). The latter distinction is not clear-cut as the examples in the question show.

(a) MgO: oxide of a metal, so is basic.

(b) SnO: oxide of a metal, but is amphoteric (see end of Section 13.9 in H&S).

(c) $CO_2$: oxide of a non-metal, so is acidic.

(d) $P_2O_5$: oxide of a non-metal, so is acidic.

(e) $Sb_2O_5$: oxide of a non-metal adjacent to the 'diagonal line' and is amphoteric.

(f) $SO_2$: oxide of a non-metal, so is acidic.

(g) $Al_2O_3$: oxide of a metal adjacent to the 'diagonal line' and is amphoteric.

(h) BeO: oxide of the first member of group 2, an atypical member of this group (see Chapter 10) and is amphoteric.

6.10   (a) Upon the addition of an ionic solid, MX, to water, the following equilibrium is established:

$$MX(s) \rightleftharpoons M^{n+}(aq) + X^{n-}(aq)$$

When equilibrium is reached, the solution is *saturated*.

(b) The solubility of the solid (stated at a given temperature) is the mass of solid which dissolves when equilibrium is reached in the presence of an excess of solid, divided by the mass of the solvent. Solubility is expressed in several ways, e.g. mass of solute (in g) per 100 g of water, or moles of solute per unit volume of water.

(c) Some salts dissolve to an extremely small extent; such a salt is *sparingly soluble*.

(d) Solubility products refer to sparingly soluble salts, and are equilibrium constants ($K_{sp}$) describing the dissolution of the salt, e.g. for $PbI_2$, the equilibrium is:

➤

In answering this question, it is best to give an example – but make sure that you choose a sensible, i.e. a *sparingly soluble*, salt!

$$PbI_2(s) \rightleftharpoons Pb^{2+}(aq) + 2I^-(aq)$$

and the expression for the solubility product is:

$$K_{sp} = \frac{[Pb^{2+}][I^-]^2}{[PbI_2]} = [Pb^{2+}][I^-]^2$$

By convention, activity of all solids taken as 1; at very *low* concentrations, activities of ions approximate to their concentrations.

6.11   (a)    $AgCl(s) \rightleftharpoons Ag^+(aq) + Cl^-(aq)$

$$K_{sp} = \frac{[Ag^+][Cl^-]}{\underset{=1}{\cancel{[AgCl]}}} = [Ag^+][Cl^-]$$

(b)    $CaCO_3(s) \rightleftharpoons Ca^{2+}(aq) + [CO_3]^{2-}(aq)$

$$K_{sp} = \frac{[Ca^{2+}][CO_3^{2-}]}{\cancel{[CaCO_3]}} = [Ca^{2+}][CO_3^{2-}]$$

(c)    $CaF_2(s) \rightleftharpoons Ca^{2+}(aq) + 2F^-(aq)$

$$K_{sp} = \frac{[Ca^{2+}][F^-]^2}{\cancel{[CaF_2]}} = [Ca^{2+}][F^-]^2$$

6.12    (a) Solubility can be expressed in moles of dissolved AgCl per $dm^3$.

Since 1 mole of dissolved AgCl gives 1 mole of $Ag^+$ and 1 mole of $Cl^-$ ions, we can write:

$$\text{Solubility} = [Ag^+] \qquad \text{and} \qquad [Ag^+] = [Cl^-]$$

Since:

$$K_{sp} = [Ag^+][Cl^-] = [Ag^+]^2$$

it follows that:

$$\text{Solubility} = [Ag^+] = \sqrt{K_{sp}}$$

> Remember that '[ ]' is used in two ways: in the formula of an ion, e.g. $[CO_3]^{2-}$, and for concentration, e.g. $[CO_3^{2-}]$

(b)  1 mole of dissolved $CaCO_3$ gives 1 mole of $Ca^{2+}$ and 1 mole of $[CO_3]^{2-}$ ions.

$$\therefore \quad \text{Solubility} = [Ca^{2+}] \qquad \text{and} \qquad [Ca^{2+}] = [CO_3^{2-}]$$

$$K_{sp} = [Ca^{2+}][CO_3^{2-}] = [Ca^{2+}]^2$$

$$\therefore \quad \text{Solubility} = [Ca^{2+}] = \sqrt{K_{sp}}$$

(c) 1 mole of dissolved $CaF_2$ gives 1 mole of $Ca^{2+}$ and 2 moles of $F^-$ ions.

$$\therefore \quad \text{Solubility} = [Ca^{2+}] \qquad \text{and} \qquad [F^-] = 2[Ca^{2+}]$$

$$K_{sp} = [Ca^{2+}][F^-]^2 = 4[Ca^{2+}]^3$$

$$\therefore \quad \text{Solubility} = [Ca^{2+}] = \sqrt[3]{\frac{K_{sp}}{4}}$$

6.13    The equilibrium is:

$$BaSO_4(s) \rightleftharpoons Ba^{2+}(aq) + [SO_4]^{2-}(aq)$$

$$\therefore \quad K_{sp} = [Ba^{2+}][SO_4^{2-}] = [Ba^{2+}]^2$$

$$K_{sp} = 1.07 \times 10^{-10}$$

$$\therefore \quad [Ba^{2+}] = \sqrt{K_{sp}} = \sqrt{1.07 \times 10^{-10}} = 1.03 \times 10^{-5} \, mol \, dm^{-3}$$

The solubility of the salt can be found directly from the concentration of $Ba^{2+}$ ions. It is reasonable to assume that the calculation above gives $[Ba^{2+}]$ in units of mol $dm^{-3}$ – the question asks for the solubility of $BaSO_4$ in g per 100 g of water.

> The problem here is that $K_{sp}$ is dimensionless; one must assume that the units of concentration are mol $dm^{-3}$; these are conventional (SI) units

Mass of 1 mole $BaSO_4$ = 137.34 + 32.06 + 4(16.00) = 233.40 g
Mass of $1.03 \times 10^{-5}$ moles $BaSO_4$ = 233.40 × $1.03 \times 10^{-5}$ = $2.40 \times 10^{-3}$ g

This is the mass dissolved in 1 $dm^3$ = 1000 g of water.

$$\therefore \quad \text{Mass per 100 g} = 2.40 \times 10^{-4} \, g \qquad \text{(to 3 sig. fig.)}$$

6.14    (a) Solid NaF is an ionic salt with an NaCl lattice. When NaF dissolves in water, the lattice collapses, and the $Na^+$ and $F^-$ ions are hydrated by solvent (water) molecules. The ion⋯⋯ion interactions present in the ionic lattice are lost, but are replaced by ion⋯⋯dipole interactions; diagram **6.11** shows the first hydration shell

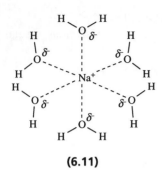

**(6.11)**

F⁻ ⅢⅢ H⎯O

**(6.12)**

Physical constants: see back
inside cover of this book

of an $Na^+$ ion – this is a coordination complex with the metal ion in an octahedral environment. The $F^-$ ion is also hydrated, interacting with the $H(\delta^+)$ atoms of the water molecules – hydrogen bond formation (**6.12**).

(b) In bulk water, $H_2O$ molecules interact with each other through hydrogen bonding (see Section 9.6 in H&S). Dissolution of NaF causes reorganization of the $H_2O$ molecules; ordered hydration shells are formed around each $Na^+$ and $F^-$ ion. Further hydration shells also form by hydrogen-bonded interactions with the inner shells.

Entropy: points to mention:
- opposing effects must be considered;
- NaF lattice going to ions in solution implies increase in entropy (positive $\Delta S$);
- $H_2O$ molecules become more ordered when ions enter the solution, so for the solvent itself, $\Delta S$ is negative;
- qualitatively, it is difficult to assess which process 'wins' and to predict the *overall* sign of $\Delta S$.

6.15    (a) In Appendix 11 in H&S, look up $E^o$ values for:

$$K^+(aq) + e^- \rightleftharpoons K(s) \qquad E^o = -2.93 \text{ V} \qquad (1)$$
$$\tfrac{1}{2}F_2(g) + e^- \rightleftharpoons F^-(aq) \qquad E^o = +2.87 \text{ V} \qquad (2)$$

Find $\Delta G^o$ for each process using: $\Delta G^o = -nFE^o$ where $\Delta G^o$ is in J mol⁻¹.
For reaction (1):
$$\Delta G^o = -nFE^o = -(1)(96\,485)(-2.93) \times 10^{-3} = +282.7 \text{ kJ mol}^{-1}$$
For reaction (2):
$$\Delta G^o = -nFE^o = -(1)(96\,485)(+2.87) \times 10^{-3} = -276.9 \text{ kJ mol}^{-1}$$

The value of $\Delta G^o$ for reaction (2) corresponds to $\Delta_f G^o$ for $F^-(aq)$
$$\therefore \Delta_f G^o(F^-,aq) = -276.9 \text{ kJ mol}^{-1}$$

The formation of $K^+(aq)$ is the *reverse* of reaction (1), so a sign change is needed:
$$\therefore \Delta_f G^o(K^+,aq) = -282.7 \text{ kJ mol}^{-1}$$

(b) To find $\Delta_{sol} G^o(KF,s)$, write down the appropriate equation and apply Hess cycle:

$$KF(s) \xrightarrow{\Delta_{sol}G^o(KF,s)} K^+(aq) + F^-(aq)$$

$$\Delta_{sol} G^o(KF,s) = \Delta_f G^o(K^+,aq) + \Delta_f G^o(F^-,aq) - \Delta_f G^o(KF,s)$$

$$= -282.7 - 276.9 - (-537.8) = -21.8 \text{ kJ mol}^{-1}$$

(c) $\Delta_{sol} G^o(KF,s)$ is significantly negative, showing that dissolution is favourable, and so KF is relatively soluble in water. Remember that $\Delta G^o = -RT\ln K$ and so a large, negative $\Delta_{sol} G^o(KF,s)$ corresponds to a large value of the equilibrium constant.

6.16    Given $\Delta_f G^o(PbS,s)$ means that one reaction for which data are available is:

$$Pb(s) + S(s) \rightarrow PbS(s) \qquad \text{(at 298 K)}$$

while $K_{sp}$ refers to the equilibrium: $\qquad PbS(s) \rightleftharpoons Pb^{2+}(aq) + S^{2-}(aq)$

These can be combined in the following thermochemical cycle:

$$Pb(s) + S(s) \xrightarrow{\Delta_f G^\circ(PbS,s)} PbS(s)$$

$$\Delta G^\circ_1 + \Delta G^\circ_2 \searrow \qquad \swarrow \Delta_{sol}G^\circ(PbS,s)$$

$$Pb^{2+}(aq) + S^{2-}(aq)$$

➤

Strictly, these are all equilibria, but single arrows are shown to give the thermochemical cycle a sense of 'direction'

$\Delta G^\circ_1 + \Delta G^\circ_2$ can be found from $E^\circ$ values. From Appendix 11 in H&S:

$$Pb^{2+}(aq) + 2e^- \rightleftharpoons Pb(s) \qquad\qquad E^\circ = -0.13 \text{ V} \qquad (1)$$
$$S(s) + 2e^- \rightleftharpoons S^{2-}(aq) \qquad\qquad E^\circ = -0.45 \text{ V} \qquad (2)$$

Find $\Delta G^\circ$ for each process using:   $\Delta G^\circ = -nFE^\circ$   where $\Delta G^\circ$ is in J mol$^{-1}$.
For reaction (1):
$$\Delta G^\circ = -nFE^\circ = -(2)(96\,485)(-0.13) \times 10^{-3} = +25 \text{ kJ mol}^{-1}$$
This refers to the *reverse* reaction to that in the cycle, and $\Delta G^\circ_1 = -25$ kJ mol$^{-1}$.
For reaction (2):
$$\Delta G^\circ = -nFE^\circ = -(2)(96\,485)(-0.45) \times 10^{-3} = +87 \text{ kJ mol}^{-1} = \Delta G^\circ_2$$

Now apply a Hess cycle :
$$\Delta_{sol}G^\circ(PbS,s) = \Delta G^\circ_1 + \Delta G^\circ_2 - \Delta_f G^\circ(PbS,s)$$
$$= -25 + 87 - (-99)$$
$$= 161 \text{ kJ mol}^{-1}$$

For $K_{sp}$:

$$\ln K_{sp} = -\frac{\Delta_{sol}G^\circ}{RT} \qquad\qquad \text{or} \qquad\qquad K_{sp} = e^{-\frac{\Delta_{sol}G^\circ}{RT}}$$

$$\therefore \quad K_{sp} = e^{-\frac{\Delta_{sol}G^\circ}{RT}} = e^{-\frac{161}{8.314\times10^{-3}\times298}} = 6.0 \times 10^{-29}$$

6.17    (a) Points to include in this answer:
• Construct thermochemical cycle shown here.

$$KCl(s) \xrightarrow{-\Delta_{lattice}G^\circ(KCl,s)} K^+(g) + Cl^-(g)$$

$$\Delta_{sol}G^\circ(KCl,s) \searrow \qquad \swarrow \begin{array}{l}\Delta_{hyd}G^\circ(K^+,g) + \\ \Delta_{hyd}G^\circ(Cl^-,g)\end{array}$$

$$K^+(aq) + Cl^-(aq)$$

• KCl is readily soluble, so $\Delta_{sol}G^\circ(KCl,s)$ must be large and negative.
• Sign and magnitude of $\Delta_{sol}G^\circ(KCl,s)$ depend on balance between $\Delta_{lattice}G^\circ(KCl,s)$ and hydration energies of ions ($\Delta_{hyd}G^\circ(KCl,s)$).
• Discuss energy losses and gains of lattice disruption versus solvation of ions.
(b) You can calculate a value for $\Delta_{sol}H^\circ(KCl,s)$:

$$\Delta_{sol}H^\circ(KCl,s) = -\Delta_{lattice}H^\circ(KCl,s) + \Delta_{hyd}H^\circ(K^+,g) + \Delta_{hyd}H^\circ(Cl^-,g) = +15 \text{ kJ mol}^{-1}$$

Since KCl is readily soluble, $\Delta_{sol}G^\circ(KCl,s)$ must be large and negative, meaning that $T\Delta_{sol}S^\circ$ term must be large and positive because:

$$\Delta_{sol}G^\circ = \Delta_{sol}H^\circ - T\Delta_{sol}S^\circ$$

6.18    Titration of NaCl against $AgNO_3$:
$$NaCl(aq) + AgNO_3(aq) \rightarrow NaNO_3(aq) + AgCl(s)$$
In the titration, $AgNO_3$ is *added to* NaCl solution. End point is when AgCl (white) stops being precipitated preferentially, and $Ag_2CrO_4$ (red) precipitates:
$$K_2CrO_4(aq) + 2AgNO_3(aq) \rightarrow 2KNO_3(aq) + Ag_2CrO_4(s)$$
From Table 6.4 in H&S, look up values of $K_{sp}$ for the two sparingly soluble salts; these are $1.77 \times 10^{-10}$ for AgCl and $1.12 \times 10^{-12}$ for $Ag_2CrO_4$.

$$\text{Solubility of AgCl} = \sqrt{1.77 \times 10^{10}} = 1.33 \times 10^{-5} \text{ mol dm}^{-3}$$

$$\text{Solubility of Ag}_2\text{CrO}_4 = \sqrt[3]{\frac{1.12 \times 10^{-12}}{4}} = 6.54 \times 10^{-5} \text{ mol dm}^{-3}$$

Hence, $Ag_2CrO_4$ is more soluble than AgCl. While there are $Cl^-$ ions in solution, AgCl is precipitated and $Ag^+$ ions are removed from solution as they are added. At the end point, the $[Ag^+]$ in solution rises, resulting in the precipitation of $Ag_2CrO_4$.

6.19    Buffer of $MeCO_2H$ and $Na[MeCO_2]$ involves the reactions:

$$MeCO_2H(aq) + H_2O(l) \;\rightleftharpoons\; [H_3O]^+(aq) + [MeCO_2]^-(aq) \qquad \text{(i)}$$

$$Na[MeCO_2](aq) \xrightarrow{\text{water}} Na^+(aq) + [MeCO_2]^-(aq) \qquad \text{(ii)}$$

A buffer withstands addition of small amounts of acid or base without significant change in pH. If a small amount of acid is added to the buffer, it reacts with $[MeCO_2]^-$ to give $MeCO_2H$. If a small amount of $[OH]^-$ is added, it reacts with $[H_3O]^+$ and is neutralized; consumption of some $[H_3O]^+$ causes equilibrium (i) to shift to right-hand side, and pH is therefore unaffected by the addition of the base.

6.20    (a) The equilibrium is:        $AgBr(s) \rightleftharpoons Ag^+(aq) + Br^-(aq)$
1 mole of dissolved AgBr gives 1 mole of $Ag^+$ and 1 mole of $Br^-$ ions.
∴   Solubility = $[Ag^+]$        and        $K_{sp} = [Ag^+][Br^-] = [Ag^+]^2$
In aqueous solution:
$$\text{Solubility} = \sqrt{K_{sp}} = \sqrt{5.35 \times 10^{-13}} = 7.31 \times 10^{-7} \text{ mol dm}^{-3}$$

Converting this to g per 100 g of water:        $M_r$ for AgBr = 187.77
Solubility of AgBr in 100 g of water = $7.31 \times 10^{-7} \times 187.77 \times 0.1$
$$= 1.37 \times 10^{-5} \text{ g / 100g H}_2\text{O}$$

(b) In 0.5 M KBr solution, $[Br^-]$ ions = 0.5 mol dm$^{-3}$ (fully dissociated salt)

|  | AgBr(s) ⇌ | Ag⁺(aq) | + | Br⁻(aq) |
|---|---|---|---|---|
| Initial aqu. ion concentrations / mol dm$^{-3}$ | | 0 | | 0.5 |
| Equilibrium concentrations / mol dm$^{-3}$ | | $x$ | | $0.5 + x$ |

> This question concerns the *common-ion effect*: see Section 6.10 in H&S

$$K_{sp} = 5.35 \times 10^{-13} = [Ag^+][Br^-] = x(0.5 + x)$$

Since $x \ll 0.5$, this simplifies to:   $5.35 \times 10^{-13} \approx 0.5x$
$$x \approx 1.07 \times 10^{-12} \text{ mol dm}^{-3}$$
∴ Solubility = $1.07 \times 10^{-12}$ mol dm$^{-3}$
Solubility of AgBr in 100 g 0.5 M KBr soln. = $1.07 \times 10^{-12} \times 187.77 \times 0.1$
$$= 2.01 \times 10^{-11} \text{ g / 100g solution}$$

6.21   MgO is sparingly soluble in water; in terms of the common-ion effect, the presence of $Mg^{2+}$ ions in the $MgCl_2$ solution might be expected to suppress dissolution of MgO. The observation in *increased* solubility suggests complex formation is aiding dissolution, e.g. formation of a species such as $[MgOMg]^{2+}$ or a hydrate thereof.

6.22   See Section 13.9 in H&S, and equations 13.52 and 13.53 in H&S.

6.23   The thermochemical cycle that is relevant is:

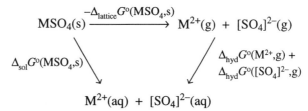

Sulfate is a common factor, so can be ignored. Variable factors are as follows.

- $\Delta_{lattice}G^\circ(MSO_4,s)$ becomes less negative on going from $Ca^{2+}$ to $Sr^{2+}$ to $Ba^{2+}$ salts (ionic radii increase), and so $\{-\Delta_{lattice}G^\circ(MSO_4,s)\}$ becomes less positive down group, although values will be similar since dealing with salts of a large cation (see also answers 5.17 and 6.24).
- From Table 6.6 in H&S, $\Delta_{hyd}G^\circ(M^{2+},g)$ becomes less negative down the group (as expected, since ion size increases).

Now look at $\Delta_{sol}G^\circ(MSO_4,s)$:

$$\Delta_{sol}G^\circ(MSO_4,s) = \{-\Delta_{lattice}G^\circ(MSO_4,s)\} + \Delta_{hyd}G^\circ(M^{2+},g) + \Delta_{hyd}G^\circ([SO_4]^{2+},g)$$

*Less positive down group,*   *Less negative*   *Constant*

*but values quite similar*   *down group*   *factor*

The two variable terms oppose each other, but $\Delta_{hyd}G^\circ(M^{2+},g)$ should be the deciding factor. Thus, $\Delta_{sol}G^\circ$ is expected to become less negative down the group; the most soluble salt will have the most negative $\Delta_{sol}G^\circ$, and this is $CaSO_4$.

6.24   A thermochemical cycle that is appropriate is:

$$PH_4X(s) \xrightarrow{\Delta_{decomp}H^\circ} PH_3(g) + HX(g)$$

$-\Delta_{lattice}H^\circ(PH_4X,s)$ ↓        ↓ $D(HX,g)$

$$[PH_4]^+(g) + X^-(g) \qquad PH_3(g) + H(g) + X(g)$$

↓ $IE(H,g) + \Delta_{EA}H^\circ(X,g)$

$\Delta_{PA}H^\circ$ = proton affinity of $PH_3(g)$

$$PH_3(g) + H^+(g) + X^-(g)$$

Applying a Hess cycle:

$$-\Delta_{lattice}H^\circ(PH_4X,s) = \Delta_{decomp}H^\circ + D(HX,g) + IE(H,g) + \Delta_{EA}H^\circ(X,g) + \Delta_{PA}H^\circ(PH_3,g)$$

$$\Delta_{decomp}H^\circ = -\Delta_{lattice}H^\circ(PH_4X,s) - D(HX,g) - IE(H,g) - \Delta_{EA}H^\circ(X,g) - \Delta_{PA}H^\circ(PH_3,g)$$

$\Delta_{PA}H(PH_3,g)$ is a constant factor for all the halides; $IE(H,g)$ is also constant. The variables that affect the value of $\Delta_{decomp}H^\circ$ are:

- $\Delta_{EA}H(X,g)$ which becomes less negative down group 17 (see Table 1.5 in H&S): F, −328; Cl, −349; Br, −325; I, −295 kJ $mol^{-1}$.

• $D(HX,g)$ which becomes less positive down group 17 (see Table 16.2 in H&S): HF, 570; Cl, 432; Br, 366; I, 298 kJ mol$^{-1}$.

• $\Delta_{lattice}H°(PH_4X,s)$ becomes less negative down the group because the anion size increases, and lattice energy is inversely proportional to the ion separation (assume the lattice type remains constant). The term $-\Delta_{lattice}H°(PH_4X,s)$ therefore becomes less positive down the group.

Now see to what extent $-\Delta_{lattice}H°(PH_4X,s)$ offsets $\{-D(HX,g) - \Delta_{EA}H°(X,g)\}$. The term $\{-D(HX,g) - \Delta_{EA}H°(X,g)\}$ becomes significantly less negative down the group: F, –242; Cl, –83; Br, –41; I, –3 kJ mol$^{-1}$, and $D(HX,g)$ shows a greater variation than $\Delta_{EA}H°(X,g)$. Although the lattice energies do decrease down the group, for a cation as large as $[PH_4]^+$ (similar in size to Cs$^+$) the difference between the lattice energies of the halides is relatively small. The dominant term is therefore $\{-D(HX,g) - \Delta_{EA}H°(X,g)\}$, and so you can predict that the trend in $\Delta_{decomp}H°$ will be to become less negative on going from PH$_4$F to PH$_4$I, i.e. the iodide is the most stable with respect to decomposition into PH$_3$ and HX.

6.25    (a) The equilibria for the question are the 2nd and 4th ligand substitutions:

$$[M(H_2O)_5L]^{z+}(aq) + L(aq) \overset{K_2}{\rightleftharpoons} [M(H_2O)_4L_2]^{z+}(aq) + H_2O(l)$$

$$[M(H_2O)_3L_3]^{z+}(aq) + L(aq) \overset{K_4}{\rightleftharpoons} [M(H_2O)_2L_4]^{z+}(aq) + H_2O(l)$$

A *stepwise* stability constant refers to an individual step. Therefore, write down an expression for the equilibrium constant for each step, remembering that $[H_2O]$ is taken as 1; (strictly this is the activity of water, but in dilute solutions activity approximates to the concentration).

$$K_2 = \frac{[M(H_2O)_4L_2{}^{z+}]}{[M(H_2O)_5L^{z+}][L]} \qquad K_4 = \frac{[M(H_2O)_2L_4{}^{z+}]}{[M(H_2O)_3L_3{}^{z+}][L]}$$

(b) The overall stability constants $\beta_2$ and $\beta_4$ refer to the equilibria:

$$[M(H_2O)_6]^{z+}(aq) + 2L(aq) \overset{\beta_2}{\rightleftharpoons} [M(H_2O)_4L_2]^{z+}(aq) + 2H_2O(l)$$

$$[M(H_2O)_6]^{z+}(aq) + 4L(aq) \overset{\beta_4}{\rightleftharpoons} [M(H_2O)_2L_4]^{z+}(aq) + 4H_2O(l)$$

The expressions for $\beta_2$ and $\beta_4$ are:

$$\beta_2 = \frac{[M(H_2O)_4L_2{}^{z+}]}{[M(H_2O)_6{}^{z+}][L]^2} \qquad \beta_4 = \frac{[M(H_2O)_2L_4{}^{z+}]}{[M(H_2O)_6{}^{z+}][L]^4}$$

6.26    (a) Ligand [acac]$^-$ is shown in structure **6.13**; it is a didentate ligand ($O,O'$-donor). Complex formation in aqueous solution takes place in 3 steps:

$$[Al(H_2O)_6]^{3+}(aq) + [acac]^-(aq) \overset{K_1}{\rightleftharpoons} [Al(H_2O)_4(acac)]^{2+}(aq) + 2H_2O(l)$$

$$[Al(H_2O)_4(acac)]^{2+}(aq) + [acac]^-(aq) \overset{K_2}{\rightleftharpoons} [Al(H_2O)_2(acac)_2]^+(aq) + 2H_2O(l)$$

$$[Al(H_2O)_2(acac)_2]^+(aq) + [acac]^-(aq) \overset{K_3}{\rightleftharpoons} [Al(acac)_3](aq) + 2H_2O(l)$$

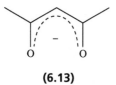

**(6.13)**

The log $K$ values refer to the steps shown. Stepwise stability constants usually *decrease* from $K_1$ to $K_n$, so it is expected that log $K_1 >$ log $K_2 >$ log $K_3$.

(b) Equation to use is:

$$\Delta G^o(303 \text{ K}) = -RT \ln K$$

For the 1st step:
$$\Delta G_1^o(303 \text{ K}) = -RT \ln K_1$$

$$\log K_1 = 8.6 \qquad \therefore K_1 = 10^{8.6}$$

$$\Delta G_1^o(303 \text{ K}) = -(8.314 \times 10^{-3})(303) \ln 10^{8.6} = -50 \text{ kJ mol}^{-1}$$

For the 2nd step:
$$\log K_2 = 7.9 \qquad \therefore K_2 = 10^{7.9}$$

$$\Delta G_2^o(303 \text{ K}) = -(8.314 \times 10^{-3})(303) \ln 10^{7.9} = -46 \text{ kJ mol}^{-1}$$

For the 3rd step:
$$\log K_3 = 5.8 \qquad \therefore K_3 = 10^{5.8}$$

$$\Delta G_3^o(303 \text{ K}) = -(8.314 \times 10^{-3})(303) \ln 10^{5.8} = -34 \text{ kJ mol}^{-1}$$

Values of $\Delta G^o$ become less negative along the series showing that the first ligand displacement step occurs more readily than the second, which in turn occurs more readily than the third.

6.27    The structures of the ligands are shown in **6.14-6.18**.

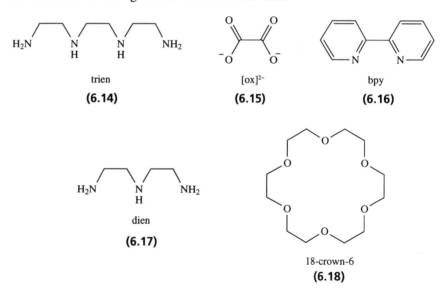

trien
**(6.14)**

$[ox]^{2-}$
**(6.15)**

bpy
**(6.16)**

dien
**(6.17)**

18-crown-6
**(6.18)**

**(6.19)**

(a) $[Cu(trien)]^{2+}$: tetradentate ligand; 4-coordinate complex with 3 chelate rings (**6.19**).
(b) $[Fe(ox)_3]^{3-}$: didentate ligand; 6-coordinate complex with 3 chelate rings (**6.20**).
(c) $[Ru(bpy)_3]^{2+}$: didentate ligand; 6-coordinate complex with 3 chelate rings (**6.21**).
(d) $[Co(dien)_2]^{3+}$: tridentate ligand; 6-coordinate complex with 4 chelate rings (**6.22**).
(d) $[K(18\text{-crown-6})]^+$: hexadentate ligand; 6-coordinate complex with 6 chelate rings (**6.23**).

(6.20)

(6.21)

(6.22)

(6.23)

# 7 Reduction and oxidation

7.1 In each answer, assign oxidation states in the order shown, because for some elements, more than one oxidation state is possible. In part (d), for example, it is impossible to assign the oxidation state of Fe (which has several possible oxidation states) before considering the oxidation state of Cl. *The sum of the oxidation states in a neutral compound equals 0, and in an ion, equals the overall charge.*

(a) CaO:      Ca    group 2     oxidation state +2

            O     group 16    oxidation state −2

(b) $H_2O$:      O     group 16    oxidation state −2

          H    oxidation state +1

(c) HF:       F     group 17    oxidation state −1

         H    oxidation state +1

(d) $FeCl_2$:    Cl    group 17    oxidation state −1

          Fe    oxidation state +2

(e) $XeF_6$:    F     group 17    oxidation state −1

          Xe    oxidation state +6

(f) $OsO_4$:    O     group 16    oxidation state −2

          Os    oxidation state +8

(g) $Na_2SO_4$:    Na    group 1     oxidation state +1

          O     group 16    oxidation state −2

          S     oxidation state +6

(h) $[PO_4]^{3-}$:    O     group 16    oxidation state −2

          P     oxidation state +5

(i) $[PdCl_4]^{2-}$:    Cl    group 17    oxidation state −1

          Pd    oxidation state +2

(j) $[ClO_4]^{-}$:    O     group 16    oxidation state −2

          Cl    oxidation state +7

(k) $[Cr(H_2O)_6]^{3+}$:    $H_2O$ is a neutral ligand (see part (b))

          Cr    oxidation state +3

7.2 Only the oxidation states of the metals are to be assigned:

(a) $[Cr_2O_7]^{2-}$ containing Cr oxidation state +6 $\rightarrow$ $Cr^{3+}$ oxidation state +3

(b) K oxidation state 0 $\rightarrow$ KOH containing $K^+$ oxidation state +1

(c) $Fe_2O_3$ containing Fe oxidation state +3 $\rightarrow$ Fe oxidation state 0

     Al oxidation state 0 $\rightarrow$ $Al_2O_3$ containing Al oxidation state +3

(d) $[MnO_4]^-$ with Mn oxidation state +7 $\rightarrow$ $MnO_2$ with Mn oxidation state +4

7.3 In a redox reaction, both oxidation and reduction must occur, and the oxidation state changes must balance.

(a)          $N_2 + 3Mg \rightarrow Mg_3N_2$

Ox. state:    0     0     +2  −3      N reduced; Mg oxidized

(b)          $N_2 + O_2 \rightarrow 2NO$

Ox. state:    0     0     +2  −2      N oxidized; O reduced

(c)          $2NO_2 \rightarrow N_2O_4$

Not a redox reaction: N is in ox. state +4 on each side, O remains in ox. state −2.

(d)    $SbF_3 + F_2 \rightarrow SbF_5$

Ox. state:    +3 –1    0    +5 –1    Sb oxidized; F in $F_2$ reduced

(e)    $6HCl + As_2O_3 \rightarrow 2AsCl_3 + 3H_2O$

Not a redox reaction; ox. states on both sides are H, +1; Cl, –1; As, +3; O, –2.

(f)    $2CO + O_2 \rightarrow 2CO_2$

Ox. state:    +2 –2    0    +4 –2    C oxidized; O in $O_2$ reduced

(g)    $MnO_2 + 4HCl \rightarrow MnCl_2 + Cl_2 + 2H_2O$

Ox. state:    +4 –2    +1 –1    +2 –1    0    +1 –2

Mn reduced; Cl going to $Cl_2$ oxidized

(h)    $[Cr_2O_7]^{2-} + 2[OH]^- \rightleftharpoons 2[CrO_4]^{2-} + H_2O$

Not a redox reaction; ox. states on both sides are Cr, +6; H, +1; O, –2.

**7.4**    (a)  Increase in ox. state for Mg = 3(+2) = +6
Decrease in ox. state for N = 2(–3) = –6

(b)  Increase in ox. state for N = 2(+2) = +4
Decrease in ox. state for O  = 2(–2) = –4

(c), (e), (h) ≠ redox reactions

(d)  Increase in ox. state for Sb = +2
Decrease in ox. state for F  = 2(–1) = –2

(f)  Increase in ox. state for C = 2(+2) = +4
Decrease in ox. state for O  = 2(–2) = –4

(g)  Increase in ox. state for Cl = 2(+1) = +2
Decrease in ox. state for Mn  = –2

**7.5**    (a)    $Zn^{2+}(aq) + 2e^- \rightleftharpoons Zn(s)$    $E^o = -0.76$ V
$Ag^+(aq) + e^- \rightleftharpoons Ag(s)$    $E^o = +0.80$ V

More positive $E^o$ corresponds to the half-cell which undergoes reduction, so the cell reaction is:

$2Ag^+(aq) + Zn(s) \rightarrow 2Ag(s) + Zn^{2+}(aq)$    $E^o_{cell} = +0.80 - (-0.76)$
$= 1.56$ V

No sign is needed for $E^o_{cell}$ because for a spontaneous reaction, it is always positive.

For physical constants including $F$: see back inside cover of the book

$$\Delta G^o = - zFE^o_{cell} = - (2)(96\ 485)(1.56) \times 10^{-3} = - 301 \text{ kJ mol}^{-1}$$

Factor to convert
J mol$^{-1}$ to kJ mol$^{-1}$

(b)    $Br_2(aq) + 2e^- \rightleftharpoons 2Br^-(aq)$    $E^o = +1.09$ V
$Cl_2(aq) + 2e^- \rightleftharpoons 2Cl^-(aq)$    $E^o = +1.36$ V

More positive $E^o$ corresponds to the half-cell which undergoes reduction, so the cell reaction is:

$Cl_2(aq) + 2Br^-(aq) \rightarrow 2Cl^-(aq) + Br_2(aq)$    $E^o_{cell} = +1.36 - (+1.09)$
$= 0.27$ V

[*Check*: result makes sense since $Cl_2$ displaces $Br_2$ from aqueous soln. of Br$^-$].

$$\Delta G^o = - zFE^o_{cell} = - (2)(96\ 485)(0.27) \times 10^{-3} = - 52.1 \text{ kJ mol}^{-1}$$

(c)    $Fe^{3+}(aq) + e^- \rightleftharpoons Fe^{2+}(aq)$                                                    $E^\circ = +0.77$ V
        $[Cr_2O_7]^{2-}(aq) + 14H^+(aq) + 6e^- \rightleftharpoons 2Cr^{3+}(aq) + 7H_2O(l)$      $E^\circ = +1.33$ V
More positive $E^\circ$ corresponds to the half-cell which undergoes reduction, so the
cell reaction is:

$$[Cr_2O_7]^{2-}(aq) + 14H^+(aq) + 6Fe^{2+}(aq) \rightarrow 2Cr^{3+}(aq) + 7H_2O(l) + 6Fe^{3+}(aq)$$

$$E^\circ_{cell} = +1.33 - (+0.77) = 0.56 \text{ V}$$

$$\Delta G^\circ = -zFE^\circ_{cell} = -(6)(96\ 485)(0.56) \times 10^{-3} = -324 \text{ kJ mol}^{-1}$$

7.6    Write out the half-equations involved, and then look up the relevant $E^\circ$ values:

(a)    $Mg^{2+}(aq) + 2e^- \rightleftharpoons Mg(s)$        $E^\circ = -2.37$ V
        $2H^+(aq) + 2e^- \rightleftharpoons H_2(g)$        $E^\circ = 0$ V

$$E^\circ_{cell} = 0 - (-2.37) = 2.37 \text{ V}$$

$$\Delta G^\circ = -zFE^\circ_{cell} = -(2)(96\ 485)(2.37) \times 10^{-3} = -457 \text{ kJ mol}^{-1}$$

This large, negative value of $\Delta G^\circ$ corresponds to the spontaneous reaction:

$$Mg(s) + 2H^+(aq) \rightarrow Mg^{2+}(aq) + H_2(g)$$

For Cu:
        $2H^+(aq) + 2e^- \rightleftharpoons H_2(g)$        $E^\circ = 0$ V
        $Cu^{2+}(aq) + 2e^- \rightleftharpoons Cu(s)$        $E^\circ = +0.34$ V

$$E^\circ_{cell} = 0 - (+0.34) = -0.34 \text{ V}$$

$$\Delta G^\circ = -zFE^\circ_{cell} = -(2)(96\ 485)(-0.34) \times 10^{-3} = +65.6 \text{ kJ mol}^{-1}$$

Positive value of $\Delta G^\circ$ means that reaction of Cu with $H^+$ is thermodynamically
unfavourable. You should also look at the possibility of Cu being oxidized to $Cu^+$,
but again this gives a positive $\Delta G^\circ$.

(b)    $I_2(aq) + 2e^- \rightleftharpoons 2I^-(aq)$        $E^\circ = +0.54$ V
        $Br_2(aq) + 2e^- \rightleftharpoons 2Br^-(aq)$        $E^\circ = +1.09$ V

$$E^\circ_{cell} = 1.09 - (+0.54) = 0.55 \text{ V}$$

$$\Delta G^\circ = -zFE^\circ_{cell} = -(2)(96\ 485)(0.55) \times 10^{-3} = -106 \text{ kJ mol}^{-1}$$

Large, negative value of $\Delta G^\circ$ corresponds to the spontaneous reaction:

$$Br_2(aq) + 2I^-(aq) \rightarrow 2Br^-(aq) + I_2(aq)$$

Answer 7.5(b) showed $Cl_2$ displacing $Br_2$. $\therefore$ The reverse reaction is not spontaneous.

(c) Consider $Fe^{3+}$ reduced to $Fe^{2+}$.
The half-equation:        $Fe^{3+}(aq) + e^- \rightleftharpoons Fe^{2+}(aq)$        $E^\circ = +0.77$ V
refers to the *hexaaqua* ions.

(7.1)

(7.2)

i.e.     $[Fe(H_2O)_6]^{3+}(aq) + e^- \rightleftharpoons [Fe(H_2O)_6]^{2+}(aq)$          $E^o = +0.77$ V

Now look in Appendix 11 for other half-equations involving reduction of $Fe^{3+}$ to $Fe^{2+}$ :

$[Fe(CN)_6]^{3-}(aq) + e^- \rightleftharpoons [Fe(CN)_6]^{4-}(aq)$          $E^o = +0.36$ V
$[Fe(bpy)_3]^{3+}(aq) + e^- \rightleftharpoons [Fe(bpy)_3]^{2+}(aq)$          $E^o = +1.03$ V
$[Fe(phen)_3]^{3+}(aq) + e^- \rightleftharpoons [Fe(phen)_3]^{2+}(aq)$          $E^o = +1.12$ V

The ligands bpy and phen are shown in **7.1** and **7.2**. Ligand coordination clearly affects the ease of reduction of $Fe^{3+}$, since $E^o$ values vary significantly; $[Fe(phen)_3]^{3+}$ is the most easily reduced.
(d) This reaction was discussed in answer 7.5(a), and shown to be spontaneous. Hence, immersing a Zn foil in an aqueous solution containing $Ag^+$ ions will result in the formation of Ag solid – Ag crystals form on the surface of the Zn foil.

7.7          $[MnO_4]^-(aq) + 8H^+(aq) + 5e^- \rightleftharpoons Mn^{2+}(aq) + 4H_2O(l)$   $E^o = +1.51$ V

Nernst equation:

$$E = E^o - \left\{ \frac{RT}{zF} \times \left( \ln \frac{[\text{reduced form}]}{[\text{oxidized form}]} \right) \right\}$$

$$= +1.51 - \left\{ \frac{8.314 \times 298}{5 \times 96\,485} \times \left( \ln \frac{[Mn^{2+}]}{[MnO_4^-][H^+]^8} \right) \right\}$$

$$= +1.51 - \left\{ (5.136 \times 10^{-3}) \times \ln \frac{1}{100[H^+]^8} \right\}$$

$$= +1.51 + \left\{ (5.136 \times 10^{-3}) \times \ln 100[H^+]^8 \right\}$$

(a) At pH = 0.5:     pH = –lg [H⁺]          $[H^+] = 10^{-pH} = 0.30$ mol dm⁻³
     $E = +1.51 + \{(5.136 \times 10^{-3}) \times \ln (100)(0.30)^8\}$
          $= +1.51 - 0.03$
          $= +1.48$ V

(b) At pH = 2.0:     $[H^+] = 10^{-2.0} = 0.010$ mol dm⁻³
     $E = +1.51 + \{(5.136 \times 10^{-3}) \times \ln (100)(0.010)^8\}$
          $= +1.51 - 0.17$
          $= +1.34$ V

(c) At pH = 3.5:     $[H^+] = 10^{-3.5} = 3.2 \times 10^{-4}$ mol dm⁻³
     $E = +1.51 + \{(5.136 \times 10^{-3}) \times \ln (100)(3.2 \times 10^{-4})^8\}$
          $= +1.51 - 0.31$
          $= +1.20$ V
As pH changes from 0.5 to 3.5, $[MnO_4]^-$ becomes a poorer oxidizing agent (less positive $E$):

$I_2(aq) + 2e^- \rightleftharpoons 2I^-(aq)$          $E^o = +0.54$ V
$Br_2(aq) + 2e^- \rightleftharpoons 2Br^-(aq)$          $E^o = +1.09$ V
$Cl_2(aq) + 2e^- \rightleftharpoons 2Cl^-(aq)$          $E^o = +1.36$ V

At pH 0.5, $[MnO_4]^-$ will oxidize all three halides; at pH 2.0 and 3.5, it will not oxidize $Cl^-$, but will still oxidize $Br^-$ and $I^-$.

7.8    (a) Half-equations needed are:

$$O_2(g) + 2H^+(aq) + 2e^- \rightleftharpoons H_2O_2(aq) \qquad E^\circ = +0.70 \text{ V}$$
$$H_2O_2(aq) + 2H^+(aq) + 2e^- \rightleftharpoons 2H_2O(l) \qquad E^\circ = +1.78 \text{ V}$$

$$E^\circ_{cell} = +1.78 - (+0.70) = 1.08 \text{ V}$$

Overall reaction is:    $2H_2O_2(aq) \rightarrow 2H_2O(l) + O_2(g)$

(b)    $\Delta G^\circ = -zFE^\circ_{cell} = -(2)(96\ 485)(1.08) \times 10^{-3} = -208 \text{ kJ mol}^{-1}$

(c) $H_2O_2$ is thermodynamically unstable with respect to disproportionation, but is kinetically stable (high activation barrier). The reaction is therefore slow, but in the presence of a catalyst, disproportionation occurs rapidly.

7.9    From the Nernst equation:

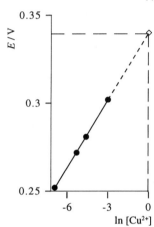

Activity of solid = 1

$$E = E^\circ - \left\{ \frac{RT}{zF} \times \left( \ln \frac{[\text{reduced form}]}{[\text{oxidized form}]} \right) \right\} = E^\circ - \left\{ \frac{RT}{2F} \times \left( \ln \frac{[Cu(s)]}{[Cu^{2+}]} \right) \right\}$$

$$\therefore \; E = E^\circ + 0.0128 \ln [Cu^{2+}] \qquad \text{(at 298 K)}$$

Plot values of $E$ against $\ln [Cu^{2+}]$ — the four ● points in Figure 7.1.
Extrapolate the line to $\ln [Cu^{2+}] = 0$ (i.e. $[Cu^{2+}] = 1$) to find $E^\circ$ as shown in Figure 7.1.

$$E^\circ = +0.34 \text{ V}$$

Data points are subject to experimental error, therefore *all* data should be used.

**Figure 7.1** Plot of data for answer 7.9. The dashed line is an extrapolation to find $E^\circ$.

7.10    (a)    $$E = E^\circ - \left\{ \frac{RT}{zF} \times \left( \ln \frac{[\text{reduced form}]}{[\text{oxidized form}]} \right) \right\} = E^\circ - \left\{ \frac{RT}{F} \times \left( \ln \frac{[Ag(s)]}{[Ag^+]} \right) \right\}$$

$= 1$

$$\therefore \; E = +0.80 - \left( \frac{8.314 \times 298}{96\ 485} \right) \left( \ln \frac{1}{0.1} \right)$$

$$= +0.74 \text{ V}$$

(b)    $Zn^{2+}(aq) + 2e^- \rightleftharpoons Zn(s) \qquad E^\circ = -0.76 \text{ V}$
       $Ag^+(aq) + e^- \rightleftharpoons Ag(s) \qquad E = +0.74 \text{ V}$ (0.1 mol dm$^{-3}$)
       $\qquad\qquad\qquad\qquad\qquad\qquad E^\circ = +0.80 \text{ V}$ (1.0 mol dm$^{-3}$)

Under standard conditions:    $E^\circ_{cell} = +0.80 - (-0.76) = 1.56 \text{ V}$
$\qquad\qquad\qquad\qquad\qquad \Delta G^\circ = -(2)(96\ 485)(1.56) \times 10^{-3} = -301 \text{ kJ mol}^{-1}$

In 0.1 M solution:    $E_{cell} = +0.74 - (-0.76) = 1.50 \text{ V}$
$\qquad\qquad\qquad\qquad \Delta G = -(2)(96\ 485)(1.50) \times 10^{-3} = -289 \text{ kJ mol}^{-1}$

Less negative $\Delta G$ in 0.1 M solution. $\therefore$ Less easily reduced (but still spontaneous).

7.11   For AgI, $K_{sp} = 8.51 \times 10^{-17}$ and this refers to the equilibrium:
$$AgI(s) \rightleftharpoons Ag^+(aq) + I^-(aq)$$
For:
$$Ag^+(aq) + e^- \rightleftharpoons Ag(s) \qquad E^o = +0.80 \text{ V}$$

Set up a thermochemical cycle that combines these equilibria with the reduction step in the question, and let $\Delta_r G^o$ be the free energy change for this step:

Strictly, these are all equilibria, but single arrows are shown to give the thermochemical cycle a sense of 'direction'

Applying Hess's Law:    $\Delta_r G^o = \Delta_{sol} G^o(AgI,s) + \Delta G^o(Ag^+/Ag)$

Find $\Delta_{sol} G^o(AgI,s)$ from $K_{sp}$:

$$\Delta_{sol} G^o(AgI,s) = -RT\ln K_{sp} = -(8.314)(298)\ln (8.51 \times 10^{-17}) \times 10^{-3}$$
$$= +91.7 \text{ kJ mol}^{-1}$$

Find $\Delta G^o(Ag^+/Ag)$ from $E^o$:

$$\Delta G^o(Ag^+/Ag) = -zFE^o = -(1)(96\,485)(+0.80) \times 10^{-3} = -77.2 \text{ kJ mol}^{-1}$$

$$\therefore \; \Delta_r G^o = \Delta_{sol} G^o(AgI,s) + \Delta G^o(Ag^+/Ag)$$

$$= 91.7 - 77.2 = +14.5 \text{ kJ mol}^{-1}$$

To find $E^o$:    $E^o = -\dfrac{\Delta_r G^o}{zF} = -\dfrac{14.5}{96\,485 \times 10^{-3}} = -0.15 \text{ V}$

For the reduction of AgCl, equation 7.31 in H&S gives $\Delta_r G^o = -21.6 \text{ kJ mol}^{-1}$, and so, the reduction of AgI (for which $\Delta_r G^o = +14.5 \text{ kJ mol}^{-1}$) is thermodynamically less favourable than reduction of AgCl.

7.12   Ag(s) does not liberate $H_2$ from dilute mineral acids:

$$H^+(aq) + e^- \rightleftharpoons \tfrac{1}{2}H_2(g) \qquad E^o = 0 \text{ V}$$
$$Ag^+(aq) + e^- \rightleftharpoons Ag(s) \qquad E^o = +0.80 \text{ V}$$

In the presence of excess $I^-$, (i.e. present in conc. HI solution), $Ag^+$ forms the very stable complex $[AgI_3]^{2-}$. So, the half-equations to consider are:

See equations 7.32 and 7.33 in H&S, and related discussion

$$[AgI_3]^{2-}(aq) + e^- \rightleftharpoons Ag(s) + 3I^-(aq) \qquad E^o = -0.03 \text{ V}$$
$$H^+(aq) + e^- \rightleftharpoons \tfrac{1}{2}H_2(g) \qquad E^o = 0 \text{ V}$$

For the overall reaction:    $Ag(s) + H^+(aq) + 3I^-(aq) \rightleftharpoons [AgI_3]^{2-}(aq) + \tfrac{1}{2}H_2(g)$
$E^o_{cell} = 0.03$ V, and $\Delta_r G^o \approx -3 \text{ kJ mol}^{-1}$. Use of conc. HI means $[H^+]$ is very large, and so $\Delta_r G$ will be much more negative than $\Delta_r G^o$. Therefore, reaction will be thermodynamcially favoured at 298 K; heating removes $H_2$ from solution.

7.13    Construct a thermochemical cycle using the half-equations given in the question, and equations for the formation of the complex ions:

➤

Strictly, these are all equilibria, but single arrows are shown to give the thermochemical cycle a sense of 'direction'

$$Fe^{3+}(aq) + e^- \xrightarrow{\Delta G^\circ_2} Fe^{2+}(aq)$$

$$\Delta G^\circ_1 \downarrow \qquad\qquad\qquad \downarrow \Delta G^\circ_3$$

$$[Fe(CN)_6]^{3-}(aq) + e^- \xrightarrow{\Delta G^\circ_4} [Fe(CN)_6]^{4-}(aq)$$

Values of $\Delta G^\circ_2$ and $\Delta G^\circ_4$ can be found from the $E^\circ$ values in the question:

$$\Delta G^\circ_2 = -zFE^\circ = -(1)(96\,485)(+0.77) \times 10^{-3} = -74 \text{ kJ mol}^{-1}$$
$$\Delta G^\circ_4 = -zFE^\circ = -(1)(96\,485)(+0.36) \times 10^{-3} = -35 \text{ kJ mol}^{-1}$$

For the formation of $[Fe(CN)_6]^{4-}$, $K \approx 10^{35}$:

$$\therefore \ \Delta G^\circ_3 = -RT \ln K = -(8.314)(298)\ln (10^{35}) \times 10^{-3} = -200 \text{ kJ mol}^{-1}$$

From the cycle above:    $\Delta G^\circ_1 = \Delta G^\circ_2 + \Delta G^\circ_3 - \Delta G^\circ_4 = -74 - 200 - (-35)$
$$= -239 \text{ kJ mol}^{-1}$$

For the overall formation constant of $[Fe(CN)_6]^{3-}$, use:

$$\ln K = -\frac{\Delta G^\circ}{RT} \qquad \text{or} \quad K = e^{-\frac{\Delta G^\circ}{RT}}$$

$$K = e^{-\frac{\Delta G^\circ}{RT}} = e^{-\frac{(-239)}{(8.314 \times 10^{-3})(298)}}$$

$$\therefore \quad K = 7.8 \times 10^{41} \approx 10^{42}$$

7.14    (a) In Appendix 11, look for half-equations containing $Fe^{2+}$ in aqueous solution:

$$Fe^{2+}(aq) + 2e^- \rightleftharpoons Fe(s) \qquad E^\circ = -0.44 \text{ V}$$
$$Fe^{3+}(aq) + e^- \rightleftharpoons Fe^{2+}(aq) \qquad E^\circ = +0.77 \text{ V}$$

Disproportionation of $Fe^{2+}$:        $3Fe^{2+}(aq) \rightarrow Fe(s) + 2Fe^{3+}(aq)$

for which $E^\circ_{cell}$ is negative and $\Delta G^\circ$ is positive; the reaction is *not* thermodynamically favoured and $Fe^{2+}$ is stable with respect to disproportionation.

(b) Half-equations containing $Sn^{2+}$ in aqueous solution:

$$Sn^{2+}(aq) + 2e^- \rightleftharpoons Sn(s) \qquad E^\circ = -0.14 \text{ V}$$
$$Sn^{4+}(aq) + 2e^- \rightleftharpoons Sn^{2+}(aq) \qquad E^\circ = +0.15 \text{ V}$$

Disproportionation of $Sn^{2+}$:        $2Sn^{2+}(aq) \rightarrow Sn(s) + Sn^{4+}(aq)$

for which $E^\circ_{cell}$ is negative and $\Delta G^\circ$ is positive; the reaction is *not* thermodynamically favoured and $Sn^{2+}$ is stable with respect to disproportionation.

(c) Half-equations containing $[ClO_3]^-$ in aqueous solution:

$$[ClO_4]^-(aq) + H_2O(l) + 2e^- \rightleftharpoons [ClO_3]^-(aq) + 2[OH]^-(aq)$$
$$E^o = +0.36 \text{ V}$$
$$[ClO_3]^-(aq) + 6H^+(aq) + 6e^- \rightleftharpoons Cl^-(aq) + 3H_2O(l) \qquad E^o = +1.45 \text{ V}$$
$$2[ClO_3]^-(aq) + 12H^+(aq) + 10e^- \rightleftharpoons Cl_2(g) + 6H_2O(l) \qquad E^o = +1.47 \text{ V}$$

Two reactions are possible for the disproportionation of Cl in ox. state +5, i.e. $[ClO_3]^-$.

(i)

$$4Cl(+5) \rightarrow Cl(-1) + 3Cl(+7)$$

i.e.    $4[ClO_3]^-(aq) \rightarrow Cl^-(aq) + 3[ClO_4]^-(aq)$

➤ See equation 16.65 in H&S and accompanying discussion

for which $E^o_{cell}$ is positive and $\Delta G^o$ is negative; the reaction is thermodynamically favoured and $[ClO_3]^-$ is unstable with respect to disproportionation.

(ii)

$$7Cl(+5) \rightarrow 2Cl(0) + 5Cl(+7)$$

i.e.    $7[ClO_3]^-(aq) + 2H^+(aq) \rightarrow Cl_2(g) + H_2O(l) + 5[ClO_4]^-(aq)$

for which $E^o_{cell}$ is positive and $\Delta G^o$ is negative, showing that this reaction is also thermodynamically favourable.

7.15    Construct an appropriate thermochemical cycle:

$$2CuCl(s) \xrightarrow{\Delta G^o_1} Cu^{2+}(aq) + 2Cl^-(aq) + Cu(s)$$

$\Delta G^o_2 \searrow \qquad \nearrow \Delta G^o_3$

$$2Cu^+(aq) + 2Cl^-(aq)$$

Find $\Delta G^o_2$ from $K_{sp}$, noting that the equation above concerns 2 moles of CuCl:

$$\Delta G^o_2 = 2(-RT\ln K_{sp}) = 2 \times \{-(8.314)(298)\ln (1.72 \times 10^{-7}) \times 10^{-3}\}$$
$$= 77.2 \text{ kJ mol}^{-1}$$

Find $\Delta G^o_3$ from $K = 1.81 \times 10^6$ :

$$\Delta G^o_3 = -RT\ln K_{sp} = -(8.314)(298)\ln (1.81 \times 10^6) \times 10^{-3}$$
$$= -35.7 \text{ kJ mol}^{-1}$$

Hence, to find $\Delta G^o_1$ :

$$\Delta G^o_1 = \Delta G^o_2 + \Delta G^o_3 = +77.2 - 35.7 = +41.5 \text{ kJ per mole of reaction}$$

The positive value of $\Delta G^o$ means that the reaction is thermodynamically unfavourable, and precipitated CuCl is stable with respect to disproportionation.

7.16    The data available are:

$$Mn^{2+}(aq) + 2e^- \rightleftharpoons Mn(s) \qquad\qquad E^\circ = -1.19 \text{ V}$$
$$[MnO_4]^-(aq) + e^- \rightleftharpoons [MnO_4]^{2-}(aq) \qquad\qquad E^\circ = +0.56 \text{ V}$$
$$MnO_2(s) + 4H^+(aq) + 2e^- \rightleftharpoons Mn^{2+}(aq) + 2H_2O(l) \qquad E^\circ = +1.23 \text{ V}$$
$$[MnO_4]^-(aq) + 8H^+(aq) + 5e^- \rightleftharpoons Mn^{2+}(aq) + 4H_2O(l) \quad E^\circ = +1.51 \text{ V}$$
$$Mn^{3+}(aq) + e^- \rightleftharpoons Mn^{2+}(aq) \qquad\qquad E^\circ = +1.54 \text{ V}$$

and the equation for which $E^\circ$ is to be found is:

$$[MnO_4]^-(aq) + 4H^+(aq) + 3e^- \rightleftharpoons MnO_2(s) + 2H_2O(l)$$

The best way to proceed is to construct a potential diagram from the data which refer to acidic solution:

You cannot find $E^\circ$ directly, but must use appropriate $\Delta G^\circ$ values.

$$\Delta G^\circ_2 = -zFE^\circ = -(2)(96\,485)(1.23) \times 10^{-3} = -237 \text{ kJ mol}^{-1}$$
$$\Delta G^\circ_3 = -zFE^\circ = -(5)(96\,485)(1.51) \times 10^{-3} = -728 \text{ kJ mol}^{-1}$$

$$\Delta G^\circ_1 = \Delta G^\circ_3 - \Delta G^\circ_2 = -491 \text{ kJ mol}^{-1}$$

$$\therefore E^\circ = -\frac{\Delta G^\circ}{zF} = -\frac{(-491)}{(3)(96\,485) \times 10^{-3}} = 1.69 \text{ V}$$

The value in equation 7.47 in H&S is confirmed. By looking at the above calculation, you will see that $F$ could be carried through the calculation without evaluation, and the factor $10^{-3}$ also cancels out and can be ignored. So, a short cut method is:

$$\Delta G^\circ_2 = -zFE^\circ = -(2)(1.23)F = -2.46F \text{ J mol}^{-1}$$
$$\Delta G^\circ_3 = -zFE^\circ = -(5)(1.51)F = -7.55F \text{ J mol}^{-1}$$

$$\Delta G^\circ_1 = \Delta G^\circ_3 - \Delta G^\circ_2 = -5.09F \text{ J mol}^{-1}$$

$$\therefore E^\circ = -\frac{\Delta G^\circ}{zF} = -\frac{(-5.09)F}{3F} = 1.69 \text{ V}$$

7.17    Potential diagrams for Mn are drawn out in Figure 7.2 and each equilibrium is coded with an equation letter; the corresponding equilibria are:
In acidic solution:
(a)    $[MnO_4]^-(aq) + H^+(aq) + e^- \rightleftharpoons [HMnO_4]^-(aq)$
(b)    $[HMnO_4]^-(aq) + 3H^+(aq) + 2e^- \rightleftharpoons MnO_2(s) + 2H_2O(l)$
(c)    $MnO_2(s) + 4H^+(aq) + e^- \rightleftharpoons Mn^{3+}(aq) + 2H_2O(l)$
(d)    $Mn^{3+}(aq) + e^- \rightleftharpoons Mn^{2+}(aq)$
(e)    $Mn^{2+}(aq) + 2e^- \rightleftharpoons Mn(s)$

**Acidic solution (pH 0)**

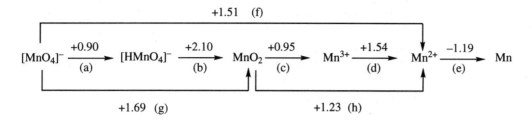

$$+1.69 \ (g) \qquad\qquad\qquad +1.23 \ (h)$$

**Alkaline solution (pH 14)**

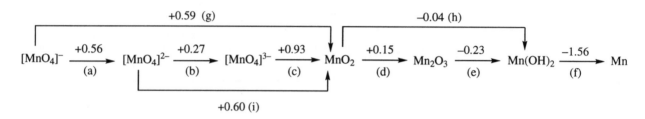

$$+0.60 \ (i)$$

**Figure 7.2**   Potential diagrams for Mn (values in V) in acidic (pH 0) and alkaline (pH 14) solutions.

(f)    $[MnO_4]^-(aq) + 8H^+(aq) + 5e^- \rightleftharpoons Mn^{2+}(aq) + 4H_2O(l)$
(g)    $[MnO_4]^-(aq) + 4H^+(aq) + 3e^- \rightleftharpoons MnO_2(s) + 2H_2O(l)$
(h)    $MnO_2(s) + 4H^+(aq) + 2e^- \rightleftharpoons Mn^{2+}(aq) + 2H_2O(l)$

In alkaline solution:
(a)    $[MnO_4]^-(aq) + e^- \rightleftharpoons [MnO_4]^{2-}(aq)$
(b)    $[MnO_4]^{2-}(aq) + e^- \rightleftharpoons [MnO_4]^{3-}(aq)$
(c)    $[MnO_4]^{3-}(aq) + 2H_2O(l) + e^- \rightleftharpoons MnO_2(s) + 4[OH]^-(aq)$
(d)    $2MnO_2(s) + H_2O(l) + 2e^- \rightleftharpoons Mn_2O_3(s) + 2[OH]^-(aq)$
(e)    $Mn_2O_3(s) + 3H_2O(l) + 2e^- \rightleftharpoons 2Mn(OH)_2(s) + 2[OH]^-(aq)$
(f)    $Mn(OH)_2(s) + 2e^- \rightleftharpoons Mn(s) + 2[OH]^-(aq)$
(g)    $[MnO_4]^-(aq) + 2H_2O(l) + 3e^- \rightleftharpoons MnO_2(s) + 4[OH]^-(aq)$
(h)    $MnO_2(s) + 2H_2O(l) + 2e^- \rightleftharpoons Mn(OH)_2(s) + 2[OH]^-(aq)$
(i)    $[MnO_4]^{2-}(aq) + 2H_2O(l) + 2e^- \rightleftharpoons MnO_2(s) + 4[OH]^-(aq)$

7.18    (a) Approach the question by first identifying all the half-equations (Appendix 11 in H&S) which involve vanadium-containing species in acidic solution. Then, determine the oxidation state of V in each – in the potential diagram, the species are arranged in order of decreasing oxidation state. Each step in the diagram corresponds to one reduction half-equation:

$$[VO_2]^+ \xrightarrow{+0.99} [VO]^{2+} \xrightarrow{+0.34} V^{3+} \xrightarrow{-0.26} V^{2+} \xrightarrow{-1.18} V$$

(b) The only species which need to be considered with respect to possible disproportionation are $[VO]^{2+}$, $V^{3+}$ and $V^{2+}$. By inspection, you can see no species is unstable with respect to disproportionation because the values of $E^\circ$ become sequentially more negative on going from left to right across the diagram. If you are unable to see why this logic follows, consider all possible pairs of half-reactions that could lead to disproportionation, and work out $E^{\circ}_{cell}$ for each resultant reaction. $E^{\circ}_{cell}$ must be positive (and, thus, $\Delta G^{\circ}$ negative) for a reaction to be spontaneous.

7.19    The potential diagram for the question refers to conditions of pH 0:

$$[UO_2]^{2+} \xrightarrow{+0.06} [UO_2]^{+} \xrightarrow{+0.61} U^{4+} \xrightarrow{-0.61} U^{3+} \xrightarrow{-1.80} U$$

$$+0.33$$

Points to include in your answer:
- oxidation states accessible are +6, +5, +4, +3 and 0;
- strongest oxidant among the species listed is $[UO_2]^{+}$;
- both $[UO_2]^{+}$ and $[UO_2]^{2+}$ are mild oxidizing agents;
- strongest reducing agent of species listed is U metal, and the large negative $E^{\circ}$ shows it to be a strong reducing agent;
- $U^{3+}$ is a relatively strong reducing agent;
- $[UO_2]^{+}$ is unstable with respect to disproportionation into $[UO_2]^{2+}$ and $U^{4+}$;
- U should react with aqueous $H^{+}$ to liberate $H_2$ and form $U^{4+}$ (thermodynamically more favourable than forming $U^{3+}$);
- $O_2$ should oxidize all forms of U at pH 0 to give $[UO_2]^{2+}$;
- half-equations involving $[UO_2]^{2+}$ and $[UO_2]^{+}$ will involve $H^{+}$ and so the $E^{\circ}$ values are pH dependent.

7.20    The potential diagram for the question is:

$$[ClO_3]^{-} \xrightarrow{+1.15} ClO_2 \xrightarrow{+1.28} HClO_2$$

$$E^{\circ}$$

(a) Follow a rigorous method to ensure no error is made with numbers of electrons, although you can omit the J to kJ conversion, and also work with the constant $F$ without evaluation as explained in answer 7.16.

For $[ClO_3]^{-}$ to $ClO_2$:    $\Delta G^{\circ}_1 = -zFE^{\circ} = -(1)(1.15)F = -1.15F$ J mol$^{-1}$

For $ClO_2$ to $HClO_2$:    $\Delta G^{\circ}_2 = -zFE^{\circ} = -(1)(1.28)F = -1.28F$ J mol$^{-1}$

For $[ClO_3]^{-}$ to $HClO_2$:    $\Delta G^{\circ} = \Delta G^{\circ}_1 + \Delta G^{\circ}_2 = -2.43F$ J mol$^{-1}$

$$\therefore E^{\circ} = -\frac{\Delta G^{\circ}}{zF} = -\frac{(-2.43)F}{2F} = 1.215 \approx 1.22 \text{ V}$$

(b) In this example, each separate step is a one-electron reduction, and the overall reaction is a two-electron reduction. Hence, it is possible to take a short cut and take the mean of +1.15 and +1.28 V to find $E^{\circ}$. Such short cuts should only be made with caution – it is all too easy to make a mistake!

7.21    An appropriate thermochemical cycle for the alkali metals, M, is:

$$\begin{array}{ccc} & \Delta H^\circ(M^+/M) & \\ M^+(aq) + e^- & \xrightarrow{\hspace{2cm}} & M(s) \\ \Big\downarrow {\scriptstyle -\Delta_{hyd}H^\circ(M^+,g)} & & \Big\uparrow {\scriptstyle -\Delta_a H^\circ} \\ M^+(g) & \xrightarrow[\hspace{1cm}-IE_1\hspace{1cm}]{} & M(g) \end{array}$$

Using enthalpy values (and the usual assumption that $IE \approx$ associated $\Delta H$ value), we can estimate $\Delta H^\circ(M^+,M)$ from:

$$\Delta H^\circ(M^+/M) = -\Delta_{hyd}H^\circ(M^+,g) - IE_1 - \Delta_a H^\circ$$

Trends in each of $-\Delta_{hyd}H^\circ(M^+,g)$ (from Table 6.6 in H&S), $-IE_1$ (from Appendix 8 in H&S) and $-\Delta_a H^\circ$ (from Appendix 10 in H&S) for the group 1 metals are plotted in Figure 7.3. The trend in the *sum* of these quantities is plotted in Figure 7.4 and is not regular. This trend agrees quite well with that for values of $\Delta G^\circ(M^+/M)$ (shown in Figure 7.5) calculated from the $E^\circ$ values (from Appendix 11 in H&S); the trend in $E^\circ$ values is shown in Figure 7.6.

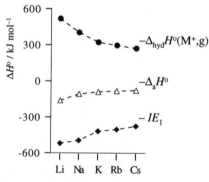

**Figure 7.3**   Plots of $-\Delta_{hyd}H^\circ(M^+,g)$, $-\Delta_a H^\circ$ and $-IE_1$ for the group 1 metals.

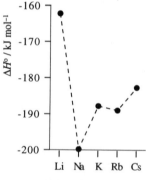

**Figure 7.4**   Plot of the sum $\{-\Delta_{hyd}H^\circ(M^+,g) - \Delta_a H^\circ - IE_1\}$ for the group 1 metals.

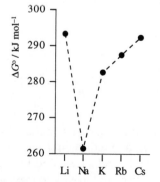

**Figure 7.5**   The trend in values of $\Delta G^\circ(M^+/M)$ for the group 1 metals.

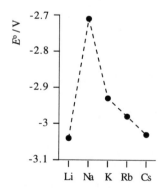

**Figure 7.6**   The trend in values of $E^\circ(M^+/M)$ for the group 1 metals.

# 8 Non-aqueous media

8.1 (a) Many possible solvents for this answer, e.g.
- hydrocarbons such as hexane
- $CH_2Cl_2$
- ethers such as $Et_2O$, THF and diglyme
- acetonitrile

Many reactions that you could choose, or choose a general type, e.g. Grignard reagents in ether solvents.

(b) Points to include in your answer:
- aqueous media widely used in inorganic preparative chemistry, limitations being for compounds that react with water, e.g. organometallic complexes;
- non-aqueous media include use of common solvents such as those in part (a) as well as 'exotic' solvents such as $BrF_3$, $NH_3$;
- in organic synthesis, aqueous media not so commonly used.

8.2 For 2 unit electronic charges, the coulombic potential energy is $\dfrac{e^2}{4\pi\varepsilon_0 r}$

where $\varepsilon_0$ is the absolute permittivity of a vacuum, $e$ is the charge on the electron and $r$ is the separation between the point charges. Putting a substance between the

point charges changes the coulombic potential energy to $\dfrac{e^2}{4\pi\varepsilon_0\varepsilon_r r}$

where $\varepsilon_r$ is the relative permittivity (dielectric constant) of the substance; $\varepsilon_r$ is dimensionless. In terms of solvents, one with high dielectric constant, e.g. $H_2O$, has the effect of significantly reducing the interaction between point charges (e.g. ions in solution) relative to the interaction between them in a vacuum. Large values of $\varepsilon_r$ are observed for polar solvents (e.g. $H_2O$ and $NH_3$) or those that are readily polarized. For solvents with related structures, the trend in values of $\varepsilon_r$ follows the trend in dipole moments ($\mu$). In general, solvents of high $\varepsilon_r$ favour ionic reactions, while reactions which do not involve ions are favoured by solvents of low $\varepsilon_r$. However, solvation often involves *specific* interactions rather than bulk properties, so predictions on the basis of $\varepsilon_r$ are not always reliable.

8.3 Consider each structure, looking for electronegative atoms which may make the molecule polar: for answers (a) to (f), and (h) to (j), see diagrams **8.1** to **8.9**. (g) and (k): hexane and benzene are non-polar.

**(8.1)**  **(8.2)**  **(8.3)**  **(8.4)**  **(8.5)**

**(8.6)**  **(8.7)**  **(8.8)**  **(8.9)**

8.4    In each, consider that $[NH_2]^-$ can act as a base in liquid $NH_3$, and $[NH_4]^+$ can act as an acid.

(a) $ZnI_2 + 2KNH_2 \rightarrow 2KI + Zn(NH_2)_2$

(b) $Zn(NH_2)_2 + 2KNH_2 \rightarrow K_2[Zn(NH_2)_4]$    (compare to the dissolution of $Zn(OH)_2$ in aqueous KOH)

(c) $Mg_2Ge + 4NH_4Br \rightarrow GeH_4 + 2MgBr_2 + 4NH_3$    (action of an acid on metal germide)

(d) $CH_3CO_2H + NH_3 \rightarrow [NH_4]^+ + [CH_3CO_2]^-$    ($NH_3$ acting as proton acceptor)

(e) $O_2 \xrightarrow{\text{Na in liquid } NH_3} Na_2O_2$ or $NaO_2$    (Na in liq. $NH_3$ = reducing agent forming $[O_2]^{2-}$ or $[O_2]^-$)

(f) $HC{\equiv}CH + KNH_2 \rightarrow K[HC{\equiv}C] + NH_3$    ($[NH_2]^-$ deprotonates terminal alkyne)

In aqueous solution, $CH_3CO_2H$ only partially dissociates, but fully ionizes in liquid $NH_3$.

8.5    $NH_3$ acts as an acid (compare Zn liberating $H_2$ from a mineral acid) giving $Zn(NH_2)_2$ which then reacts with excess $[NH_2]^-$ provided by sodium amide:

$$Zn + 2NaNH_2 + 2NH_3 \rightarrow Na_2[Zn(NH_2)_4] + H_2$$

$[NH_4]^+$ reacts with $[NH_2]^-$ (acid-base reaction) and as a result, $Zn(NH_2)_2$ precipitates:

$$[Zn(NH_2)_4]^{2-} + 2[NH_4]^+ \rightarrow Zn(NH_2)_2 + 4NH_3$$

Acid-base reaction again, followed by coordination of $NH_3$ to $Zn^{2+}$:

$$Zn(NH_2)_2 + 2NH_4I \rightarrow [Zn(NH_3)_4]^{2+}[I^-]_2$$

(b) In water:   $2K + 2H_2O \rightarrow 2KOH + H_2$

In liquid $NH_3$, at low concentrations:

$$K \xrightarrow{\text{Liquid } NH_3} K^+(NH_3) + e^-(NH_3)$$    i.e. solvated ion and electron

On standing:   $2NH_3 + 2e^- \rightarrow 2[NH_2]^- + H_2$

8.6    In each case, make appropriate replacements, e.g. $[NH_2]^-$ for $[OH]^-$, $NH_3$ for $H_2O$: As a starting point, structures of each oxygen-containing species (excluding the metal oxide HgO) are drawn in **8.10** to **8.14**.
(a) $H_2NNH_2$; (b) a nitride: $Hg_3N_2$; (c) $O_2NNH_2$; (d) $MeNH_2$; (e) $OC(NH_2)_2$; (f) $[Cr(NH_3)_6]Cl_3$.

(8.10)    (8.11)    (8.12)    (8.13)    (8.14)

8.7    AlF$_3$ is a Lewis acid and, in the presence of F$^-$, forms the adduct [AlF$_4$]$^-$ :

$$AlF_3 + NaF \rightarrow Na[AlF_4]$$

and the ionic salt Na[AlF$_4$] is soluble in liquid HF.
BF$_3$ is a stronger fluoride acceptor than AlF$_3$, and accepts F$^-$ from [AlF$_4$]$^-$ :

$$Na[AlF_4] + BF_3 \rightarrow AlF_3 + Na[BF_4]$$

In the absence of excess F$^-$, AlF$_3$ precipitates.

For acid-base chemistry, the self-ionization you need to consider is:

$$3HF \rightleftharpoons [H_2F]^+ + [HF_2]^-$$

8.8    Some species in the question act as donors or acceptors of F$^-$ with respect to HF.
(a) ClF$_3$ donates F$^-$ to HF:

$$ClF_3 + HF \rightarrow [ClF_2]^+ + [HF_2]^-$$

(b) Acid-base reaction with HF acting as the acid:

$$MeOH + 2HF \rightarrow [MeOH_2]^+ + [HF_2]^-$$

(c) Acid-base reaction with HF acting as the acid:

$$Et_2O + 2HF \rightarrow [Et_2OH]^+ + [HF_2]^-$$

(d) CsF donates F$^-$ to HF:

$$CsF + HF \rightarrow Cs^+ + [HF_2]^-$$

(e) SrF$_2$ donates F$^-$ to HF:

$$SrF_2 + 2HF \rightarrow Sr^{2+} + 2[HF_2]^-$$

(f) Acid-base reaction, with HClO$_4$ acting as the acid:

$$HClO_4 + HF \rightarrow [H_2F]^+ + [ClO_4]^-$$

8.9    (a) Monobasic, therefore donates one proton:

$$H_2S_2O_7 + H_2SO_4 \rightleftharpoons [H_3SO_4]^+ + [HS_2O_7]^-$$

(b) A value of $K = 1.4 \times 10^{-2}$ corresponds to significant dissociation, and so H$_2$S$_2$O$_7$ acts as a relatively strong acid.

8.10    Need to consider relative H$^+$ donor/acceptor abilities of species with respect to H$_2$SO$_4$.
(a) H$_2$O will act as a base (H$_2$SO$_4$ as an acid):

$$H_2O + H_2SO_4 \rightarrow [H_3O]^+ + [HSO_4]^-$$

(b) $NH_3$ will act as a base ($H_2SO_4$ as an acid):

$$NH_3 + H_2SO_4 \rightarrow [NH_4]^+ + [HSO_4]^-$$

(c) $HCO_2H$ will act as a base, and acceptance of $H^+$ triggers the decomposition:

$$HCO_2H + H_2SO_4 \rightarrow CO + [H_3O]^+ + [HSO_4]^-$$

For parts (d) and (e):

> For more details of $v$ and $\gamma$,
> see Section 8.8 in H&S

$v$ = total number of particles produced per molecule of solute
$\gamma$ = number of $[H_3SO_4]^+$ or $[HSO_4]^-$ ions produced per molecule of solute

(d) $H_3PO_4$ will act as a base ($H_2SO_4$ as an acid), and since $v = 2$ and $\gamma = 1$:

$$H_3PO_4 + H_2SO_4 \rightarrow [H_4PO_4]^+ + [HSO_4]^-$$

(e) HCl will act as an acid ($H_2SO_4$ as a base), and since $v = 3$ and $\gamma = 1$:

$$HCl + 2H_2SO_4 \rightarrow HOSO_2Cl + [H_3O]^+ + [HSO_4]^-$$

8.11    Points that should be included:
  • aqueous solution, strong acid, fully dissociated:

$$HNO_3 + H_2O \rightarrow [H_3O]^+ + [NO_3]^-$$

  • in $H_2SO_4$, acts as a base but reaction involves N–O cleavage:

$$HNO_3 + H_2SO_4 \rightleftharpoons [H_2NO_3]^+ + [HSO_4]^-$$

$$[H_2NO_3]^+ \rightleftharpoons [NO_2]^+ + H_2O$$

$$H_2O + H_2SO_4 \rightleftharpoons [H_3O]^+ + [HSO_4]^-$$

Overall:
$$HNO_3 + 2H_2SO_4 \rightarrow [NO_2]^+ + [H_3O]^+ + 2[HSO_4]^-$$

Inorganic reaction examples: any in which dilute aqueous $HNO_3$ is used as a mineral acid, e.g.

$$Mg + 2HNO_3 \rightarrow Mg(NO_3)_2 + H_2$$

$$CaCO_3 + 2HNO_3 \rightarrow Ca(NO_3)_2 + H_2O + CO_2$$

$$NaOH + HNO_3 \rightarrow NaNO_3 + H_2O$$

Organic nitration reactions use $HNO_3$ in $H_2SO_4$, with nitrating agent being $[NO_2]^+$, e.g.

$$C_6H_6 \xrightarrow{HNO_3,\ H_2SO_4} C_6H_5NO_2$$

8.12    (a) In liquid HCl, HCl acts as an acid:

$$Ph_2C=CH_2 + HCl \rightleftharpoons [Ph_2CCH_3]^+ + Cl^-$$

This equilibrium is then upset by the addition of $BCl_3$ which acts as a chloride acceptor, removing $Cl^-$ from the right-hand side and causing complete reaction to occur:

$$Cl^- + BCl_3 \rightarrow [BCl_4]^-$$

(b) Self-ionization of $N_2O_4$ is:

$$N_2O_4 \rightleftharpoons [NO]^+ + [NO_3]^-$$

Using the values of $v = 6$ and $\gamma = 3$, reactions that follow in $H_2SO_4$ are:

$$[NO_3]^- + H_2SO_4 \rightleftharpoons [NO_2]^+ + [HSO_4]^- + [OH]^-$$

$$[OH]^- + 2H_2SO_4 \rightleftharpoons [H_3O]^+ + 2[HSO_4]^-$$

giving an overall reaction of:

$$N_2O_4 + 3H_2SO_4 \rightleftharpoons [NO]^+ + [NO_2]^+ + [H_3O]^+ + 3[HSO_4]^-$$

8.13    $[BrF_2]^+$

> More about VSEPR theory in Chapter 1 (answer 1.29)

Central atom is Br
Br is in group 17, so number of valence electrons = 7
Subtract one extra electron from the positive charge
Number of bonding pairs (2 Br–F bonds) = 2
Number of lone pairs = 2
Total number of electron pairs = 4 = 2 bonding and 2 lone pairs
'Parent' shape = tetrahedral, see structure **8.15**
*Molecular* shape = bent.

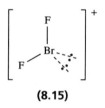

**(8.15)**

$[BrF_4]^-$

Central atom is Br
Br is in group 17, so number of valence electrons = 7
Add one extra electron from the negative charge
Number of bonding pairs (4 Br–F bonds) = 4
Number of lone pairs = 2
Total number of electron pairs = 6 = 4 bonding and 2 lone pairs
'Parent' shape = octahedral, see structure **8.16**
*Molecular* shape = square planar.

**(8.16)**

Compare the structures **8.15** and **8.16** in this book with **8.16** and **8.17** in H&S. Using the VSEPR model, the observed structures can be rationalized.

8.14    Prepare salts containing $[AsCl_2]^+$ and $[AsCl_4]^-$, e.g.

$$AsCl_3 + AlCl_3 \rightarrow [AsCl_2]^+[AlCl_4]^-$$

$$AsCl_3 + Me_4NCl \rightarrow [Me_4N]^+[AsCl_4]^-$$

and show that they give conducting solutions in $AsCl_3$, and that on mixing, the conductivity is a minimum at a 1:1 ratio.

8.15    (a) The $[Al_2Cl_7]^-$ ion is shown here in diagram **8.17**. $AlCl_3$ is a Lewis acid, accepting $Cl^-$, and the formation of $[Al_2Cl_7]^-$ can be considered in terms of:

$$AlCl_3 + Cl^- \rightleftharpoons [AlCl_4]^-$$
$$AlCl_3 + [AlCl_4]^- \rightleftharpoons [Al_2Cl_7]^-$$

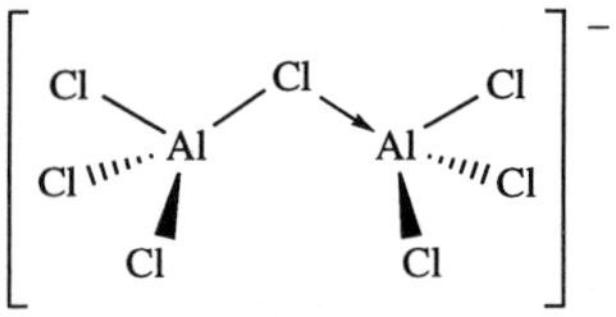

with each Al–Cl interaction being localized 2c-2e.

(b) Formation of $[Al_3Cl_{10}]^{2-}$ is a continuation of the series of equilibria:

$$AlCl_3 + [Al_2Cl_7]^- \rightleftharpoons [Al_3Cl_{10}]^{2-}$$

Suggested structure is shown in **8.18**.

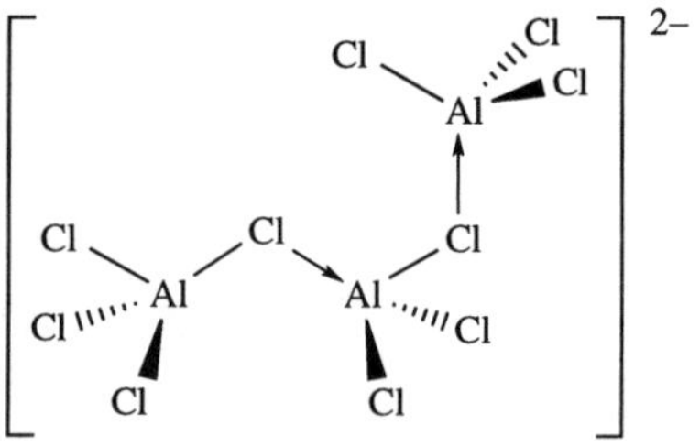

**(8.17)**

**(8.18)**

8.16    Use VSEPR theory for $[BiCl_5]^{2-}$:

Bi is in group 15, so number of valence electrons = 5
  Add 2 extra electrons from the negative charge
  Number of bonding pairs (5 Bi–F bonds) = 5
  Number of lone pairs = 1
  Total number of electron pairs = 6 = 1 lone and 5 bonding pairs
  'Parent' shape = octahedral,  see structure **8.19**
  *Molecular* shape = square-based pyramidal.

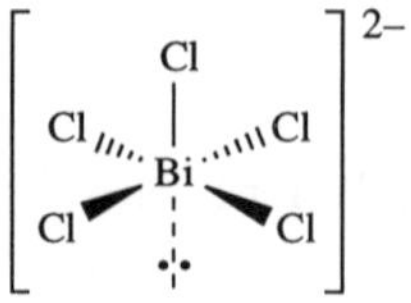

**(8.19)**

$[Bi_2Cl_8]^{2-}$ will involve bridging Cl. A possible structure would be **8.20** but this is less symmetrical than allowing two Cl atoms to be in bridging modes. Thus, **8.21** is a more realistic structure (and is that observed). Note that **8.21** has a *trans* arrangement of lone pairs; a *cis* arrangement is possible but is less favourable in terms of electron-electron repulsions than the *trans* arrangement.

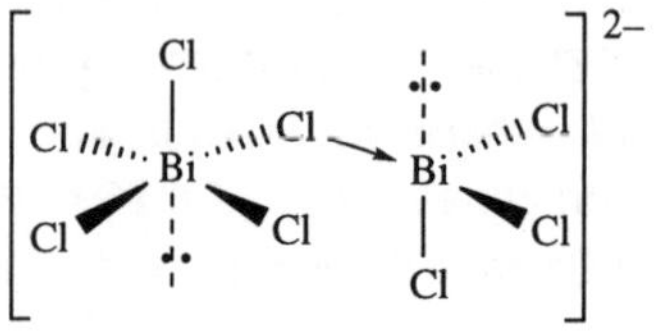

**(8.20)**

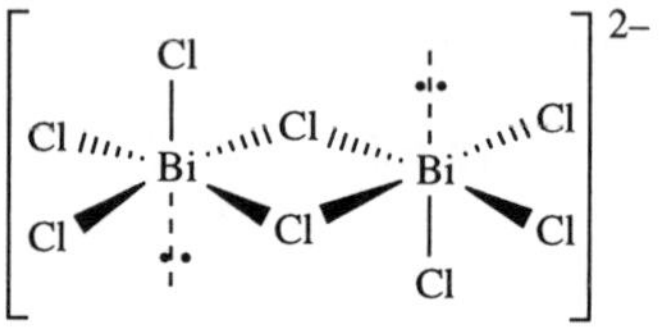

**(8.21)**

# 9 Hydrogen

9.1 From Table 9.2 in H&S, the values of $\bar{v}_1$ are 3657 and 2671 cm$^{-1}$. Equation needed is:

$$\bar{v} \propto \sqrt{\frac{1}{\mu}}$$

where $\mu$ is the reduced mass and: $\qquad \dfrac{1}{\mu} = \dfrac{1}{m_1} + \dfrac{1}{m_2}$

Since we are looking at a ratio, we can leave out the factors of $1.67 \times 10^{-27}$ in the calculation; in the steps below, the factors are left in to show why they cancel. For an O–H bond:

$$\frac{1}{\mu_{O-H}} = \frac{1}{(16 \times 1.67 \times 10^{-27})} + \frac{1}{(1 \times 1.67 \times 10^{-27})} = \frac{1.06}{1.67 \times 10^{-27}}$$

$$\mu_{O-H} = 0.94(1.67 \times 10^{-27}) \text{ kg}$$

For an O–D bond:

$$\frac{1}{\mu_{O-D}} = \frac{1}{(16 \times 1.67 \times 10^{-27})} + \frac{1}{(2 \times 1.67 \times 10^{-27})} = \frac{0.56}{1.67 \times 10^{-27}}$$

$$\mu_{O-D} = 1.78(1.67 \times 10^{-27}) \text{ kg}$$

For the ratio of the wavenumbers:

$$\frac{\bar{v}_{O-H}}{\bar{v}_{O-D}} = \sqrt{\frac{\mu_{O-D}}{\mu_{O-H}}} = \sqrt{\frac{1.78(1.67 \times 10^{-27})}{0.94(1.67 \times 10^{-27})}} = 1.37$$

Now compare this ratio with that obtained from the experimental values:

$$\frac{\bar{v}_{O-H}}{\bar{v}_{O-D}} = \frac{3657}{2671} = 1.37$$

Therefore the difference in values in the stretching frequencies is consistent with the isotopic masses of H and D.

9.2 (a) Points to include:
   • the signal from the deuterated solvent is used to lock the field – essential in an FT experiment where data are summed together;
   • when observing $^1$H NMR spectra, the solvent should be ≥99.5% deuterated otherwise proton signals due to the non-deuterated solvent dominate the spectrum, masking signals from the sample;
   • in $^1$H NMR spectra, ≤0.5% of the solvent which is non-deuterated gives $^1$H signals which can be used as an internal reference, e.g. residual CHCl$_3$ in ≥99.5% deuterated CDCl$_3$ gives a signal in the $^1$H NMR spectrum at $\delta$ 7.25.
   (b) See structures **9.1** and **9.2**.

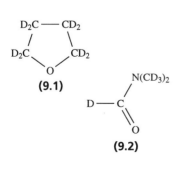
(9.1)

(9.2)

**9.3**

*Remember*: D = $^2$H

For $^2$H with $I = 1$, there are 3 possible spin states: +1, 0 and −1. In CDCl$_3$ (100% labelled), the $^{13}$C nucleus couples to one $^2$H nucleus.

Multipicity of the signal = $2nI + 1$
$$= 2(1)(1) + 1 = 3$$

Since each spin state is equally probable, a 1:1:1 three-line signal with *equal intensity* lines is observed (Figure 9.1).

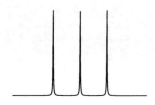

**Figure 9.1** Simulated $^{13}$C NMR spectrum of CDCl$_3$.

**9.4**

The solvent is deuterated to an extent of 99.6%. It contains small amounts of CD$_2$HCN, CDH$_2$CN and CH$_3$CN (where H represents $^1$H), although on probability grounds, CDH$_2$CN and CH$_3$CN can be assumed to be present in negligible amounts. The $^1$H NMR spectrum arises, therefore, from the presence of CD$_2$HCN. $^1$H nucleus couples to two $^2$H nuclei, each with $I = 1$.

Multiplicity of the signal = $2nI + 1$
$$= 2(2)(1) + 1 = 5$$

This is *not* a binomial quintet, but is a 1:2:3:2:1 five-line signal which arises as follows:

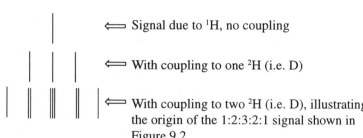

⟸ Signal due to $^1$H, no coupling

⟸ With coupling to one $^2$H (i.e. D)

⟸ With coupling to two $^2$H (i.e. D), illustrating the origin of the 1:2:3:2:1 signal shown in Figure 9.2.

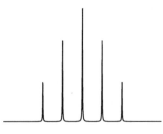

**Figure 9.2** Simulated $^1$H NMR spectrum of CD$_2$HCN.

**9.5**

Need to start with a commercially available deuterium-containing compound, e.g. D$_2$O with >99.9% label. Treat D$_2$O with AlCl$_3$ to prepare DCl:

$$AlCl_3 + 3D_2O \rightarrow Al(OD)_3 + 3DCl$$

Then treat Li[AlH$_4$] with DCl which will liberate HD (but not H$_2$ or D$_2$).

The gaseous product can be collected and a mass spectrum recorded: accurate masses of $^1$H = 1.008 and of $^2$H = 2.014. In the mass (accurate mass) spectrum, a parent ion at *m/z* 3.022 is expected for HD. Any contamination with H$_2$ or D$_2$ will be seen from peaks at *m/z* 2.016 and 4.028 respectively. Fragmentation of HD, H$_2$ and D$_2$ leads to peaks at *m/z* 1.008 and 2.014.

Alternatively, combustion of HD will give HDO – accurate density measurement distinguishes this from H$_2$O or D$_2$O.

**9.6**

In dilute solutions (e.g. 0.01 mol dm$^{-3}$), *tert*-butanol (Me$_3$COH) is essentially monomeric, and the absorption at 3610 cm$^{-1}$ is assigned to the stretch $\nu$(OH) in an isolated molecule. The 1.0 mol dm$^{-3}$ solution is concentrated enough that there will be intermolecular hydrogen bonding. Structure **9.3** illustrates this between two molecules, but hydrogen bonding can be more extensive, leading to small aggregates in solution. The result of hydrogen bonding is that the covalent O–H bond is weakened, and hence a shift in $\nu$(OH) to lower wavenumber (3610 to 3330 cm$^{-1}$). The broadening of the band is characteristic of hydrogen-bond formation since $\nu$(OH) is no longer at one diagnostic value.

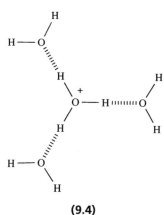

**(9.3)**

9.7    As HCl is absorbed, the equilibrium:

$$MCl(s) + HCl(g) \rightleftharpoons M[HCl_2](s)$$

is established. The position of equilibrium is determined by the relative lattice energies of MCl and M[HCl$_2$]. The relevant thermochemical cycle is:

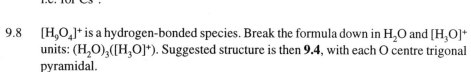

From this cycle:

$$\Delta_r H^\circ(1) = \Delta_r H^\circ(2) + \Delta_{lattice} H^\circ(M[HCl_2],s) - \Delta_{lattice} H^\circ(MCl,s)$$

$\Delta_r H^\circ(2)$ is the enthalpy change for the reaction:    $HCl(g) + Cl^-(g) \rightarrow [HCl_2]^-(g)$ and is independent of the metal.

∴ $\Delta_r H^\circ(1)$ depends on the difference between the two lattice energies. Both lattice energies will be negative, but $\Delta_{lattice} H^\circ(MCl,s)$ will be more negative than $\Delta_{lattice} H^\circ(M[HCl_2],s)$. Therefore, difference $\{\Delta_{lattice} H^\circ(M[HCl_2],s) - \Delta_{lattice} H^\circ(MCl,s)\}$ is always positive, meaning that the absorption of HCl is endothermic. However, the difference will be smallest (and so $\Delta_r H^\circ(1)$ least positive) for the larger M$^+$ ion, i.e. for Cs$^+$.

9.8    [H$_9$O$_4$]$^+$ is a hydrogen-bonded species. Break the formula down in H$_2$O and [H$_3$O]$^+$ units: (H$_2$O)$_3$([H$_3$O]$^+$). Suggested structure is then **9.4**, with each O centre trigonal pyramidal.

9.9    Points to include:
- Hydrogen bond (X–H$\cdots$Y) is formed between an H atom attached to an electronegative atom X, and an electronegative atom Y which possesses a lone pair of electrons.
- Evidence for hydrogen bonding comes from physical data and structural features in the solid state.
- Typical strength of a 'normal' hydrogen bond ≈20-25 kJ mol$^{-1}$, but some are much stronger, e.g. 165 kJ mol$^{-1}$ in [HF$_2$]$^-$.
- The 'normal' hydrogen bond is considered as a weak electrostatic interaction (**9.5**) between H ($\delta^+$) and Y (lone pair). Some hydrogen bonds

**(9.4)**

**(9.5)**

(e.g. that in $[HF_2]^-$) are more covalent in nature; a 3c-2e bond description is appropriate.

• Hydrogen bonds may be asymmetrical (e.g. in ice or $[H_9O_4]^+$, see **9.4**) or symmetrical (e.g. in $[HF_2]^-$).

• Formation of X–H⋯Y hydrogen bond results in lengthening and weakening of X–H covalent bond; this can be observed by a shift in the IR spectroscopic band for $v(XH)$ (see answer 9.6).

• Hydrogen bonding may tend to cause association, e.g. may be formation of dimers (e.g. $RCO_2H$, **9.6**) or may lead to polymeric (e.g. HF, **9.7**) or 3D-lattice (e.g. ice).

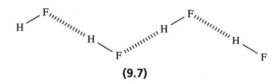

**(9.7)**

• Boiling points of hydrogen-bonded liquids (e.g. HF, $H_2O$, $NH_3$) are anomalously high compared to those of later group congeners (e.g. HF as compared to HCl, HBr, HI); similarly, values of $\Delta_{vap}H$ are high; melting points are also affected.

• Biological examples are crucial to life, e.g. formation of double-helical DNA through Watson-Crick or Hoogsteen hydrogen-bonded pairs of complementary bases.

9.10    (a)    $KH + NH_3 \rightarrow KNH_2 + H_2$

$KH + EtOH \rightarrow KOEt + H_2$

(b) KH is acting as a base; its conjugate acid is $H_2$; the other conjugate acid-base pairs are $NH_3$ with $[NH_2]^-$, and EtOH with $[EtO]^-$.

9.11    (a)    $2H_2O \rightarrow 2H_2 + O_2$    ($O_2$ formed at anode, $H_2$ at cathode; condition is that an electrolyte is present)

(b)    $2LiH \rightarrow 2Li + H_2$    ($H_2$ formed at anode, Li at cathode)

(c)    $CaH_2 + H_2O \rightarrow Ca(OH)_2 + H_2$    (occurs at 298 K)

(d)    $Mg + 2HNO_3 \rightarrow Mg(NO_3)_2 + H_2$    (occurs at 298 K)

(e)    $2H_2 + O_2 \rightarrow 2H_2O$    (radical reaction that needs initiating with e.g. a spark)

(f)    $CuO + H_2 \rightarrow Cu + H_2O$    (requires heat)

9.12    Although $H_2O_2$ is *thermodynamically* unstable with respect to the reaction:

$$H_2O_2 \rightarrow H_2O + \tfrac{1}{2}O_2 \qquad \Delta G° = -116.7 \text{ kJ mol}^{-1}$$

it is *kinetically* stable; $E_a$ is high, and a catalyst (e.g. $MnO_2$) is needed to speed up decomposition.

9.13    (a) See Figure 9.3.

(b) Magnesium hydride has the formula $MgH_2$; compare this to $TiO_2$ (rutile). $Mg^{2+}$ ions will take the sites of $Ti^{4+}$ (see Figure 9.3), therefore $Mg^{2+}$ is 6-coordinate, octahedral. $H^-$ ions will take the sites of $O^{2-}$ ions (see Figure 9.3), therefore will be 3-coordinate, trigonal planar.

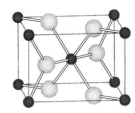

**Figure 9.3** A unit cell of rutile, $TiO_2$; Ti(IV) centres are in the very central and the corner sites.

9.14    The text description: 'AlH$_3$ consists of an infinite lattice, in which each Al(III) centre is in an AlH$_6$-octahedral site; H atoms bridge pairs of Al centres.'

From this, sketch part of the lattice, as in **9.8.**
The ratio of the coordination numbers is:   Al : H = 6 : 2.
It follows (see p. 46) that the stoichiometry is: Al : H = 2 : 6
$\qquad\qquad\qquad\qquad\qquad\qquad\qquad\qquad$ = 1 : 3 , i.e.  AlH$_3$

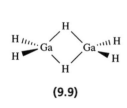

**(9.8)**

9.15    Figure 9.4 shows the structure of $BeH_2$. Each Be is tetrahedrally coordinated, each H atom bridges 2 Be.
An $sp^3$ hybridization scheme is appropriate for Be; Be–H–Be bridges formed by overlap of $sp^3$ hybrid and H 1$s$ orbitals. Each Be has 2 valence electrons, each H, one. This gives enough electrons to form 3c-2e bridges:

**Figure 9.4** Part of the chain structure of polymeric BeH$_2$.

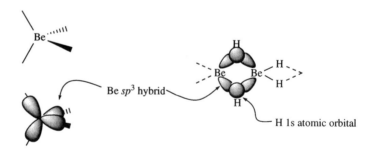

Be $sp^3$ hybrid

H 1s atomic orbital

**(9.9)**

$Ga_2H_6$ (**9.9**) is isostructural with $B_2H_6$ (see Section 4.7 in H&S); the bridging H atoms are involved in 3c-2e interactions related to those in BeH$_2$.

9.16    (a) The trend in bond angles can be rationalized in terms of VSEPR theory.

CH$_4$: C (group 14) uses all 4 valence electrons to form 4 C–H bonds, so no lone pairs. Tetrahedral molecule with ideal H–C–H bond angles of 109°.

NH$_3$: N (group 15) uses 3 valence electrons for 3 N–H bonds, with one lone pair remaining. 'Parent' shape is tetrahedron; molecular shape is trigonal pyramidal (**9.10**). Lone pair-bonding pair > bonding pair-bonding pair repulsions, leading to H–N–H bond angles less than the ideal value. Consistent with observed 106.7°.

**(9.10)**

**(9.11)**

**(9.12)**

**(9.13)**

$H_2O$: O (group 16) uses 2 valence electrons for 2 O–H bonds, with 2 lone pairs remaining. 'Parent' shape is tetrahedral; molecular shape is bent (**9.11**). Lone pair-lone pair > lone pair-bonding pair > bonding pair-bonding pair repulsions, leading to H–O–H bond angles less than the ideal value, and less than in $NH_3$. Consistent with observed 104.5°.

(b) For $NH_3$, the dipole will act as shown in **9.12**, being determined by the greater electronegativity of N versus H, and the lone pair of electrons on N. In $NH_2OH$ (**9.13**), it is difficult to predict the direction of the dipole, but take into account the following:
- electronegativities, $\chi^P$, of N and O are 3.0 and 3.4 respectively;
- N bears 1 lone pair, O has two;
- the shape of $NH_2OH$ leads to the lone pairs of O facing away from the lone pair of N.

The direction of the molecular dipole moment is the resultant of the bond dipole moments and those due to the lone pairs. This will not mimic that in $NH_3$, nor will it be of the same magnitude.

(c) Trouton's rule:

For liquid $\rightleftharpoons$ vapour: $\quad \Delta S_{vap} = \dfrac{\Delta H_{vap}}{bp} \approx 88 \text{ J K}^{-1} \text{ mol}^{-1}$

$PH_3$, $P_2H_4$, $SiH_4$, $Si_2H_6$ show ratios close to this value and obey Trouton's rule. $NH_3$ and $N_2H_4$ have higher values of $\Delta_{vap}S$ due to effects of hydrogen bonding; hydrogen-bonded interactions are destroyed as the liquid vaporizes.

The low value of $\Delta_{vap}S$ for $HCO_2H$ implies vaporization is accompanied by *more* ordering of the system. Value lower than $\approx 88$ J K$^{-1}$ mol$^{-1}$ implies more (or at least stronger) hydrogen bonding in vapour than in liquid, the vapour contains dimers.

# 10 Group 1: the alkali metals

10.1    (a)    Li     lithium
            Na     sodium
            K      potassium
            Rb     rubidium
            Cs     caesium          ('cesium' in the US)
            Fr      francium

        (b) $ns^1$

10.2    Removal of first electron is from a half-filled $ns$ atomic orbital:

➤

See Section 1.10 in H&S for
further discussion of trends
in ionization energies

$$M(g) \rightarrow M^+(g) + e^- \qquad \text{involves:} \quad [X]ns^1 \rightarrow [X] \quad \text{where } X = \text{noble gas}$$

Removal of the second electron is from a filled atomic orbital – for Li, from the $1s^2$ orbital, for the heavier group 1 metals, from an $np^6$ level:

$$M^+(g) \rightarrow M^{2+}(g) + e^-$$

10.3    (a) Points to include:
- all metals have body-centred cubic lattices;
- sketch a unit cell (Figure 10.1);
- each atom is 8-coordinate (Figure 10.1);
- lattice is *not* close-packed.

(b) Points to include:
- at 298 K, LiCl, NaCl, KCl, RbCl adopt NaCl lattice (see p. 46, Figure 5.5a);
- in NaCl lattice, each $Na^+$ is 6-coordinate (octahedral) with respect to $Cl^-$, and each $Cl^-$ is 6-coordinate (octahedral) with respect to $Na^+$ ;
- NaCl lattice can be considered as interpenetrating fcc $Na^+$ and fcc $Cl^-$ ;
- for CsCl, coordination number of each ion is 8 in a lattice related to a bcc structure (see p. 46, Figure 5.5b).

See also Box 5.3 in H&S for discussion of radius ratio rules which are relevant to this question – but use the rules with caution!

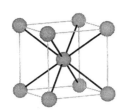

**Figure 10.1**   Unit cell of a bcc lattice.

10.4    (a) Melting points (Figure 10.2). Points to include:
- melting the solid refers to:      $M(s) \xrightarrow{\Delta} M(l)$

➤

Values for the question are
listed in Table 10.1 in H&S

- all structures are the same type (see answer 10.3a) so a direct comparison of melting points is valid down the group;
- all melting points are relatively low;
- low values can be related to relatively weak M–M bonding; only one valence electron per M centre;
- bcc lattice is non-close-packed – contributes to low melting points;
- down the group, M–M bonding becomes weaker as $ns$–$ns$ overlap becomes less effective (more diffuse orbitals as $n$ increases).

(b) Cation radii (Figure 10.3). Points to include:
- general trend is an increase in cation size down the group;
- charge is constant ($M^+$), so a direct comparison down the group is valid;
- going from Li to Na, extra quantum shell is added; similarly from Na to K, K to Rb, and Rb to Cs.

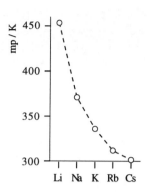

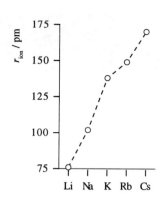

**Figure 10.2** Trend in the melting points of the alkali metals.

**Figure 10.3** Trend in the ionic radii of the alkali metals.

10.5    (a) Take Li as representative; bonding in all $M_2$ species is similar as each M atom has one valence electron. Lewis structure for $Li_2$ (**10.1**) indicates an Li–Li single bond. Valence bond theory gives a set of resonance structures **10.2**, but the ionic structures will contribute very little.

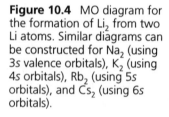

**(10.1)**

$$Li \longrightarrow Li \longleftrightarrow Li^+ \quad Li^- \longleftrightarrow Li^- \quad Li^+$$

**(10.2)**

MO theory: construct an MO diagram using LCAO approach – consider only the valence atomic orbitals and electrons (Figure 10.4).

**Figure 10.4** MO diagram for the formation of $Li_2$ from two Li atoms. Similar diagrams can be constructed for $Na_2$ (using 3s valence orbitals), $K_2$ (using 4s orbitals), $Rb_2$ (using 5s orbitals), and $Cs_2$ (using 6s orbitals).

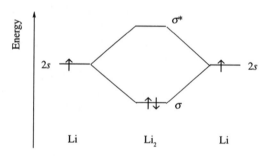

The conclusion from the MO treatment is that there is one, fully occupied bonding MO, a bond order of 1, and the molecule is diamagnetic.

(b) Bond dissociation energies decrease down the group: 110 kJ mol$^{-1}$ for $Li_2$, to 44 kJ mol$^{-1}$ for $Cs_2$. Each molecule has one filled $\sigma$ MO, but $ns$–$ns$ overlap depends on $n$ (more diffuse orbitals as $n$ increases): $2s$–$2s$ > $3s$–$3s$ > $4s$–$4s$ > $5s$–$5s$ > $6s$–$6s$. Therefore, the Li–Li bond is the strongest, and the Cs–Cs, the weakest.

10.6    (a)    $^{40}_{19}K \longrightarrow {}^{40}_{18}Ar + \beta^+$

(b)    1.0 g $^{40}K$ = 0.025 moles

∴ Amount of Ar formed = 0.025 moles

Assume ideal gas at standard pressure (1 bar = $10^5$ Pa) and temperature (273 K). Since the volume of 1 mole of gas under these conditions is 22.7 dm$^3$:

Volume of 0.025 moles = 0.025 × 22.7 = 0.57 dm$^3$ (at 273 K, 1 bar) or, use:    $PV = nRT$

to obtain a volume of gas at 298 K:

$$V = \frac{nRT}{P} = \frac{(0.025)(8.314)(298)}{10^5}$$

$$= 6.2 \times 10^{-4} \text{ m}^3$$

$$= 0.62 \text{ dm}^3 \text{ (at 298 K, 1 bar)}$$

(c) K occurs in, e.g. some silicates. $^{40}$K decays: half-life $1.26 \times 10^9$ yr. Measure radioactivity of mineral containing 1g of K and compare with that of any current material containing 1 g of K. Apply integrated rate equation as in worked example 2.5 in H&S.

10.7    (a) Consider the Born-Haber cycle:

$$\Delta_f H^\circ(M_3N,s) = 3\Delta_a H^\circ(M,s) + 3IE_1 + \Delta_a H^\circ(N,g) + \Sigma\Delta_{EA}H^\circ + \Delta_{lattice}H(M_3N,s)$$

Formation of $N^{3-}$ is independent of metal, but requires a lot of energy ($\Sigma\Delta_{EA}H^\circ = 2100 \text{ kJ mol}^{-1}$) – major factor for which lattice energy must compensate. This is seen from Figure 10.5; values relating to N are constant as metal varies. Sum $\{3\Delta_a H^\circ(M,s) + 3IE_1 + \Delta_a H^\circ(N,g) + \Sigma\Delta_{EA}H^\circ\}$ is always large and positive. Lattice energy is inversely proportional to internuclear separation; cation radii increase substantially down the group (see Figure 10.2) and so the most negative lattice energy is for $Li_3N$. ∴ This compound is predicted to be the most thermodynamically viable of alkali metal nitrides.
(b) Assume all the $M^+$ ions have a primary solvation number of 6, i.e. all form $[M(H_2O)_6]^+$ in aqueous solution. Smallest hydrated ion with highest charge density has the greatest attraction for further $H_2O$ solvent molecules, increasing effective hindrance to ion movement through the solution. Cation radii (see Figure 10.2) increase down the group; charge density is highest for $Li^+$. This rationalizes the order of the ionic mobilities: $Li^+ < Na^+ < K^+ < Rb^+ < Cs^+$.
(c) See answer 7.21, p. 74.

**Figure 10.5** Trends in the values of the contributions in the Born-Haber cycle for the alkali metal nitrides (see text).

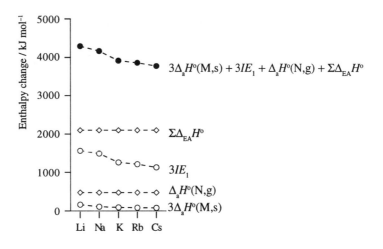

10.8    From Table 10.2 in H&S:

$\Delta_f H^\circ(\text{LiI,s}) = -270 \text{ kJ mol}^{-1}$      $\Delta_f H^\circ(\text{NaI,s}) = -288 \text{ kJ mol}^{-1}$

$\Delta_f H^\circ(\text{LiF,s}) = -616 \text{ kJ mol}^{-1}$      $\Delta_f H^\circ(\text{NaF,s}) = -577 \text{ kJ mol}^{-1}$

The possible reaction that could occur is:

$$\text{LiI(s)} + \text{NaF(s)} \xrightarrow{\Delta} \text{LiF(s)} + \text{NaI(s)}$$

for which:

$$\Delta_r H^\circ(298 \text{ K}) = \Delta_f H^\circ(\text{LiF,s}) + \Delta_f H^\circ(\text{NaI,s}) - \Delta_f H^\circ(\text{LiI,s}) - \Delta_f H^\circ(\text{NaF,s})$$

$$= -57 \text{ kJ mol}^{-1}$$

One should really be looking at the value of $\Delta_r G^\circ$, but since $\Delta_r S^\circ$ is negligible (solid state reaction), a negative value of $\Delta_r H^\circ$ can be taken to indicate that reaction is thermodynamically favourable.

10.9    Grinding a metal halide or halo-anion with an alkali metal halide may lead to halide exchange (see question 10.8, for example). Grinding $[\text{PtCl}_4]^{2-}$ with KBr or KI may lead to anions such as $[\text{PtCl}_3\text{Br}]^{2-}$ or $[\text{PtCl}_3\text{I}]^{2-}$. IR spectra that are recorded may therefore be of these salts (or of a mixture of anions), and may not represent IR spectroscopic data for $[\text{PtCl}_4]^{2-}$.

10.10    Consider the replacement of the organic Cl by F in terms of thermochemical cycle:

The only variable when M changes is the difference between the lattice energies of MCl and MF. Use the lattice energy data in Table 10.2 in H&S:

For KF:  Difference in lattice energies $= \Delta U(\text{KCl}) - \Delta U(\text{KF})$
$$= -701 - (-808) = +107 \text{ kJ mol}^{-1}$$

For NaF:  Difference in lattice energies $= \Delta U(\text{NaCl}) - \Delta U(\text{NaF})$
$$= -769 - (-910) = +141 \text{ kJ mol}^{-1}$$

Both oppose the reaction (i.e. positive enthalpy change), but for the larger metal ion ($\text{K}^+$), the difference in lattice energies is smaller and so is less unfavourable.

10.11    The phase of the solid in equilibrium with the dissolved salt must alter at 305 K, for example:

$$\text{hydrate} \rightleftharpoons \text{anhydrous salt}$$

The solubility of the lower temperature phase increases as the temperature is increased, and so (by Le Chatelier's Principle) the dissolution must be an endothermic process. The solubility of the higher temperature phase decreases as the temperature increases, and so must be an exothermic process.

**10.12** Consider ion sharing between the units of the lithium nitride structure shown in Figure 10.6 (or Figure 10.3 in H&S).

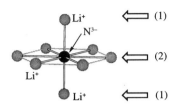

**Figure 10.6** Part of the layer structure of lithium nitride.

(a)    $N^{3-}$ :                this lies wholly in the unit

$Li^+$, layer (2): each is shared between 3 units as shown for the central $Li^+$ atom in the diagram on the right:

Thus, in layer (2):
Number of $N^{3-}$ = 1
Number of $Li^+$ = $6 \times \frac{1}{3}$ = 2
∴  Ratio $Li^+$ : $N^{3-}$ = 2 : 1

(b) $Li^+$ in layer (1) is shared between two layers, therefore:
Number of $Li^+$ = $2 \times \frac{1}{2}$ = 1
For layers (1) and (2), the ratio $Li^+$ : $N^{3-}$ = 3 : 1, so stoichiometry = $Li_3N$.

**10.13** Figure 10.7 shows the general MO diagram that you should construct (using the

**Figure 10.8** MO diagram for the formation of $[O_2]^-$ and $[O_2]^{2-}$ (see text for occupation of MOs).

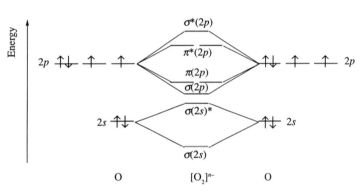

LCAO approach, see Chapter 1). Each O atom provides 6 valence electrons and so in $O_2$, the MOs are filled (*aufbau* principle) to give a configuration:
$\sigma(2s)^2\sigma^*(2s)^2\sigma(2p)^2\pi(2p)^4\pi^*(2p)^2$
In $[O_2]^-$, there is an extra electron to be accommodated giving:
$\sigma(2s)^2\sigma^*(2s)^2\sigma(2p)^2\pi(2p)^4\pi^*(2p)^3$
and this rationalizes why $[O_2]^-$ is paramagnetic.
On going from $[O_2]^-$ to $[O_2]^{2-}$, one more electron is added giving:
$\sigma(2s)^2\sigma^*(2s)^2\sigma(2p)^2\pi(2p)^4\pi^*(2p)^4$
and hence, MO theory is consistent with $[O_2]^{2-}$ being diamagnetic.

**10.14** Equation 10.21 in H&S is:        $2Na \rightleftharpoons Na^+ + Na^-$
This is disproportionation since it is the combination of:

$\quad\quad\quad\quad Na \rightleftharpoons Na^+ + e^-$        *oxidation*
and    $Na + e^- \rightleftharpoons Na^-$        *reduction*

**10.15**    (a) $[O_2]^-$ ; (b) $[O_2]^{2-}$ ; (c) $[O_3]^-$ ; (d) $[N_3]^-$ ; (e) $N^{3-}$ ; (f) $Na^-$

**10.16** Sources of information for this answer are Section 10.2, Boxes 10.1-10.4 and Box 8.4 in H&S.

$$\left[ C \equiv N \right]^-$$

**(10.3)**

$$: \overset{-}{C} \equiv N : \quad \longleftrightarrow \quad : C \overset{-}{\underset{\cdot\cdot}{=}} \overset{\cdot\cdot}{N} \cdot$$

**(10.4)**

10.17 (a) Structure of [CN]⁻ can be represented as in **10.3**; **10.4** shows resonance structures. Using MO approach, bonding in [CN]⁻ can be described in the same way as that in CO since [CN]⁻ is isoelectronic with CO; see Section 1.17 and Figure 1.28 in H&S – on the left hand side of Figure 1.28a and 1.28b in H&S, replace O by N⁻.
(b) Consider unit cell of NaCl (Figure 5.5a, p. 46). If NaCN possesses this same lattice, then Cl⁻ ions in NaCl must be replaced by [CN]⁻. This ion can only be treated as a sphere if it is either rotating or occupies the lattice sites with random orientations; see Figure 5.17 in H&S which illustrates how [NH₄]⁺ can be treated as spherical. Structure of KOH was described similarly at the end of Section 10.6 in H&S.

10.18 Points to include (using information from Section 8.6 in H&S):
• Dilute solutions of Na in NH₃ are bright blue; process is:
$$M \rightarrow M^+(NH_3) + e^-(NH_3)$$
where '(NH₃)' represents solvated species in liquid NH₃.
• Very dilute solutions are paramagnetic; magnetic susceptibility corresponds to one free electron per Na.
• Saturated solutions are bronze and diamagnetic.
• Increasing the concentration of Na in liquid NH₃ leads to an initial decrease in molar conductivity, followed by an increase; molar conductivity of a saturated solution is similar to that of Na metal.
• Explanation of conductivity data:
– at low concentrations:    $M \rightarrow M^+(NH_3) + e^-(NH_3)$
– at concentrations ≈ 0.05 mol dm⁻³, association of M⁺(NH₃) and e⁻(NH₃)
– metal-like behaviour at higher concentrations.
• Magnetic susceptibility data at higher concentrations explained by equilibria:
$$2M^+(NH_3) + 2e^-(NH_3) \rightleftharpoons M_2(NH_3)$$
$$M(NH_3) + e^-(NH_3) \rightleftharpoons M^-(NH_3)$$
• Blue solutions of Na in liquid NH₃ decompose slowly:
$$2NH_3 + 2e^- \rightarrow 2[NH_2]^- + H_2$$

10.19 (a)    $NaH + H_2O \rightarrow NaOH + H_2$    (NaH is source of H⁻, so H₂ formed by $H^+ + H^- \rightarrow H_2$ )

(b)    $KOH + CH_3CO_2H \rightarrow K[CH_3CO_2] + H_2O$    (Brønsted base-acid)

(c)    $2NaN_3 \overset{\Delta}{\longrightarrow} 2Na + 3N_2$    (thermal decomposition of an azide)

(d)    $K_2O_2 + 2H_2O \rightarrow 2KOH + H_2O_2$
but presence of base will catalyse the decomposition of H₂O₂:
$$H_2O_2 \rightarrow H_2O + \tfrac{1}{2}O_2$$

(e)    $NaF + BF_3 \rightarrow Na[BF_4]$    (NaF is a source of F⁻; BF₃ is a Lewis acid and accepts F⁻ to give [BF₄]⁻)

(f) *Molten* KBr, and so only K⁺ and Br⁻ ions present:

At the cathode:    $K^+ + e^- \rightarrow K$
At the anode:    $2Br^- \rightarrow Br_2 + 2e^-$

(g) *Aqueous solution* of the salt and so preferential release of H₂ at the cathode:

At the cathode:    $2H_2O + 2e^- \rightarrow 2[OH]^- + H_2$
At the anode:    $2Cl^- \rightarrow Cl_2 + 2e^-$

# 11 The group 2 metals

11.1 (a)

| | |
|---|---|
| Be | beryllium |
| Mg | magnesium |
| Ca | calcium |
| Sr | strontium |
| Ba | barium |
| Ra | radium |

Alkaline earth metals

(b) $ns^2$

11.2 1 mole of dissolved $Ca(OH)_2$ gives 1 mole of $Ca^{2+}$ and 2 moles of $[OH]^-$ ions.

$$\therefore \quad \text{Solubility} = [Ca^{2+}] \qquad \text{and} \qquad [OH^-] = 2[Ca^{2+}]$$

$$K_{sp} = [Ca^{2+}][OH^-]^2 = 4[Ca^{2+}]^3$$

$$\therefore \quad \text{Solubility} = [Ca^{2+}] = \sqrt[3]{\frac{K_{sp}}{4}} = \sqrt[3]{\frac{4.68 \times 10^{-6}}{4}} = 1.05 \times 10^{-2} \text{ mol dm}^{-3}$$

Similarly:
$$\text{Solubility of } Mg(OH)_2 = \sqrt[3]{\frac{K_{sp}}{4}} = \sqrt[3]{\frac{5.61 \times 10^{-12}}{4}} = 1.12 \times 10^{-4} \text{ mol dm}^{-3}$$

Relative solubilities of $Ca(OH)_2 : Mg(OH)_2 \approx 94 : 1$.

In extraction of Mg from seawater, $Mg^{2+}$ is first precipitated as $Mg(OH)_2$ by addition of $Ca(OH)_2$. From the calculated solubilities, $Mg(OH)_2$ is about 90 times *less* soluble than $Ca(OH)_2$, and addition of $Ca(OH)_2$ will provide $[OH]^-$ ions which then combine with $Mg^{2+}$ and precipitate $Mg(OH)_2$.

11.3 (a) Formation of magnesium nitride containing $Mg^{2+}$ and $N^{3-}$ (but see Section 14.1 in H&S):

$$3Mg + N_2 \xrightarrow{\Delta} Mg_3N_2$$

(b) Nitrides of group 2 (and group 1) metals liberate ammonia when they react with $H_2O$ (see Sections 14.5 and 14.6 in H&S):

$$Mg_3N_2 + 6H_2O \rightarrow 2NH_3 + 3Mg(OH)_2$$

11.4 (a) Consider the unit cell of NaCl (Figure 5.5a, p. 46). In $MgC_2$, $Mg^{2+}$ ions replace $Na^+$ ions in the NaCl lattice, and $[C_2]^{2-}$ ions replace $Cl^-$. A $Cl^-$ ion is spherical, but a $[C_2]^{2-}$ ion (**11.1**) is not. Start with an NaCl lattice – the unit cell dimensions along the three axes are the same. Now, replace the $Cl^-$ ions by $[C_2]^{2-}$ ions so that all the $[C_2]^{2-}$ ions are aligned in the *same* direction, coincident with one of the axes. The effect of packing the ions together in this way will be to elongate the unit cell in one direction – the direction along which the $[C_2]^{2-}$ ions are aligned.

(b) Free rotation of $[CN]^-$ or random orientations of the anions in NaCN means $[CN]^-$ is pseudo-spherical (see answer 10.17b).

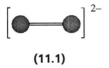

**(11.1)**

11.5    (a) Reaction *could* be driven by increase in entropy:

$$[NH_4]_4[BeF_4] \xrightarrow{\Delta} BeF_2 + 2NH_4F$$

(b) NaCl acts as a source of Cl$^-$; BeCl$_2$ is Cl$^-$ acceptor, giving [BeCl$_4$]$^{2-}$:

$$2NaCl + BeCl_2 \rightarrow Na_2[BeCl_4]$$

(c) Dissolving BeF$_2$ in water leads to hydrated ions:

$$BeF_2(s) \xrightarrow{water} [Be(H_2O)_4]^{2+}(aq) + 2F^-(aq)$$

11.6    (a) BeCl$_2$ monomer is linear and Be atom has two vacant $2p$ atomic orbitals; left hand diagram below shows one of these vacant orbitals – with axis set shown, the other vacant orbital is $2p_y$. Dimerization occurs by Cl lone pair donation into vacant orbital on Be:

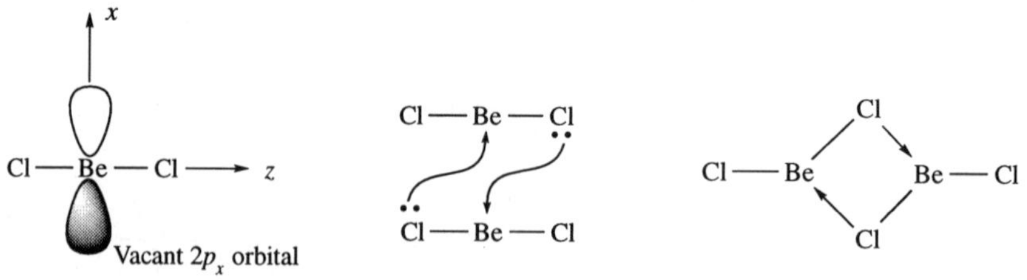

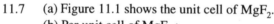

Dimer contains trigonal planar Be; $sp^2$ hybridization scheme for Be is appropriate. Each Be atom still has one vacant $2p$ atomic orbital.

(b) Et$_2$O is a Lewis base and BeCl$_2$ is a Lewis acid; adduct formation occurs. BeCl$_2$ has 2 vacant $2p$ orbitals (see above) and can accept 2 pairs of electrons:

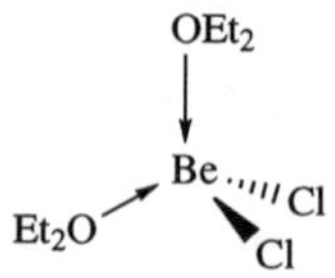

**(11.2)**

$$BeCl_2 + 2Et_2O \rightarrow BeCl_2 \cdot 2Et_2O$$

In BeCl$_2$·2Et$_2$O (**11.2**), Be is tetrahedrally coordinated and $sp^3$ hybridization is an appropriate description. Four localized $\sigma$-interactions.

11.7    (a) Figure 11.1 shows the unit cell of MgF$_2$.
(b) Per unit cell of MgF$_2$ :

| Site | Number of Mg$^{2+}$ | Number of F$^-$ |
|---|---|---|
| Within unit cell | 1 | 2 |
| Corner site | $8 \times \frac{1}{8} = 1$ | 0 |
| In face of unit cell | 0 | $4 \times \frac{1}{2} = 2$ |
| Total | 2 | 4 |

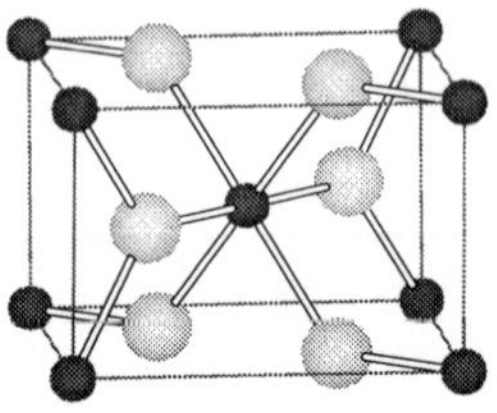

**Figure 11.1** Unit cell of MgF$_2$ i.e. rutile lattice. Mg$^{2+}$ ions are shown as the small spheres

Stoichiometry is Mg$^{2+}$ : F$^-$ = 1:2, or MgF$_2$.

11.8    The data in Table 11.2 in H&S give values of $\Delta_f H^\circ$(298 K) for group 2 metal halides; the trends are represented in Figure 11.2. Points to note from data:
  • With a constant metal, the trend is always that $\Delta_f H^\circ$(MF$_2$,s) is more negative than $\Delta_f H^\circ$(MCl$_2$,s), than $\Delta_f H^\circ$(MBr$_2$,s), than $\Delta_f H^\circ$(MI$_2$,s).
  • With a constant halide (X = Cl, Br and I), the trend is for $\Delta_f H^\circ$(MX$_2$,s) to become more negative as group 2 is descended.
  • For MF$_2$, $\Delta_f H^\circ$(MF$_2$,s) is least negative for M = Mg, and is ≈constant for the heavier metals.

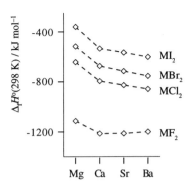

**Figure 11.2** Trends in values of $\Delta_f H^\circ$ for group 2 metal halides.

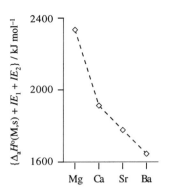

**Figure 11.3** Trend in $\{\Delta_a H^\circ(M,s) + IE_1 + IE_2\}$ for the group 2 metals.

Consider the Born-Haber cycle:

For a diatomic $X_2$:
$$2\Delta_a H^\circ(X,s) = D(X_2,g)$$

$$\Delta_f H^\circ(MX_2,s) = \Delta_a H^\circ(M,s) + IE_1 + IE_2 + D(X_2,g) + 2\Delta_{EA}H^\circ(X,g) + \Delta_{lattice}H(MX_2,s)$$

***For constant metal:*** Factors that influence $\Delta_f H^\circ(MX_2,s)$ with varying X are $D(X_2,g)$, $2\Delta_{EA}H^\circ(X,g)$ and $\Delta_{lattice}H(MX_2,s)$. From Appendices 9 and 10 in H&S, the sum $\{D(X_2,g) + 2\Delta_{EA}H^\circ(X,g)\}$ becomes *less negative* from F to I. Values of $\Delta_{lattice}H(MX_2,s)$ become *less negative* on going from F to I for a given M. Therefore, on going from F to I, values of $\Delta_f H^\circ(MX_2,s)$ become *less negative* for a given M.
***With constant halide:*** Factors that influence $\Delta_f H^\circ(MX_2,s)$ with varying M are $\Delta_a H^\circ(M,s)$, $IE_1$, $IE_2$ and $\Delta_{lattice}H(MX_2,s)$. From Appendices 8 and 10 in H&S, the sum $\{\Delta_a H^\circ(M,s) + IE_1 + IE_2\}$ becomes *less positive*. The decrease from Mg to Ca is significantly larger than from Ca to Sr, or Sr to Ba (Figure 11.3). This general trend corresponds to the observed trends in Figure 11.2 for X = Cl, Br and I; values of $\Delta_{lattice}H(MX_2,s)$ become *less negative* from $MgCl_2$ to $BaCl_2$, from $MgBr_2$ to $BaBr_2$ and from $MgI_2$ to $BaI_2$ – this offsets the sum $\{\Delta_a H^\circ(M,s) + IE_1 + IE_2\}$, but the trend set by $\{\Delta_a H^\circ(M,s) + IE_1 + IE_2\}$ remains reflected in the trend in values of $\Delta_f H^\circ(MX_2,s)$. For X = F, there is substantial variation in $\Delta_{lattice}H(MF_2,s)$ on going from M = Mg to Ba; this offsets $\{\Delta_a H^\circ(M,s) + IE_1 + IE_2\}$ leaving $\Delta_f H^\circ(MX_2,s)$ varying only by a small amount.

11.9   (a) Anhydrous $CaCl_2$ is hygroscopic, and forms a hydrate, probably $CaCl_2 \cdot 2H_2O$. In the presence of a lot of water, hydrate becomes liquid (it is deliquescent). Action as a drying agent depends on hygroscopic nature of anhydrous $CaCl_2$. Anhydrous $CaH_2$ reacts with water liberating $H_2$:

$$CaH_2 + H_2O \rightarrow Ca(OH)_2 + H_2$$

$CaH_2$ is used to remove $H_2O$ from, e.g., a solvent (suitable only if solvent does not react with $CaH_2$). ***Caution!*** Do not expose $CaH_2$ to large amounts of water – it reacts vigorously.

(b) Points to include:

- $BeCl_2$, colourless, deliquescent solid; $CaCl_2$, colourless solid, hygroscopic, forming hydrate which is deliquescent.
- $BeCl_2$ is covalent solid with infinite chains (**11.3**); each Be can be taken as being $sp^3$ hybridized; localized bonding.
- Anhydrous $CaCl_2$ is ionic solid; distorted rutile lattice ($MCl_2$, $MBr_2$ and $MI_2$ for M = Ca, Sr and Ba typically have layer structures or distorted lattice-types).
- $BeCl_2$ and $CaCl_2$ soluble in water forming $[Be(H_2O)_4]^{2+}$ and $[Ca(H_2O)_6]^{2+}$.
- $BeCl_2$ acts as Lewis acid; used as Friedel-Crafts catalyst; adducts illustrating Lewis acidity include $[BeCl_4]^{2-}$ and $BeCl_2 \cdot 2Et_2O$ (see **11.2**).
- $CaCl_2$ readily forms a hydrate – application as drying agent (see above).
- Solubility of chlorides in polar solvents (e.g. $Et_2O$) due to adduct formation.

(11.3)

11.10  (a)    $MgCl_2 + Mg \rightarrow 2MgCl$

$Mg^{2+}$ being reduced to $Mg^+$, *and* Mg being oxidized to $Mg^+$, so is the reverse of a disproportionation reaction. Set up a suitable thermochemical cycle (see Section 5.16 in H&S):

$$Mg(s) + MgCl_2(s) \xrightarrow{\Delta_r H^\circ} 2MgCl(s)$$

$$\Delta_{lattice}H^\circ(MgCl_2,s) - \Delta_a H^\circ(Mg,s) \uparrow \qquad \uparrow 2\Delta_{lattice}H^\circ(MgCl,s)$$

$$Mg(g) + Mg^{2+}(g) + 2Cl^-(g) \xrightarrow{IE_1 - IE_2 (Mg,g)} 2Mg^+(g) + 2Cl^-(g)$$

From the cycle:

$$\Delta_r H^\circ = IE_1 - IE_2 + 2\Delta_{lattice}H^\circ(MgCl,s) - \Delta_{lattice}H^\circ(MgCl_2,s) + \Delta_a H^\circ(Mg,s)$$

From Appendices 8 and 10 in H&S:

$$IE_1 - IE_2 \text{ (for Mg)} = 737.7 - 1451 = -713.3 \text{ kJ mol}^{-1}$$
$$\Delta_a H^\circ(Mg,s) = 146 \text{ kJ mol}^{-1}$$

$$\therefore \quad IE_1 - IE_2 + \Delta_a H^\circ = -567.3 \text{ kJ mol}^{-1}$$

Now consider the term $\{2\Delta_{lattice}H^\circ(MgCl,s) - \Delta_{lattice}H^\circ(MgCl_2,s)\}$. The lattice energy for $MgCl_2$ will greatly exceed that of MgCl because:

➤ Look at the Born-Landé equation in Section 5.13 in H&S

- $|z^+|$ for $Mg^{2+}$ is twice that of $Mg^+$
- $r_0$ for $Mg^{2+}$ is smaller than that of $Mg^+$
- Madelung constants for $MX_2$ structures are $\approx 1.5$ times those of MX lattices (see Table 5.4 in H&S).

Thus $\Delta_{lattice}H^\circ(MgCl_2,s)$ will be more negative than $2\Delta_{lattice}H^\circ(MgCl,s)$, and the term $\{2\Delta_{lattice}H^\circ(MgCl,s) - \Delta_{lattice}H^\circ(MgCl_2,s)\}$ will be significantly positive (this can be quantified using the Born-Landé equation), sufficiently so to offset the −567.3 kJ mol$^{-1}$ of the $\{IE_1 - IE_2 + \Delta_a H^\circ\}$ term.

$\therefore \quad MgCl_2 + Mg \rightarrow 2MgCl$ is predicted to be endothermic, and an estimate of $\Delta_r H^\circ$ can be obtained using Born-Landé equation and cycle above.

(b) To estimate $\Delta_r H^\circ$ for the reaction:

$$CaCO_3(calcite) \rightarrow CaCO_3(aragonite)$$

dissolve each in dilute HCl, measure $\Delta_{sol}H^\circ$ for each, and apply a Hess cycle. In solution and after reaction, there will no longer be a distinction between the two phases of $CaCO_3$, a common set of products being obtained.

$$CaCO_3(calcite) + 2HCl(aq) \xrightarrow{\Delta_rH^\circ} CaCO_3(aragonite) + 2HCl(aq)$$

$\Delta_{sol}H^\circ(1)$ $\searrow$ $\swarrow$ $\Delta_{sol}H^\circ(2)$

$$Ca^{2+}(aq) + 2Cl^-(aq) + H_2O(l) + CO_2$$

From this cycle:     $\Delta_rH^\circ = \Delta_{sol}H^\circ(1) - \Delta_{sol}H^\circ(2)$

11.11    (a) The reaction to be considered is:        $SrO_2 + 2HCl \rightarrow SrCl_2 + H_2O_2$
On the left-hand side, $SrO_2$ acts as base, and HCl as acid. In the reverse direction. $H_2O_2$ acts as acid (forming $[O_2]^{2-}$ and $H^+$). Thus the conjugate acid-base pairs are:

$$SrO_2 \ + \ 2HCl \ \rightarrow \ SrCl_2 \ + \ H_2O_2$$

conjugate    conjugate    conjugate    conjugate
base(2)        acid (1)        base (1)        acid (2)

(b)     $BaO_2 + 2H_2O \rightarrow Ba(OH)_2 + H_2O_2$         (reaction of base with a weak acid)

11.12    (a)     $MO(s) + H_2O(l) \rightarrow M(OH)_2(s)$

Data:
$\Delta_fH^\circ(SrO,s) = -592.0 \ kJ \ mol^{-1}$
$\Delta_fH^\circ(BaO,s) = -553.5 \ kJ \ mol^{-1}$
$\Delta_fH^\circ(Sr(OH)_2,s)$
$\qquad = -959.0 \ kJ \ mol^{-1}$
$\Delta_fH^\circ(Ba(OH)_2,s)$
$\qquad = -944.7 \ kJ \ mol^{-1}$
$\Delta_fH^\circ(H_2O,l) = -285.5 \ kJ \ mol^{-1}$

Only enough water is available for the stated reaction, so no dissolution of hydroxide; $Sr(OH)_2$ and $Ba(OH)_2$ dissolve in water and are strong bases.

$$\Delta_rH^\circ = \Delta_fH^\circ(M(OH)_2,s) - \Delta_fH^\circ(MO,s) - \Delta_fH^\circ(H_2O,l)$$

For M = Sr:    $\Delta_rH^\circ = -959.0 - (-592.0) - (-285.5) = -81.5 \ kJ \ mol^{-1}$
For M = Ba:    $\Delta_rH^\circ = -944.7 - (-553.5) - (-285.5) = -105.7 \ kJ \ mol^{-1}$

(b) For:        $CaO(s) + H_2O(l) \rightarrow Ca(OH)_2(s)$        $\Delta_rH^\circ = -65 \ kJ \ mol^{-1}$

On going down group 2 from Ca to Ba, $\Delta_rH^\circ$ for :  $MO(s) + H_2O(l) \rightarrow M(OH)_2(s)$ becomes more negative. Set up an appropriate thermochemical cycle:

$$MO(s) + H_2O(l) \xrightarrow{\Delta_rH^\circ} M(OH)_2(s)$$

$\Delta_{lattice}H^\circ(MO,s)$ $\uparrow$ $\uparrow$ $\Delta_{lattice}H^\circ(M(OH)_2,s)$

$$M^{2+}(g) + O^{2-}(g) + H_2O(l) \xrightarrow{\Delta_rH^\circ(1)} M^{2+}(g) + 2[OH]^-(g)$$

The value of $\Delta_rH^\circ(1)$ is independent of metal and so there is no need to consider this part of the cycle in more detail. As M varies, trend in $\Delta_rH^\circ$ depends on the difference between the lattice energies (or, associated enthalpy changes) of $M(OH)_2$ and MO :

$$\Delta_rH^\circ = \Delta_rH^\circ(1) + \Delta_{lattice}H^\circ(M(OH)_2,s) - \Delta_{lattice}H^\circ(MO,s)$$

**11.13**  (a) Bubble $CO_2$ through 'limewater' – slaked lime is $Ca(OH)_2$.
(b) Reaction:

$$Ca(OH)_2(aq) + CO_2(g) \rightarrow CaCO_3(s) + H_2O(l)$$

(c) Positive result for $CO_2$ is the appearance of a white precipitate ($CaCO_3$ formation), commonly referred to as a 'milky' appearance.

**11.14**  Data to be discussed for the formation of the complexes $[M(crypt-222)]^{n+}$ :

| $M^{n+}$ | $Na^+$ | $K^+$ | $Rb^+$ |
|---|---|---|---|
| log $K$ | 4.2 | 5.9 | 4.9 |
| $K$ | $1.6 \times 10^4$ | $7.9 \times 10^5$ | $7.9 \times 10^4$ |

| $M^{n+}$ | $Mg^{2+}$ | $Ca^{2+}$ | $Sr^{2+}$ | $Ba^{2+}$ |
|---|---|---|---|---|
| log $K$ | 2.0 | 4.1 | 13.0 | >15 |
| $K$ | $10^2$ | $1.3 \times 10^4$ | $10^{13}$ | $>10^{15}$ |

> See Section 6.12 in H&S for discussion of stability constants

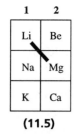

**(11.4)**

Ligand is **11.4**, a *cryptand* that encapsulates $M^{n+}$. Values of $K$ refer to equilibrium:

$$M^{n+}(aq) + crypt\text{-}222(aq) \rightleftharpoons [M(crypt\text{-}222)]^{n+}(aq)$$

Trends to consider:
- $K$ for group 2 metal ions > for group 1 metal ions;
- for group 1 $M^+$, $K$ varies within one order of magnitude – pattern irregular;
- for group 2 $M^{2+}$, $K$ varies significantly; increases down group;

Point (1): rationalized in terms of the charge on the ion; higher charge gives more stable complex. For points (2) and (3), consider metal ion sizes:

| $M^{n+}$ | $Na^+$ | $K^+$ | $Rb^+$ | $Mg^{2+}$ | $Ca^{2+}$ | $Sr^{2+}$ | $Ba^{2+}$ |
|---|---|---|---|---|---|---|---|
| $r_{ion}$ / pm | 102 | 138 | 149 | 72 | 100 | 126 | 142 |

> For further detail about this type of coordination, see the suggested reading at the end of Chapter 10 in H&S, under *'Macrocyclic effect'*.

Cryptand has a characteristic cavity-size. For group 1 metals, most stable complex formed is for $K^+$, for which good ion-size/cavity-size match obtained; $Na^+$ is rather small and $Rb^+$, rather large, for crypt-222 cavity. Down group 2, $r_{cation}$ increases with $Ba^{2+}$ giving good match to crypt-222 cavity; compare $r_{ion}$ for $K^+$ and $Ba^{2+}$. All the group 2 metal ion complexes are relatively stable however – see point (1).

**11.15**  Points to include:
- Define 'diagonal relationship' in relation to periodic table (scheme **11.5**).
- Li shows 'anomalous' behaviour in group 1 – shows similarities to Mg.
- $\Delta_a H^\circ$ values similar: $\Delta_a H^\circ$ for Li and Mg = 161  and 146 kJ mol$^{-1}$ respectively; value for Li significantly higher than for Na and other group 1 metals.
- Metallic radii similar: $r_{metal}$(Li) = 157 pm; $r_{ion}$(Mg) = 160 pm.
- Cation radii similar: $r_{ion}$(Li$^+$) = 76 pm; $r_{ion}$(Mg$^{2+}$) = 72 pm.
- Pauling electronegativities similar, $\chi^P$(Li) = 1.0, $\chi^P$(Mg) = 1.3.
- Chemical examples which relate Li to Mg rather than other group 1 metals: Li and Mg combine directly with $N_2$ to form nitrides; Li and Mg do not form stable peroxides under conditions of combustion in $O_2$; formation of many (synthetically important) organometallics (see Chapter 18 in H&S).

| 1 | 2 |
|---|---|
| Li | Be |
| Na | Mg |
| K | Ca |

**(11.5)**

**11.16**  MgO is sparingly soluble in water; in terms of the common-ion effect, the presence of $Mg^{2+}$ ions in the $MgCl_2$ solution might be expected to suppress dissolution of MgO. The observation in *increased* solubility suggests complex formation is aiding dissolution, e.g. formation of a species such as $[MgOMg]^{2+}$ or a hydrate thereof.

# 12 The group 13 elements

12.1 (a) B boron
        Al aluminium (aluminum in the US)
        Ga gallium
        In indium
        Tl thallium
(b) Non-metallic: B; metallic: Al, Ga, In, Tl
(c) $ns^2np^1$

12.2 Data from Table 12.1 in H&S are:
$$E^{\circ}(Tl^{3+}/Tl) = +0.72 \text{ V}$$
$$E^{\circ}(Tl^{+}/Tl) = -0.34 \text{ V}$$

➤ The potential diagram is therefore:

**Potential diagrams:**
**see Section 7.5 in H&S**

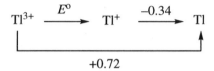

To find $E^{\circ}(Tl^{3+}/Tl^{+})$ ($E^{\circ}$ on the potential diagram), find $\Delta G^{\circ}$ for each step:

➤ For $Tl^{3+}$ to Tl: $\quad \Delta G^{\circ}_1 = -zFE^{\circ} = -(3)(0.72)F = -2.16F \text{ J mol}^{-1}$

**See answer 7.16 on p. 71 for**
**more examples of this type**
**and more details of the**
**calculation**

For $Tl^{+}$ to Tl: $\quad \Delta G^{\circ}_2 = -zFE^{\circ} = -(1)(-0.34)F = +0.34F \text{ J mol}^{-1}$

For $Tl^{3+}$ to $Tl^{+}$: $\quad \Delta G^{\circ}_3 = \Delta G^{\circ}_1 - \Delta G^{\circ}_2 = -2.16F - 0.34F = -2.50F \text{ J mol}^{-1}$

$$E^{\circ}(Tl^{3+}/Tl^{+}) = -\frac{\Delta G^{\circ}_3}{zF} = -\frac{(-2.50F)}{2F} = +1.25 \text{ V}$$

12.3 The data for the question are plotted in Figures 12.1 and 12.2.

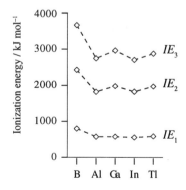

**Figure 12.1** Trends in values of the first three ionization energies for the group 13 elements.

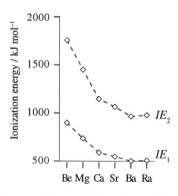

**Figure 12.2** Trends in values of the first two ionization energies for the group 2 elements.

Group 13 elements: ground state electronic configuration $ns^2np^1$. $IE_2$ refers to process:

$$M^+(g) \rightarrow M^{2+}(g)$$

and removal of one electron from an $ns^2$ pair. Group 2 elements: ground state electronic configuration is $ns^2$. $IE_2$ refers to process:

$$M^+(g) \rightarrow M^{2+}(g)$$

and removal of one electron from an $ns^1$ configuration. Both involve removal of an electron from a singly charged ion, so a direct comparison is valid on charge grounds. The lower values of $IE_2$ for the group 2 metals correspond to the fact an electron is being removed from a singly occupied level. The decrease in values of $IE_2$ on descending group 2 corresponds to increased distance between nucleus and outer electron – $ns$ orbital becomes more diffuse as $n$ increases from 2 for Be to 6 for Ba. Trend in values of $IE_2$ for the group 13 elements is less simple. Decrease from B to Al as for decrease from Be to Mg. On going from Ca to Ga, the $3d$ metals are present, from Sr to In, the $4d$ metals are present, between Ba and Tl, there are the $5d$ and $4f$ metals. Discontinuities in values of $IE_2$ are due to the failure of the $d$ and $f$-electrons (which have low screening power) to compensate for the increase in nuclear charge.

➤ Screening effects: see Section 1.7 in H&S

12.4    (a) Redox reaction: B reduced, Mg oxidized:

$$B_2O_3(s) + 3Mg(s) \rightarrow 2B(s) + 3MgO(s)$$

(b) $Al_2O_3$ is amphoteric, $Fe_2O_3$ is basic; only $Al_2O_3$ reacts, leaving solid $Fe_2O_3$:

$$Al_2O_3(s) + 3H_2O(l) + 2NaOH(aq) \rightarrow 2Na[Al(OH)_4](aq)$$

(c)    $2Na[Al(OH)_4](aq) + CO_2(g) \rightarrow Al_2O_3 \cdot 3H_2O(s) + Na_2CO_3(aq) + H_2O(l)$

12.5    Nuclear spin data:    $^1H$        $I = {}^1/_2$        100%

$^{31}P$        $I = {}^1/_2$        100%

$^{11}B$        $I = {}^3/_2$        80.4%        (ignore $^{10}B$)

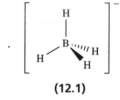

**(12.1)**

(a) $[BH_4]^-$, structure **12.1**; $^{11}B$ couples to 4 equivalent $^1H$ to give a binomial quintet (Figure 12.3). Measure $J_{BH}$ between any pair of adjacent lines.

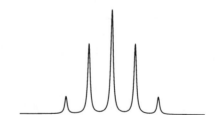

**Figure 12.3**  Simulated $^{11}B$ NMR spectrum of $[BH_4]^-$.

**Figure 12.4**  Simulated $^1H$ NMR spectrum of $[BH_4]^-$.

(b) One signal arising from the 4 equivalent $^1H$; each $^1H$ couples to the $^{11}B$ nucleus; $I = {}^3/_2$, so four possible spin states ($+{}^3/_2, +{}^1/_2, -{}^1/_2, -{}^3/_2$) giving rise to a 1:1:1:1 multiplet (Figure 12.4). **NB**: *this is not a binomial quartet.* Measure $J_{BH}$ between any pair of adjacent lines; value must be the same as that obtained from the $^{11}B$ NMR spectrum in part (a).

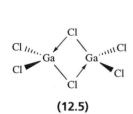

**(12.2)**

**(12.3)**

(c) $H_3B \cdot PMe_3$, structure **12.2**. $^{11}B$ couples to one $^{31}P$ nucleus and to 3 equivalent $^1H$ nuclei to give a doublet of binomial quartets (Figure 12.5). See answer 2.23b (p. 22) for a more detailed explanation for the related $H_3B \cdot PMe_2Ph$.

(d) The notation $^{11}B\{^1H\}$ means 'proton decoupled $^{11}B$'. $^{11}B$–$^1H$ spin-spin coupling is instrumentally removed from the $^{11}B$ NMR spectrum. $^{11}B\{^1H\}$ NMR spectrum of $THF \cdot BH_3$ (**12.3**) is a singlet. See answer 2.23a for the $^{11}B$ NMR spectrum of $THF \cdot BH_3$

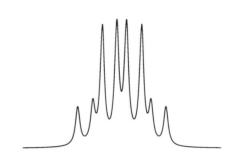

**Figure 12.5** Simulated $^{11}B$ NMR spectrum of $H_3B \cdot PMe_3$.

12.6   Thermite process is:   $2Al(s) + Fe_2O_3(s) \rightarrow Al_2O_3(s) + 2Fe(s)$

Data (at 298 K):
$\Delta_f H^\circ(Al_2O_3,s)$
   $= -1675.7$ kJ mol$^{-1}$
$\Delta_f H^\circ(Fe_2O_3,s)$
   $= -824.2$ kJ mol$^{-1}$
$\Delta_f H^\circ(Al,s) = 0$ (standard state)
$\Delta_f H^\circ(Fe,s) = 0$ (standard state)

To find $\Delta_r H^\circ$, apply a Hess cycle.

$\Delta_r H^\circ = \Delta_f H^\circ(Al_2O_3,s) + 2\Delta_f H^\circ(Fe,s) - \Delta_f H^\circ(Fe_2O_3,s) - 2\Delta_f H^\circ(Al,s)$

   $= -1675.7 - (-824.2)$

   $= -851.5$ kJ mol$^{-1}$  (per mole of $Fe_2O_3$)

$\Delta_{fus} H^\circ(Fe,s) = 13.8$ kJ mol$^{-1}$; this gives the amount of heat energy needed to melt 1 mole of Fe. Enough energy is released in the thermite process to melt the Fe formed.

12.7   Scheme **12.4** shows $BH_3$ acting:
   • as a Lewis base by donating a pair of electrons (a B–H bonding pair);
   • as a Lewis acid by accepting a pair of electrons.
The result is the formation of two 3c-2e B–H–B bridges in $B_2H_6$.

**(12.4)**

12.8   $Ga_2H_6$ is isostructural with $B_2H_6$; analogous bonding pictures are appropriate. In $Ga_2H_6$, 2c-2e localized terminal Ga–H bonds and 3c-2e Ga–H–Ga bridge bonds:

Overlap of H $1s$ orbital with two B $sp^3$ hybrid orbitals

**(12.5)**

In $Ga_2Cl_6$ (**12.5**) there are sufficient valence electrons to invoke all 2c-2e Ga-Cl interactions; analogous to $Al_2Cl_6$.

**12.9**    Order of relative stabilities of adducts $L \cdot BH_3$ increases for L:

$$Me_2O < THF < Me_2S < Me_3N < Me_3P < H^-$$

(a) Adding $Me_3N$ to $THF \cdot BH_3$ in THF solution: $Me_3N$ displaces THF to give $Me_3N \cdot BH_3$. In the $^{11}B$ NMR, the chemical shift of the signal will change. (Observed shift is from $\delta$ 0 to $\delta -12$.) Each of $THF \cdot BH_3$ and $Me_3N \cdot BH_3$ exhibits a binomial (1:3:3:1) quartet in the $^{11}B$ NMR spectrum.

(b) No displacement since $Me_2O$ is lower in the series than $PMe_3$. Monitor with either $^{11}B$ or $^{31}P$ NMR spectroscopy – no change observed.

(c) $[BH_4]^-$ is an adduct of $BH_3$ and $H^-$. Since $H^-$ is higher in the series than THF, $[BH_4]^-$ is stable in THF solution with respect to the formation of $THF \cdot BH_3$. Monitor by $^{11}B$ NMR spectroscopy – no change in the signal.

(d) $Ph_2PCH_2CH_2PPh_2$ **(12.6)** has two donor sites, each of which can coordinate to $BH_3$ giving **12.7**, the likely product if $THF \cdot BH_3$ is in excess. In $^{11}B$ NMR spectrum, see a change from binomial quartet for $THF \cdot BH_3$ to doublet of quartets (Figure 12.5). If only one P→B coordinate bond forms, the final $^{11}B$ NMR spectrum will contain both a 1:3:3:1 quartet and a doublet of quartets at different chemical shifts. $^{31}P$ NMR spectroscopy could also be used to follow the reaction.

(12.6)

(12.7)

**12.10**    (a) Can be rationalized in terms of smaller B atom being sterically protected by 4 OH groups from attack, or in terms of possible participation of Al $3d$ orbitals expanding valence shell of Al allowing formation of 5-coordinate intermediate.

(b) For hydrolysis of $B_2H_6$:

$$Rate \propto \left(P_{B_2H_6}\right)^{\frac{1}{2}} \left(P_{H_2O}\right)$$

Reaction scheme must show that the rate is order $^1/_2$ with respect to $B_2H_6$, and first order with respect to $H_2O$. Reaction steps consistent with this are:

$$B_2H_6 \underset{}{\overset{fast}{\rightleftharpoons}} 2BH_3$$

$$BH_3 + H_2O \xrightarrow{slow} products$$

(c) In water, equilibrium for $B(OH)_3$:

$$B(OH)_3 + 2H_2O \rightleftharpoons [B(OH)_4]^- + [H_3O]^+ \qquad pK_a = 9.1$$

➤ For relevant $pK_a$ values, see Sections 6.5 and 12.7 in H&S

so is neutral to bromocresol green (pH range 3.8-4.5). In water, $[HF_2]^-$ is in equilibrium:

$$[HF_2]^- + H_2O \rightleftharpoons [H_3O]^+ + 2F^- \qquad pK_a = 3.45$$

and appears acidic to bromocresol green. When excess $B(OH)_3$ is added to $K[HF_2]$:

$$B(OH)_3 + 2[HF_2]^- \rightarrow [BF_4]^- + 2H_2O + [OH]^-$$

and the solution appears alkaline to the indicator.

**12.11**    (a)    $BCl_3 + EtOH \rightarrow B(OEt)_3 + 3HCl$    (Ethoxide for chloride displacement)

(b)    $BF_3 + EtOH \rightarrow EtOH \cdot BF_3$    (Adduct formation)

(c)     $BCl_3 + 3PhNH_2 \rightarrow B(NHPh)_3 + 3HCl$     (B(NHPh)₃ stabilized by B–N π-interactions)

(d)     $BF_3 + KF \rightarrow KBF_4$     (Contains $[BF_4]^-$ formed by $BF_3$ accepting F⁻)

12.12    (a) Cryolite: $Na_3[AlF_6]$.
(b) Perovskite: $CaTiO_3$.
(c) Rewrite $Na_3[AlF_6]$ as $Na_2[NaAlF_6]$ giving a formula that is equivalent to $NaXF_3$ in which X is Na or Al. The formula $NaXF_3$ can be related to $CaTiO_3$. Cryolite possesses a perovskite lattice (Figure 12.6) with $^2/_3$ of the Na in the Ca sites , and the Al and $^1/_3$ of the Na in the Ti sites.

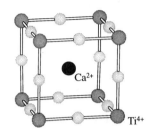

**Figure 12.6** Unit cell of perovskite, $CaTiO_3$.

12.13    (a) Group 13 element has 3 valence electrons; add extra electron(s) from negative charges.
$[MCl_6]^{3-}$
Electrons in valence shell of M = 3 + 3 = 6
Number of bonding pairs (6 M–Cl bonds) = 6
Total number of electron pairs = 6  (no lone pairs)
'Parent' shape = molecular shape = octahedral; see structure **12.8**.
$[MCl_5]^{2-}$
Electrons in valence shell of M = 3 + 2 = 5
Number of bonding pairs (5 M–Cl bonds) = 5
Total number of electron pairs = 5  (no lone pairs)
'Parent' shape = molecular shape = trigonal bipyramidal; see structure **12.9**.
$[MBr_4]^-$
Electrons in valence shell of M = 3 + 1 = 4
Number of bonding pairs (4 M–Br bonds) = 4
Total number of electron pairs = 4  (no lone pairs)
'Parent' shape = molecular shape = tetrahedral; see structure **12.10**.
(b) VSEPR suggests that structures should be trigonal bipyramidal. For 5-coordination, energy difference between trigonal bipyramidal and square-based pyramidal structures is usually small. Balance could be tipped in solid state by crystal packing effects.
(c) Compounds contain chloro anions, and can be prepared from $TlCl_3$ (Lewis acid-base reactions). A source of Cl⁻ is needed, as well as a cation:

$TlCl_3 + H_2N(CH_2)_5NH_2 + 2HCl \rightarrow [H_3N(CH_2)_5NH_3]^{2+} + [TlCl_5]^{2-}$

$2TlCl_3 + 3CsCl \rightarrow 3Cs^+ + [Tl_2Cl_9]^{3-}$

(d) Monomeric $GaCl_2$ would be paramagnetic (**12.11**); since '$GaCl_2$' is *diamagnetic*, formulation $Ga[GaCl_4]$ containing Ga⁺ and $[GaCl_4]^-$ ions is correct.

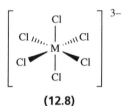

**(12.8)**

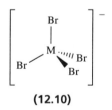

**(12.9)**

**(12.10)**

**(12.11)**

12.14    (a) Dissolution requires source of F⁻ to form complex ion (i.e. requires KF to be present in HF, see Section 8.7 in H&S for self-ionization of anhydrous HF):

$AlF_3 + 3F^- \rightarrow [AlF_6]^{3-}$

On adding $BF_3$, $[BF_4]^-$ is formed in preference to $[AlF_6]^{3-}$, i.e. $BF_3$ accepts F⁻ from $[AlF_6]^{3-}$. This regenerates $AlF_3$ which precipitates.

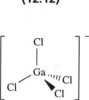

**(12.12)**

**(12.13)**

(b) The data (i.e. 4 absorptions at similar wavenumbers, close values for Ga species, and shifted a little for Ge species) indicate a common species for $GaCl_2$ and $GaCl_3$/HCl, and one which is isostructural with $GeCl_4$ (**12.12**, Ge is group 14). Gallium species must be $[GaCl_4]^-$ (**12.13**, Ga is group 13). $GaCl_3$ reacts with $Cl^-$ to give $[GaCl_4]^-$, and $GaCl_2$ exists as $Ga[GaCl_4]$, there being only one Ga–Cl containing species according to the Raman spectroscopic data.

(c) Group 1 metal triiodides are $M^+[I_3]^-$, so solid $TlI_3$ is $Tl^+[I_3]^-$. Hydrated $Tl_2O_3$ is insoluble, and oxidation of $Tl^+(aq)$ to solid $Tl_2O_3$ is very much easier than to $Tl^{3+}(aq)$ (see Section 7.3 in H&S). The oxidizing agent is $I_2$, from:

$$[I_3]^- \rightarrow I_2 + I^-$$

**12.15**  In the solid state (Figure 12.7), $[Al(BH_4)_4]^-$ has two H environments, terminal and bridging. The data in the question refer to $[Al(BH_4)_4]^-$ *in solution*.
(a) 298 K to 203 K: one broad signal (in addition to cation protons) means that the terminal and bridging H atoms are involved in a dynamic process that renders them equivalent on the NMR spectroscopic timescale.
(b) All $^{11}B$ nuclei are equivalent giving rise to one signal in the $^{11}B$ NMR spectrum. The quintet (analogous to Figure 12.3, p. 100) is due to coupling of $^{11}B$ nucleus to 4 *equivalent* $^1H$ nuclei, indicating that the dynamic process involves exchange of terminal and bridging H about a *single* B atom. Notice that each $^{11}B$ nucleus 'sees' *only* 4 $^1H$ nuclei, and so the H atoms are not fluxional over the whole molecule (compare with $[B_3H_8]^-$ in answer 12.19b).
(c) IR and NMR spectroscopic timescales are *not the same*. On NMR timescale, the anion appears fluxional, but on IR timescale, the anion appears to be static. The IR spectroscopic data are consistent with a structure like that in the solid state.

**Figure 12.7**  The structure of $[Al(BH_4)_4]^-$ in the solid state as determined for the salt $[Ph_3MeP][Al(BH_4)_4]$.

**12.16**  Points to include:
- Borazine, $(HBNH)_3$, has planar structure **12.14** with equal B–N bond lengths.
- Bonding can be represented by pair of resonance structures of which **12.15** is one; $B^-$ and $N^+$ are isoelectronic with C; $(HBNH)_3$ is isoelectronic (and isostructural) with $C_6H_6$.
- In $C_6H_6$, all C atoms carry same partial charge; in $(HBNH)_3$, difference in electronegativities of B and N makes B $\delta^+$ and N $\delta^-$; B susceptible to nucleophilic attack, and N susceptible to electrophilic attack. *NB*: compare actual partial charges with the formal charges in VB theory.
- Unlike benzene (which is kinetically inert towards attack by, for example, HCl and $H_2O$), borazine undergoes addition reactions, e.g.

144 pm

**(12.14)**

**(12.15)**

$$(HBNH)_3 + 3HCl \rightarrow [ClHBNH_2]_3$$

$$(HBNH)_3 + 3H_2O \rightarrow [H(HO)BNH_2]_3$$
**(12.16)**

- Fully hydrogenated derivative made by substitution:

$$(ClHBNH_2)_3 \xrightarrow{Na[BH_4]} (H_2BNH_2)_3$$

- Saturated products have chair-conformation of $B_3N_3$-ring; analogues of *cyclo*-$C_6H_{12}$.

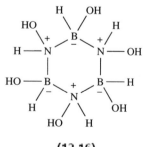

**(12.16)**

12.17    Each compound can be described in terms of localized 2c-2e bonds, utilizing a lone pair of electrons from each N and vacant orbital on each Al centre. Each Al and N centre can be considered to be *sp*³ hybridized, although bond angles in each compound do not exactly match idealized hybrid model. Diagram **12.17** shows a bonding representation using N→Al coordinate bonds. Alternative is to draw resonance structure **12.18**.

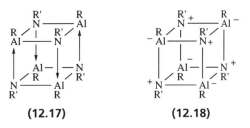

(12.17)          (12.18)

12.18    $B_5H_9$

There are 5 {BH}-units and 4 additional H atoms.

Each {BH}-unit provides 2 electrons, each H atom provides 1 electron.

Total number cage-bonding electrons $= (5 \times 2) + 4 = 14$ electrons $= 7$ pairs

$B_5H_9$ has 7 pairs of electrons with which to bond 5 {BH}-units.

This means that there are $(n + 2)$ pairs of electrons for $n$ vertices.

∴ $B_5H_9$ is a *nido*-cage, derived from a parent 6-vertex deltahedron, see **12.19**.

4 extra H atoms adopt B–H–B bridging positions around the square face.

$[B_8H_8]^{2-}$

There are 8 {BH}-units and no additional H atoms.

Each {BH}-unit provides 2 electrons.

2 electrons from the 2– charge.

Total number cage-bonding electrons $= (8 \times 2) + 2 = 18$ electrons $= 9$ pairs

$[B_8H_8]^{2-}$ has 9 pairs of electrons with which to bond 8 {BH}-units.

This means that there are $(n + 1)$ pairs of electrons for $n$ vertices.

∴ $[B_8H_8]^{2-}$ is a *closo*-cage, an 8-vertex deltahedron (dodecahedron), see **12.20**.

$C_2B_{10}H_{12}$

There are 10 {BH}-units and 2 {CH}-units.

Each {BH}-unit provides 2 electrons; each {CH}-unit provides 3 electrons.

Total number cage-bonding electrons $= (10 \times 2) + (2 \times 3) = 26$ electrons $= 13$ pairs

$C_2B_{10}H_{12}$ has 13 pairs of electrons with which to bond 12 cluster-units.

This means that there are $(n + 1)$ pairs of electrons for $n$ vertices.

∴ $C_2B_{10}H_{12}$ is a *closo*-cage, a 12-vertex deltahedron (icosahedron), see **12.21**. Structure **12.21** shows the 1,2-isomer with the C atoms adjacent to each other. Cage atom numbering follows the scheme shown in **12.21**, with atoms 2-6 being around the top pentagonal ring, and atoms 7-11 around the lower pentagonal ring. Other isomers of $C_2B_{10}H_{12}$ are 1,7- and 1,12-$C_2B_{10}H_{12}$ in which the C atoms are remote from each other.

$[B_6H_9]^-$

There are 6 {BH}-units and 3 additional H atoms.

Each {BH}-unit provides 2 valence electrons; each H provides 1 electron.

1 electron from 1– charge.

Total number cage-bonding electrons $= (6 \times 2) + 3 + 1 = 16$ electrons $= 8$ pairs

$[B_6H_9]^-$ has 8 pairs of electrons with which to bond 6 {BH}-units.

This means that there are $(n + 2)$ pairs of electrons for $n$ vertices.

∴ $[B_6H_9]^-$ is a *nido*-cage, derived from a parent 7-vertex deltahedron, see **12.22**.

3 extra H atoms adopt B–H–B bridging positions around the open face.

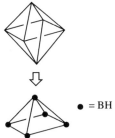

● = BH

(12.19)

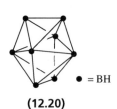

● = BH

(12.20)

1
2
7
12

● = BH
○ = CH

(12.21)

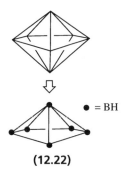

● = BH

(12.22)

In a static model, two arrangements of the bridging H atom could be suggested:

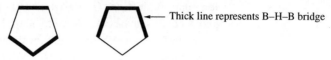

Thick line represents B–H–B bridge

In practice, the anion is fluxional in solution.

12.19 (a) Reaction to consider: $B_5H_9 \xrightarrow{2e^-} [B_5H_9]^{2-} \xrightarrow{2H^+} B_5H_{11}$

From Wade's rules, adding 2 electrons means parent deltahedron changes from being $n = 6$ (for $B_5H_9$, see answer 12.18) to $n = 7$ (for $B_5H_{11}$). Therefore, a change from a *nido-* to *arachno-*cage is predicted (and is observed).

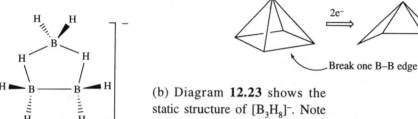

Break one B–B edge

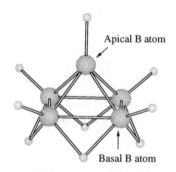

Apical B atom

Basal B atom

**Figure 12.9** The structure of $B_5H_9$.

(12.23)

(b) Diagram **12.23** shows the static structure of $[B_3H_8]^-$. Note there are 2 B environments, therefore this is inconsistent with the solution $^{11}B$ NMR spectrum. The anion must be undergoing a dynamic process in solution, with *every* B centre 'seeing' all 8 H atoms. This gives a nonet (Figure 12.8).

**Figure 12.8** Simulated $^{11}B$ NMR spectrum of $[B_3H_8]^-$. The outer lines of the binomial nonet are just visible; relative intensities of lines in the nonet are 1:8:28:56:70:56:28:8:1.

(c) Intermolecular elimination of $H_2$: $2B_5H_9 \rightarrow B_{10}H_{16} + H_2$
suggests cage-coupling by formation of inter-cage B–B bond: 2 different B sites in $B_5H_9$ (Figure 12.9). Coupling may be $B_{apical}$–$B_{apical}$, $B_{basal}$–$B_{apical}$ or $B_{basal}$–$B_{basal}$, so 3 possible isomers of $B_{10}H_{16}$:

12.20 (a) Electrophilic substitution; initial attack at apical B atom (see Figure 12.9):

$B_5H_9 + Br_2 \xrightarrow{-HBr} \text{1-BrB}_5H_8 \xrightarrow{\text{isomerization}} \text{2-BrB}_5H_8$

(b) $PF_3$ is Lewis base; attacks B (Lewis acid), displacing $H_2$; electron count for the cluster remains unchanged – no structural change for the cage:

$B_4H_{10} + PF_3 \rightarrow B_4H_8(PF_3) + H_2$

(c) KH supplies $H^-$, a base which deprotonates the cluster by removal of *bridging* $H^+$:

$\text{1-BrB}_5H_8 \xrightarrow{\text{KH, 195 K}} K^+[\text{1-BrB}_5H_7]^-$

(d) Reaction with ROH degrades cluster; B–H and B–B interactions are destroyed, but *not* the B–C bond; $H_2$ is liberated:

$\text{2-MeB}_5H_8 + 14ROH \rightarrow 4B(OR)_3 + MeB(OR)_2 + 11H_2$

> Use Figures 12.20 and 12.21 in H&S to help with this question

> In $B_5H_9$, the apical site is numbered 1, and the basal sites are 2-5.

# 13 The group 14 elements

13.1 (a)    C      carbon
            Si      silicon
            Ge     germanium
            Sn     tin
            Pb     lead

(b) Non-metallic: C; semi-metallic: Ge, Si; metallic: Sn, Pb

(c) $ns^2np^2$

13.2 Figures 13.1-13.3 show the data to be discussed in this answer.

(a) Melting points: see Figure 13.1. Data for C refer to diamond; diamond, Si and Ge are isostructural with 3D-covalent diamond-type lattice (see Figure 5.4, p. 45). Melting solid breaks down lattice; trend in melting points for C, Si and Ge follows decrease in strength of covalent bonds (see Table 13.2 in H&S). White Sn (stable allotrope at 298 K) has metallic lattice with coordination number of 6; lead has ccp structure (see Section 5.3 in H&S).

(b) Values of $\Delta_aH^\circ(298\ K)$: see Figure 13.2. Enthalpy change refers to the process:

$$E(s) \rightarrow E(g)$$

i.e. the solid state lattice for element E is *completely* disrupted to give gas phase atoms. Decrease down group 14 corresponds to decrease in strength of E–E bonding (see Table 13.2 in H&S); comparing the isostructural C, Si and Ge, the orbital overlap becomes less effective as the principal quantum number increases from 2 (for C) to 4 (for Ge). Trend continues for the metallic members of the group.

(c) Values of $\Delta_{fus}H^\circ(mp)$: see Figure 13.3. Enthalpy change refers to the process:

$$E(s) \rightarrow E(l)$$

i.e. the solid state lattice for element E collapses but is not completely disrupted. Value of $\Delta_{fus}H^\circ(mp) < \Delta_aH^\circ(298\ K)$ for a given element. Trend in $\Delta_{fus}H^\circ(mp)$ approximately mimics that for the melting points.

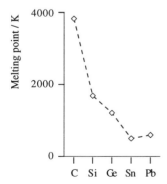

**Figure 13.1** Trend in melting points for the group 14 elements.

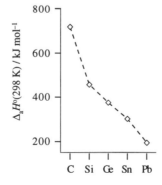

**Figure 13.2** Trend in standard enthalpies of atomization for the group 14 elements.

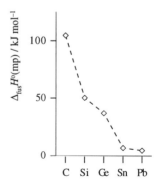

**Figure 13.3** Trend in standard enthalpies of fusion for the group 14 elements.

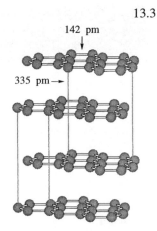

142 pm

335 pm →

**Figure 13.4** Part of the layered structure of graphite. The vertical lines are shown to emphasize which atoms lie over which; adjacent layers are staggered.

13.3    Structure of α-graphite: see Figure 13.4. Comment on bonding: 3-coordinate C atoms within layers with delocalized π-bonding within each layer.

(a) Interactions between the layers in the solid state are weak (van der Waals) enabling the layers to slide past one another easily. This slippage is the basis for the use of graphite powder as a solid lubricant.

(b) Each C atom is 3-coordinate in a layer; consider as $sp^2$ hybridized with 3 electrons used within σ-bonding framework and the fourth electron occupying a $2p$ atomic orbital involved in electronic delocalization within a layer:

Electrical conductivity of graphite is dependent on direction: high conductivity (resistivity is low) in a direction parallel to the layers, but perpendicular to this direction, conductivity is low. Use in electrodes follows from the high electrical conductivity, and the relative chemical inertness of graphite under ambient conditions.

(c) Density of diamond > graphite (3.51 versus 2.25 g cm$^{-3}$). At high pressures, graphite converts to diamond and this is a method for preparing artificial diamonds from graphite.

13.4    Look at Figure 13.7 in H&S. The fulleride anions are in an fcc arrangement. Use the method of working as for unit cells (see p. 46).

| Site | Number of $[C_{60}]^{n-}$ | Number of $K^+$ |
|---|---|---|
| Central | 0 | 9 |
| Corner | $8 \times \frac{1}{8} = 1$ | 0 |
| Face | $6 \times \frac{1}{2} = 3$ | 0 |
| Edge | 0 | $12 \times \frac{1}{4} = 3$ |
| Total | 4 | 12 |

The ratio of $[C_{60}]^{n-} : K^+ = 1 : 3$, and the charge on the fulleride ion is 3–.

13.5    Selected reaction types that illustrate the carbon-carbon double bond character include:
  • addition of halogens;
  • cycloaddition reactions;
  • formation of epoxide $C_{60}O$;
  • organometallic π-complex formation.

For details of individual reactions, refer to Section 13.4 (subsection 'Fullerenes: reactivity') and Section 23.10 (subsection 'Alkene ligands', equations 23.65 and 23.66) in H&S.

13.6    (a) Carbides belong to different families, and their reactivity with water is a distinguishing feature.

$Mg_2C_3$ and $CaC_2$ contain $[C=C=C]^{4-}$ and $[C\equiv C]^{2-}$ ions respectively; reactions with water are:

$$Mg_2C_3 + 4H_2O \rightarrow CH_3C\equiv CH + 2Mg(OH)_2$$

$$CaC_2 + 2H_2O \rightarrow HC\equiv CH + Ca(OH)_2$$

$ThC_2$ contains $[C_2]^{4-}$. Hydrolysis gives a mixture of $C_2H_2$, $C_2H_6$ and $H_2$ and distinguishes this carbide from one containing an acetylide ion, $[C_2]^{2-}$.

TiC is an interstitial carbide, i.e. cubic close-packed metal atoms with C atoms in octahedral interstitial holes. Interstitial carbides are robust materials that are inert towards the action of water.

> Review acid-base properties of non-aqueous solvents in Section 8.4 in H&S

(b) In liquid ammonia, $[NH_4]Br$ acts as an acid because it is a source of $[NH_4]^+$ ions. Magnesium silicide reacts with dilute acids to give silanes. Therefore, in liquid $NH_3$, $Mg_2Si$ reacts with $[NH_4]Br$ according to:

$$Mg_2Si + 4[NH_4]Br \rightarrow 2MgBr_2 + 4NH_3 + SiH_4$$

Reactions of $Mg_2Si$ with acids are usually non-specific, giving mixtures of silanes.

(c) If hydrolysis of compound **13.1** involves cleavage of the Si–H bond, then the rate of alkaline hydrolysis of the deuterated compound **13.2** should be slower than the rate of alkaline hydrolysis of **13.1**. In this case, the reaction would show a kinetic isotope effect (see Section 2.9 in H&S) because the bond enthalpy of the Si–D bond is greater than that of the Si–H bond. Since the rates of hydrolysis of **13.1** and **13.2** are the same, it follows that the rate-determining step is *not* the cleavage of the Si–H (or Si–D) bond.

An alternative mechanism that can be suggested is attack by $[OH]^-$ at the Si centre in the rate-determining step.

**(13.1)**

**(13.2)**

13.7    (a) An explanation of the planarity of the $NSi_3$-skeleton in **13.3** is not straightforward. One theory centres on $(p\text{-}d)\pi$-bonding as shown below:

**(13.3)**

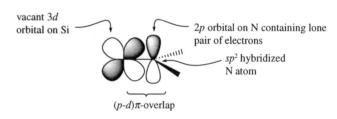

However, current arguments favour 'negative hyperconjugation' in which the N lone pair electrons are donated into the vacant Si–$C_{methyl}$ $\sigma^*$ orbitals rather than into the Si $3d$ orbitals. The difficulty with the $(p\text{-}d)\pi$-bonding scheme is that it assumes that Si $3d$ orbitals lie at low enough energy to be accessible, and this is debatable.

(b) At 298 K, $CO_2$ exists as linear molecules (**13.4**) and $SiO_2$ forms an infinite 3D-lattice (**13.5**). Both have covalent bonding.

The bonding in $CO_2$ can be described in terms of $sp$ hybridized C, and the formation of two C–O $\sigma$-bonds and two C–O $\pi$-bonds involving overlap between C $2p$ and O $2p$ orbitals.

**(13.4)**

**(13.5)**

The bonding in $SiO_2$ is best described in terms of $sp^3$ hybridized Si with Si–O $\sigma$-bonding. In terms of bond enthalpy terms:

| Bond | Bond enthalpy term / kJ mol⁻¹ |
|------|-------------------------------|
| C=O | 810 |
| C–O | 359 |
| Si=O | 642 |
| Si–O | 466 |

you can see that the $(p\text{-}p)\pi$ contribution to the C=O double bond is much greater than to the Si=O bond – this reflects the difference between $2p$-$2p$ overlap and $3p$-$2p$ overlap. To consider the balance between discrete molecular and macromolecular $CO_2$ on the grounds of bond enthalpies at 298 K, you must look at the difference between two C=O (1620 kJ mol⁻¹) and four C–O bonds (1436 kJ mol⁻¹): the discrete molecule with two C=O bonds wins. A similar exercise for $SiO_2$ shows that the macromolecular structure with four Si–O wins.

**13.8** Use VSEPR theory to suggest shapes for the molecular species in the question. For full details of the method, look back to answer 1.29 (p. 11).

(a) ClCN

    C: number of valence electrons = 4
    Number of bonding 'pairs' (Cl–C and C≡N) = 2
    No lone pairs on C
    'Parent' shape = molecular shape = linear (see **13.6**)

$$Cl\!-\!\!-\!C\!\equiv\!\!\equiv\!N$$

**(13.6)**

(b) OCS

    C: number of valence electrons = 4
    Number of bonding 'pairs' (C=O and C=S) = 2
    No lone pairs on C
    'Parent' shape = molecular shape = linear (see **13.7**)

$$O\!=\!\!=\!C\!=\!\!=\!S$$

**(13.7)**

(c) $[SiH_3]^-$

    Si: number of valence electrons = 4
    1 electron from the negative charge   ∴ 5 electrons in valence shell of Si
    Number of bonding pairs (3 Si–H) = 3
    This leaves 1 lone pair on Si
    'Parent' shape = tetrahedral
    Molecular shape = trigonal pyramidal (see **13.8**)

**(13.8)**

(d) $[SnCl_5]^-$

    Sn: number of valence electrons = 4
    1 electron from the negative charge   ∴ 5 electrons in valence shell of Sn
    Number of bonding pairs (5 Sn–Cl) = 5
    No lone pairs on Sn
    'Parent' shape = molecular shape = trigonal bipyramidal (see **13.9**)

**(13.9)**

(e) $Si_2OCl_6$

    First, recognize that the O atom bridges between the Si atoms.
    Si: number of valence electrons = 4
    Number of bonding pairs (3 Si–Cl + 1 Si–O) = 4
    No lone pairs on Si
    'Parent' shape = molecular shape = tetrahedral at each Si (see **13.10**)

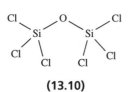

**(13.10)**

(f) $[Ge(C_2O_4)_3]^{2-}$

    First, note that you are dealing with 3 didentate oxalate ligands (see

                                   Table 6.7 in H&S)

    Ge: number of valence electrons = 4
    2 electrons from the negative charge   ∴ 6 electrons in valence shell of Ge

Number of bonding pairs (6 Ge–O) = 6
No lone pairs on Ge
'Parent' shape = molecular shape = octahedral (see **13.11**)

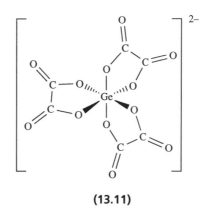

**(13.11)**

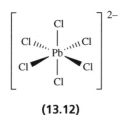

**(13.12)**

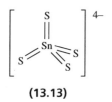

**(13.13)**

(g)  [PbCl$_6$]$^{2-}$
    Pb: number of valence electrons = 4
    2 electrons from the negative charge  ∴ 6 electrons in valence shell of Pb
    Number of bonding pairs (6 Pb–Cl) = 6
    No lone pairs on Pb
    'Parent' shape = molecular shape = octahedral (see **13.12**)

(h)  [SnS$_4$]$^{4-}$
    Sn: number of valence electrons = 4
    4 electrons from the negative charge  ∴ 8 electrons in valence shell of Sn
    Number of bonding 'pairs' (4 Sn=S) = 4
    No lone pairs on Sn
    'Parent' shape = molecular shape = tetrahedral (see **13.13**)

13.9    (a) This question is about the application of Wade's rules – discussed in detail in answer 12.18 (p. 105).

➤ Look at worked example 13.1 in H&S

Apply Wade's rules to [Sn$_9$Tl]$^{3-}$ :
    9 Sn atoms, and each provides 2 electrons (assume 1 lone pair per Sn).
    1 Tl atom provides 1 electron (assume 1 lone pair on the Tl centre).
    3 electrons from the 3– charge.
    Total number cage-bonding electrons = (9 × 2) + 1 + 3
                                   = 22 electrons = 11 pairs
    [Sn$_9$Tl]$^{3-}$ has 11 pairs of electrons with which to bond 10 cluster atoms.
    This means that there are ($n$ + 1) pairs of electrons for $n$ vertices.
    ∴ [Sn$_9$Tl]$^{3-}$ is a *closo*-cage, a 10-vertex deltahedron (bicapped square
                                         antiprism), see **13.14**.

(b) Isomers arise because the bicapped square antiprism has two different sites (see **13.4**). In principle, the Tl atom can occupy either of these sites, and 2 isomers of [Sn$_9$Tl]$^{3-}$ may be drawn.

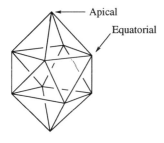

Apical
Equatorial

**(13.14)**

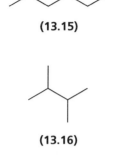

**(13.15)**

**(13.16)**

13.10    For information on the structures and chemistries of hydrides of the group 14 elements, see Section 13.6 in H&S. The hydrides of B and Al (group 12) are discussed in Section 12.5 of H&S. Points that should be included in your answer:

- Alkanes $C_nH_{2n+2}$ known for wide range of $n$; all have molecular structures with tetrahedral C; isomers with unbranched and branched chains, e.g. **13.15-13.16**.
- Silanes $Si_nH_{2n+2}$ for $1 \le n \le 10$; all have molecular structures; isomers with straight and branched chains; structural analogues of alkanes.
- Comparing straight chain silanes with alkanes with $n$ C or Si atoms: for a given $n$, boiling point of silane > alkane; silanes (but not alkanes, the difference being kinetic) are spontaneously explosively inflammable in air:

$$SiH_4 + 2O_2 \rightarrow SiO_2 + 2H_2O$$

- Germanes $Ge_nH_{2n+2}$ (straight and branched chain isomers) known for $1 \le n \le 9$; structural analogues of alkanes and silanes.
- Stannane $SnH_4$; isostructural with $CH_4$, but decomposes at 298 K:

$$SnH_4 \rightarrow Sn + 2H_2$$

- $PbH_4$ is poorly characterized – may not actually exist.
- Reactivities of hydrides of Si, Ge and Sn follow trend $SiH_4 > GeH_4 < SnH_4$; $GeH_4$ (like $SiH_4$) is highly inflammable in air.
- $CH_4$, $SiH_4$ and $GeH_4$ all insoluble in water; $CH_4$ kinetically stable with respect to hydrolysis by water, but $SiH_4$ hydrolysed, also by alkalis:

$$SiH_4 + 2NaOH + H_2O \rightarrow Na_2SiO_3 + 4H_2$$

- Reactions of $SiH_4$ and $GeH_4$ with alkali metals give synthetically useful salts of $[EH_3]^-$:

$$2GeH_4 + 2M \xrightarrow{\text{liquid NH}_3} 2M[GeH_3] + H_2$$

- $BH_3$ versus $CH_4$: hydrides of first member of groups 12 and 13 respectively. $BH_3$ is trigonal planar, and B has sextet of valence electrons – unstable with respect to dimerization (see **12.4**, p. 101). $B_2H_6$ reacts with Lewis bases, L, forming adducts $L{\rightarrow}BH_3$; $B_2H_6$ rapidly hydrolysed to $B(OH)_3$ and $H_2$. Contrasts with $CH_4$ in which C has octet of electrons; tetrahedral; no dimerization. Lewis acidity of $BH_3$ not mirrored in chemistry of $CH_4$.
- Similar comparison between $AlH_3$ and $SiH_4$ as between $BH_3$ and $CH_4$.

13.11    (a) Hydrolysis of a group 14 halide liberates hydrogen halide:

$$GeCl_4 + 2H_2O \rightarrow GeO_2 + 4HCl$$

$GeO_2$ is dimorphic, possessing rutile and quartz polymorphs (see answer 13.14).
(b) Hydrolysis with aqueous alkali gives a silicate:

$$SiCl_4 + 4NaOH \rightarrow Na_4SiO_4 + 4HCl$$

Discrete, tetrahedral $[SiO_4]^{4-}$ ions are not present, rather polymeric species form containing tetrahedral $SiO_4$ units linked by Si–O–Si bridges.

(c) CsF is a fluoride donor and $GeF_2$, a fluoride acceptor. Products of such reactions depend on stoichiometry (as well as cation); with a 1:1 ratio:

$$CsF + GeF_2 \rightarrow Cs[GeF_3]$$

The salt is expected to contain trigonal pyramidal $[GeF_3]^-$ ions. Solid state structures of related species such as salts of $[SnCl_3]^-$ are cation-dependent; discrete anions may be present, but in some salts, halide bridges connect Sn centres to give polymer.
(d) Hydrolysis of $SiH_3Cl$ liberates HCl; Si–O–Si bridge formation:

$$2SiH_3Cl + H_2O \rightarrow (H_3Si)_2O + 2HCl$$

**(13.17)**

Molecular species with structure **13.17**.
(e) Hydrolysis of $SiF_4$ gives silica and HF, but $SiF_4$ acts as a fluoride acceptor:

$$2SiF_4 + 4H_2O \rightarrow SiO_2 + 2[H_3O]^+ + [SiF_6]^{2-} + 2HF$$

$SiO_2$ has 3D-lattice (see answer 13.7, p. 109). $[SiF_6]^{2-}$ is octahedral.
(f) Stoichiometry stated as 2:1, indicating that $SnCl_4$ accepts *two* $Cl^-$:

$$2[Bu_4P]Cl + SnCl_4 \rightarrow [Bu_4P]_2[SiCl_6]$$

Product is ionic salt with tetrahedral $[Bu_4P]^+$ and octahedral $[SiCl_6]^{2-}$.

13.12    Each spectrum is $^{119}Sn$ *observed*, so all splittings are due to $^{119}Sn$-$^{19}F$ spin-spin couplings; each species is octahedral (use VSEPR theory to confirm this).
(a) $[SnCl_5F]^{2-}$, **13.18**, has only one F and gives a doublet in the $^{119}Sn$ NMR spectrum.
(b) $[SnCl_4F_2]^{2-}$ has *cis* (**13.19**) and *trans* (**13.20**) isomers; each isomer has 2 equivalent F sites, giving a triplet in the $^{119}Sn$ NMR spectrum; not possible to distinguish isomers A and B from these data alone.
(c) $[SnCl_3F_3]^{2-}$ has *fac* (**13.21**) and *mer* (**13.22**) isomers. In **13.21**, F atoms are all equivalent, and so a binomial quartet is seen in the $^{119}Sn$ NMR spectrum; isomer B is the *fac* isomer. In **13.22**, there are two F environments labelled *a* and *b*; coupling of $^{119}Sn$ to F(*a*) gives a triplet, and further coupling to F(*b*) gives an overall signal that is a doublet of triplets. Isomer A is the *mer*-isomer.
(d) $[SnCl_2F_4]^{2-}$ has *trans* (**13.23**) and *cis* (**13.24**) isomers. In **13.23**, F atoms are equivalent, so a binomial quintet is observed in the $^{119}Sn$ NMR spectrum; isomer A is the *trans*-isomer. In **13.24**, there are two F environments labelled *a* and *b*; coupling of $^{119}Sn$ to F(*a*) gives a triplet, and further coupling to F(*b*) gives an overall signal that is a triplet of triplets. Isomer B is the *cis*-isomer.
(e) $[SnClF_5]^{2-}$ (**13.25**) has two F sites (*a* and *b*). Coupling of $^{119}Sn$ to F(*a*) gives a doublet; further coupling to F(*b*) gives an overall signal that is a doublet of quintets.
(f) $[SnF_6]^{2-}$ has 6 equivalent F atoms giving rise to a binomial septet.

**(13.18)**

**(13.19)**

**(13.20)**

**(13.21)**

**(13.22)**

**(13.23)**

**(13.24)**

**(13.25)**

13.13    (a) Concentrated NaOH oxidizes Sn and also provides [OH]$^-$ for complex formation:

$$2H_2O + 2e^- \rightleftharpoons H_2 + 2[OH]^-$$

$$[Sn(OH)_6]^{2-} + 4e^- \rightleftharpoons Sn + 6[OH]^-$$

Overall:
$$Sn + 2[OH]^- + 4H_2O \rightarrow [Sn(OH)_6]^{2-} + 2H_2$$
or
$$Sn + 2NaOH + 4H_2O \rightarrow Na_2[Sn(OH)_6] + 2H_2$$

(b) SO$_2$ reduces Pb(IV) to Pb(II), and is itself oxidized:

$$SO_2 + PbO_2 \rightarrow PbSO_4$$

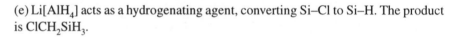

(13.26)

(c) Section 13.11 in H&S describes the reaction of H$_2$O with CS$_2$ to give H$_2$S and CO$_2$. With NaOH, reaction will give Na$_2$CS$_3$ and some Na$_2$CO$_3$. Na$_2$CS$_3$ contains the [CS$_3$]$^{2-}$ ion, **13.26**.

(d) Hydrolysis of SiCl$_4$ gives HCl and SiO$_2$. Follow on from this to suggest that SiH$_2$Cl$_2$ hydrolyses according to:

$$n\text{SiH}_2\text{Cl}_2 + n\text{H}_2\text{O} \rightarrow \text{-(SiH}_2\text{O)}_n\text{-} + 2n\text{HCl}$$

(13.27)

-(SiH$_2$O)$_n$- is polymeric; structure **13.27** shows two repeat units.

(e) Li[AlH$_4$] acts as a hydrogenating agent, converting Si–Cl to Si–H. The product is ClCH$_2$SiH$_3$.

$$4RCl_3 + 3LiAlH_4 \rightarrow 4RH_3 + 3LiAlCl_4 \qquad \text{where } R = ClCH_2Si$$

13.14    (a) To estimate $\Delta_r H^\circ$ for the reaction:

$$GeO_2(\text{quartz}) \rightarrow GeO_2(\text{rutile})$$

Compare with answer 11.10b, p. 96

dissolve each in conc. HF, measure $\Delta_{sol}H^\circ$ for each, and apply a Hess cycle. In solution and after reaction, there will no longer be a distinction between the two phases of GeO$_2$, a common set of products being obtained.
From the cycle:

$$GeO_2(\text{quartz}) + 6HF(\text{conc}) \xrightarrow{\Delta_r H^\circ} GeO_2(\text{rutile}) + 6HF(\text{conc})$$

$\Delta_{sol}H^\circ(1)$ ⟍          ⟋ $\Delta_{sol}H^\circ(2)$

$$[GeF_6]^{2-}(aq) + 2[H_3O]^+(aq)$$

apply Hess's Law:

$$\Delta_r H^\circ = \Delta_{sol}H^\circ(1) - \Delta_{sol}H^\circ(2)$$

(b) Pauling electronegativity of Si estimated using the equation:

$$\chi^P(X) - \chi^P(Si) = \sqrt{\Delta D}$$

➤
1 eV = 96.485 kJ mol$^{-1}$

where $\Delta D$ is in eV and is given by:

$$\Delta D = D(Si-X) - \tfrac{1}{2}\{D(Si-Si) + D(X-X)\} \qquad (\Delta D \text{ in kJ mol}^{-1})$$

Choose X so that compounds containing Si–Si and Si–X bonds are readily accessible, e.g. X = H. Values of $D(Si-H)$ and $D(Si-Si)$ are listed in Table 13.2 (326 and 226 kJ mol$^{-1}$ respectively), but comment on how these could be estimated:

$$D(Si-H) = \tfrac{1}{4}\Delta_a H^\circ(SiH_4, 298 \text{ K})$$

$$6D(Si-H) + D(Si-Si) = \Delta_a H^\circ(Si_2H_6, 298 \text{ K})$$

For $D(H-H)$:

➤
$\Delta_a H^\circ$(H, 298 K) from
Table 5.2 in H&S

$$D(H-H) = 2\Delta_a H^\circ(H, 298 \text{ K})$$

Values of $\Delta_a H^\circ(SiH_4, 298 \text{ K})$ and $\Delta_a H^\circ(Si_2H_6, 298 \text{ K})$ determined from Hess cycles, exemplified below for $SiH_4$. $\Delta_c H^\circ$ is an experimentally measurable quantity.

$$
\begin{array}{ccc}
SiH_4(g) & \xrightarrow{\Delta_c H^\circ} & SiO_2(s) \\
+ 2O_2(g) & & + 2H_2O(l) \\
\Delta_f H^\circ(1) \searrow & & \nearrow \Delta_f H^\circ(2) \\
& Si(s) + 2H_2(g) + 2O_2(g) &
\end{array}
\qquad
\begin{array}{ccc}
SiH_4(g) & \xrightarrow{\Delta_a H^\circ(1)} & Si(g) + 4H(g) \\
\Delta_f H^\circ(1) \searrow & & \nearrow \Delta_a H^\circ(2) \\
& Si(s) + 2H_2(g) &
\end{array}
$$

(c) Purity of samples can be assessed by using analytical methods. Analysis of $Pb(MeCO_2)_4$: determine Pb(IV) by allowing it to oxidize I$^-$ and titrating the I$_2$ formed against thiosulfate:

$$I_2 + 2[S_2O_3]^{2-} \rightarrow 2I^- + [S_4O_6]^{2-}$$

Or heat $Pb(MeCO_2)_4$ with HCl; Pb(IV) is reduced and Cl$^-$ oxidized to Cl$_2$. Pass Cl$_2$ into aqueous KI:

$$Cl_2 + 2I^- \rightarrow I_2 + 2Cl^-$$

and titrate I$_2$ formed against thiosulfate (see above).

13.15    Figure 13.5 shows the relevant part of the Ellingham diagram for this problem. The values of $\Delta_f G^\circ$ refer to the reactions:

$$C + \tfrac{1}{2}O_2 \rightarrow CO$$

$$\tfrac{1}{2}Sn + \tfrac{1}{2}O_2 \rightarrow \tfrac{1}{2}SnO_2$$

**Figure 13.5** An Ellingham diagram for CO and SnO$_2$ (also see Figure 7.4 in H&S)

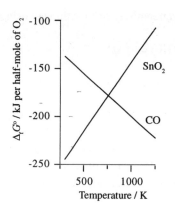

The reaction to be considered is:

$$C + \tfrac{1}{2}SnO_2 \rightarrow CO + \tfrac{1}{2}Sn$$

At 1000 K, CO is more thermodynamically stable than SnO$_2$ (CO has a more negative $\Delta_f G^\circ$) and so C reduces SnO$_2$ to Sn. At 500 and 750 K, CO is *not* more thermodynamically stable than SnO$_2$, and at these temperatures, C cannot be used to extract Sn from SnO$_2$.

13.16    (a) In the isomorphous (i.e. possess the same structures) pyroxenes CaMgSi$_2$O$_6$ and CaFeSi$_2$O$_6$, Fe$^{2+}$ and Mg$^{2+}$ occupy the same lattice positions. The ionic radii of Fe$^{2+}$ and Mg$^{2+}$ are 78 and 72 pm respectively (see Appendix 6 in H&S); the value for Fe$^{2+}$ is for the *high-spin* ion since the coordination environment consists of weak-field O$^{2-}$ centres (see Chapter 20 of H&S). The similarity in size between Fe$^{2+}$ and Mg$^{2+}$ means that ion replacement causes little structural perturbation.

*Pyroxenes* and *feldspars* are silicate minerals

(b) Consider NaAlSi$_3$O$_8$ to be the 'host' lattice. Going from this to CaAl$_2$Si$_2$O$_8$ requires that Ca$^{2+}$ replaces Na$^+$, *and at the same time* Al$^{3+}$ replaces Si$^{4+}$. These simultaneous replacements allow electrical neutrality to be retained. Consider relevant ionic radii (see Figure 13.14 in H&S):

$$r_{ion}:\quad Ca^{2+} = 100 \text{ pm}; \quad Na^+ = 102 \text{ pm}; \quad Al^{3+} = 54 \text{ pm}; \quad Si^{4+} \approx 40 \text{ pm}$$

The similarity in ion sizes of Ca$^{2+}$ and Na$^+$, and of Al$^{3+}$ and Si$^{4+}$ means ion replacement occurs with little or no structural perturbation.

(c) Quartz is a polymorph of SiO$_2$; when LiAlSi$_2$O$_6$ transforms to a quartz form, the Al$^{3+}$ ions must take lattice sites adopted by Si$^{4+}$ in SiO$_2$ – similarity in size (see above) allows this to occur. Quartz lattice is a relatively open network and Li$^+$ ions ($r_{ion} = 76$ pm) can occupy interstitial sites. Comparing LiAlSi$_2$O$_6$ with SiO$_2$:
   • rewrite LiAlSi$_2$O$_6$ as Li$^+$[AlSi$_2$O$_6$]$^-$
   • think of [AlSi$_2$O$_6$]$^-$ as 3[Al/SiO$_2$]$^{1/3-}$
   • [Al/SiO$_2$]$^{1/3-}$ compares directly with SiO$_2$
   • Li$^+$ ions provide electrical neutrality.

13.17    (a) Molecules in question are all linear ($D_{\infty h}$). 2200 cm$^{-1}$ is typical of $\nu(C\equiv N)$, and **I** is N≡C–C≡N. For CO$_2$ and CS$_2$, the lower wavenumber corresponds to the bond with lower force constant and higher reduced mass, so **II** is S=C=S, and **III** is O=C=O.
(b) All are symmetrical molecules, therefore the symmetric stretch is IR inactive, and asymmetric stretch is IR active in each.

13.18    KCN(aq) is basic (see answer 6.7, p. 53) giving [OH]$^-$ in solution. This competes with [CN]$^-$ for Al$^{3+}$. Al(OH)$_3$ forms preferentially and precipitates.

13.19    (a)    HOCN + 2H$_2$O → NH$_3$ + H$_2$CO$_3$        H$_2$CO$_3$ → CO$_2$ + H$_2$O

Cyanic acid = HOCN
Isocyanic acid = HNCO
Thiocyanic acid = HSCN

(b) HNCO reacts to give the same products as HOCN

(c)    HSCN + 2H$_2$O → NH$_3$ + H$_2$CO$_2$S        H$_2$CO$_2$S → OCS + H$_2$O

# 14 The group 15 elements

14.1 In assigning oxidation states to N and P, *first* assign oxidation states to O and H (if present) of –2 and +1 respectively. These are the most common oxidation states for these elements. *But watch H*: oxidation state of –1 is assigned when it is combined with an electropositive element.

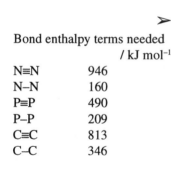

**(14.1)**

| | | |
|---|---|---|
| (a) $N_2$ | Elemental form | $\therefore$ ox. state N = 0 |
| (b) $[NO_3]^-$ | Overall charge of 1– | $\therefore$ ox. state N = +5 |
| (c) $[NO_2]^-$ | Overall charge of 1– | $\therefore$ ox. state N = +3 |
| (d) $NO_2$ | Neutral compound | $\therefore$ ox. state N = +4 |
| (e) NO | Neutral compound | $\therefore$ ox. state N = +2 |
| (f) $NH_3$ | Neutral compound | $\therefore$ ox. state N = –3 |
| (g) $NH_2OH$ | Neutral compound **14.1** | $\therefore$ ox. state N = –1 |
| (h) $P_4$ | Elemental form | $\therefore$ ox. state P = 0 |
| (i) $[PO_4]^{3-}$ | Overall charge of 3– | $\therefore$ ox. state P = +5 |
| (j) $P_4O_6$ | Neutral compound | $\therefore$ ox. state P = +3 |
| (k) $P_4O_{10}$ | Neutral compound | $\therefore$ ox. state P = +5 |

14.2 For each part of the problem, the reaction to consider is:

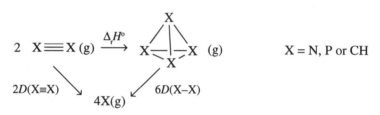

X = N, P or CH

**Bond enthalpy terms needed**
/ kJ mol$^{-1}$

| | |
|---|---|
| N≡N | 946 |
| N–N | 160 |
| P≡P | 490 |
| P–P | 209 |
| C≡C | 813 |
| C–C | 346 |

(a) X = N  $\Delta_r H^\circ = 2(946) - 6(160) = +932$ kJ mol$^{-1}$ (i.e. per mole of reaction shown)

(b) X = P  $\Delta_r H^\circ = 2(490) - 6(209) = -274$ kJ mol$^{-1}$

(c) X = CH  $\Delta_r H^\circ = 2(813) - 6(346) = -450$ kJ mol$^{-1}$

*Comments:* At 298 K, $N_2$ is thermodynamically stable with respect to $N_4$, but $P_4$ is stable with respect to $P_2$; $N_2$ and $P_4$ are elemental forms at 298 K. $C_2H_2$ does *not* spontaneously convert to $C_4H_4$; $C_2H_2$ is kinetically stable with respect to $C_4H_4$.

14.3 Allotropy: information in Section 14.4 of H&S. Points to include are:
- $N_2$; triply bonded diatomic; molecular structure; gaseous at 298 K.
- Phosphorus has wide range of allotropes; white phosphorus ($P_4$ molecules; solid at 298 K) is *defined* as standard state of the element, but is metastable; heating white phosphorus under inert conditions gives red phosphorus (complicated covalent lattice), and under pressure, white phosphorus converts to black polymorph (most stable allotrope with double-layer lattice of puckered 6-membered rings).
- At 298 K and 1 bar, As, Sb and Bi are solids with lattices comparable to black phosphorus; coordination number changes from 3 for P to 6 for Bi as layer structure distorts to a 3D-lattice.

14.4    (a) Group 2 metal phosphide, best considered as ionic; liberates $PH_3$ on hydrolysis:

$$Ca_3P_2 + 6H_2O \rightarrow 3Ca(OH)_2 + 2PH_3$$

(b) Acid-base reaction of $[NH_4]^+$ with $[OH]^-$:

$$NaOH + NH_4Cl \rightarrow NaCl + NH_3 + H_2O$$

(c) Aqueous $NH_3$ produces $[NH_4]^+$ and $[OH]^-$; $Mg^{2+}$ with $[OH]^-$ forms sparingly soluble $Mg(OH)_2$:

$$Mg(NO_3)_2 + 2NH_3 + 2H_2O \rightarrow Mg(OH)_2(s) + 2NH_4NO_3$$

(d) $I_2$ is oxidizing agent, able to oxidize As(−III) to As(V); in aqueous solution, an oxoacid of As(V) forms:

$$AsH_3 + 4I_2 + 4H_2O \rightarrow H_3AsO_4 + 8HI$$

(e) In liquid $NH_3$, $[NH_2]^-$ acts as a base and deprotonates $PH_3$

$$PH_3 + KNH_2 \xrightarrow{\text{liquid NH}_3} KPH_2 + NH_3$$

14.5    (a) HCl dissolved in $H_2O$ is fully ionized:

$$HCl(aq) + H_2O(l) \rightarrow [H_3O]^+(aq) + Cl^-(aq)$$

but aqueous $NH_3$ contains dissolved $NH_3$ which gives rise to the 'ammonia smell':

$$NH_3(aq) + H_2O(l) \rightleftharpoons [NH_4]^+(aq) + [OH]^-(aq) \qquad K = 1.8 \times 10^{-5}$$

$$[NH_4]^+(aq) + H_2O(l) \rightleftharpoons [H_3O]^+(aq) + NH_3(aq) \qquad K = 5.6 \times 10^{-10}$$

(b) Ammonium carbamate is $[NH_4][NH_2CO_2]$, the salt of a very weak acid. Add moisture from the air on opening the bottle of smelling salts:

$$[NH_4][NH_2CO_2] + H_2O \rightleftharpoons [NH_4][OH] + NH_2CO_2H$$

$$[NH_4]^+(aq) + H_2O(l) \rightleftharpoons [H_3O]^+(aq) + NH_3(aq)$$

The $NH_3$ formed results in the reviving smell.

14.6    $K_b$ for $NH_3$ refers to the equilibrium:

See Section 6.4 in H&S

$$NH_3(aq) + H_2O(l) \rightleftharpoons [NH_4]^+(aq) + [OH]^-(aq) \qquad pK_b = 4.75$$

$K_a$ for $[NH_4]^+$ refers to the equilibrium:

$$[NH_4]^+(aq) + H_2O(l) \rightleftharpoons NH_3(aq) + [H_3O]^+(aq) \qquad pK_a = ?$$

Now write out an expression for each $K$ and combine them.

$$K_b = \frac{[NH_4^+][OH^-]}{[NH_3]} \qquad\qquad K_a = \frac{[NH_3][H_3O^+]}{[NH_4^+]}$$

Combining these equations gives:

$$\frac{[NH_4^+]}{[NH_3]} = \frac{K_b}{[OH^-]} = \frac{[H_3O^+]}{K_a}$$

$$\therefore \quad K_b \times K_a = [H_3O^+] \times [OH^-]$$

In water (pH 7): $\quad [H_3O^+] \times [OH^-] = K_w = 1 \times 10^{-14}$

$$\therefore \quad K_a \times K_b = 1 \times 10^{-14}$$

$$pK_a + pK_b = 14$$

$$\therefore \quad pK_a = 14 - pK_b = 14 - 4.75 = 9.25$$

14.7   Half-equations for oxidation of $NH_2OH$ to $HNO_3$, and reduction of $[BrO_3]^-$ (use Appendix 11 in H&S to help you):

$$HNO_3(aq) + 6H^+ + 6e^- \rightleftharpoons NH_2OH(aq) + 2H_2O(l)$$

$$[BrO_3]^-(aq) + 6H^+(aq) + 6e^- \rightleftharpoons Br^-(aq) + 3H_2O(l)$$

Overall reaction is:

$$NH_2OH(aq) + [BrO_3]^-(aq) \rightarrow HNO_3(aq) + Br^-(aq) + H_2O(l)$$

14.8   (a) Preparation of sodium azide – use equation 14.40 in H&S to help you:

$$3NaNH_2 + NaNO_3 \rightarrow NaN_3 + 3NaOH + NH_3$$

➤

Na in liquid $NH_3$: see Section 8.6 in H&S

(b) Sodium metal dissolved in liquid $NH_3$ slowly liberates $H_2$ and provides a method of making sodium amide:

$$2Na + 2NH_3 \rightarrow 2NaNH_2 + H_2$$

(Compare this to Na reacting with $H_2O$ to give NaOH.)

(c) This reaction is a method of preparing lead azide:

$$2NaN_3 + Pb(NO_3)_2 \rightarrow Pb(N_3)_2 + 2NaNO_3$$

14.9   (a) For species to be strictly isoelectronic, they must possess the same *total* number of electrons, not just the same number of valence electrons. A useful starting point for the answer is to note that C and $N^+$ are isoelectronic, as are O and $N^-$. Isoelectronic species include:

$$O{=}C{=}O \qquad \overset{-}{N}{=}\overset{+}{N}{=}\overset{-}{N} \qquad \overset{-}{N}{=}C{=}\overset{-}{N} \qquad \overset{-}{O}{=}\overset{+}{N}{=}O \qquad O{=}C{=}\overset{-}{N}$$

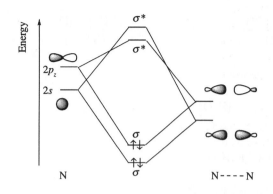

**Figure 14.1** Partial MO diagram to show $\sigma$-bond formation in $[N_3]^-$ using a ligand group orbital approach.

**Figure 14.2** Partial MO diagram to show $\pi$-bond formation in $[N_3]^-$ using a ligand group orbital approach. An analogous diagram can be constructed using $2p_x$ atomic orbitals (see text).

**(14.2)**

(b) Bonding scheme for $[N_3]^-$ is similar to that for the isoelectronic $CO_2$ (see Section 4.7 in H&S). You could begin with an *atomic orbital basis set* for each atom, but you can simplify the problem by considering the $\sigma$- and $\pi$-bonding separately. Lewis structure **14.2** is a useful starting point to allow you to keep track of electrons. It is convenient to consider each terminal N atom to be *sp* hybridized with one hybrid orbital accommodating one lone pair per N atom. Define an axis set: e.g. let the N atoms lie on the $z$ axis.

**$\sigma$-Bond formation**: Consider the interactions between in-phase and out-of-phase *sp* hybrid orbitals of the terminal N atoms (i.e. two ligand group orbitals) with the $2s$ and $2p_z$ atomic orbitals of the central N atom. This is represented in the partial MO diagram in Figure 14.1. Two $\sigma$-bonding MOs (with corresponding $\sigma^*$ MOs) result. Leave consideration of the electrons until later.

**$\pi$-Bond formation**: Consider the interactions between in-phase and out-of-phase $2p_y$ orbitals of the terminal N atoms (i.e. two ligand group orbitals) with the $2p_y$ atomic orbital of the central N atom. This is represented in the partial MO diagram in Figure 14.2. One $\pi$-bonding MO (with corresponding $\pi^*$ MO) results, as well as a non-bonding MO. An analogous partial MO diagram can be drawn using the $2p_x$ orbitals and again gives rise to one $\pi$-bonding MO (with corresponding $\pi^*$ MO) and a non-bonding MO. The $\pi_x$ and $\pi_y$ bonding MOs are degenerate; the $\pi_x^*$ and $\pi_y^*$ bonding MOs are degenerate.

**Electrons:** The total number of valence electrons available in $[N_3]^-$ is 16 (5 per N atom and 1 from the negative charge), i.e. 8 pairs.

- 2 pairs are accommodated in outward pointing *sp* hybrid orbitals on the terminal N atoms;
- 2 pairs are accommodated in the two $\sigma$-bonding MOs (Figure 14.1);
- 2 pairs occupy the $\pi$-bonding and non-bonding MOs arising from interactions of the N $2p_y$ orbitals (Figure 14.2);
- 2 pairs occupy the $\pi$-bonding and non-bonding MOs arising from interactions of the N $2p_x$ orbitals.

**Check:** There are 4 bonding ($\sigma + \pi$) pairs of electrons in $[N_3]^-$ and this agrees with Lewis structure **14.2**. There are 2 lone pairs per terminal N atom – the non-bonding electrons represent lone pairs in addition to those occupying the N *sp* hybrids.

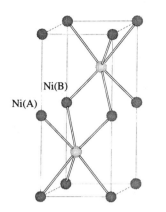

**Figure 14.3** Unit cell of NiAs; Ni centres are shown in dark grey.

14.10    (a) Figure 14.3 shows a single unit cell of NiAs. Consider the atoms labelled Ni(A) and Ni(B). *Within this unit cell*, Ni(A) has only one As neighbour, while Ni(B) has two. Ni(A) and Ni(B) lie in edge sites and each is shared between 4 unit cells. Figure 14.4 shows two adjacent unit cells – each of Ni(A) and Ni(B) is 3-coordinate. In the complete lattice, Ni(A) and Ni(B) are equivalent (to confirm this, 'grow' the structure in Figure 14.4 by adding two more unit cells in front of those shown) and are 6-coordinate.

The problem can be tackled in a similar manner by starting with a Ni centre in a corner site.

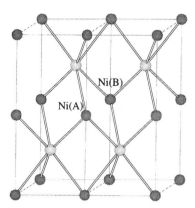

**Figure 14.4** Two unit cells of NiAs; Ni centres are shown in dark grey.

(b) Per unit cell of NiAs (Figure 14.3):

| Site | Number of $Ni^{3+}$ | Number of $As^{3-}$ |
| --- | --- | --- |
| Within unit cell | 0 | 2 |
| Corner site | $8 \times \frac{1}{8} = 1$ | 0 |
| Edge site | $4 \times \frac{1}{4} = 1$ | 0 |
| Total | 2 | 2 |

The stoichiometry is therefore $Ni^{3+} : As^{3-} = 2:2 = 1:1$, or NiAs.

14.11    (a) Gas phase structures may be investigated using electron diffraction, or vibrational spectroscopy.

(b) In the liquid phase, Raman spectroscopy could be used. IR spectroscopy is not appropriate because the symmetric stretching mode is IR inactive (no change in dipole moment).

14.12    (a) Information is found in Section 14.7 of H&S. Points to include:
- $PX_3$ (X = F, Cl, Br, I) all molecular (**14.3**);
- $PF_5$ molecular (**14.4**) in solid, liquid and gas; fluxional in solution with $F_{ax}$ and $F_{equ}$ exchanging (Berry pseudo-rotation) on NMR spectroscopic timescale;
- $PCl_5$ molecular (like **14.4**) in gas phase; solid consists of $[PCl_4]^+$ (tetrahedral) and $[PCl_6]^-$ (octahedral) ions;
- Solid $PBr_5$ is $[PBr_4]^+Br^-$; decomposes to $PBr_3$ and $Br_2$ in gas phase;
- $PI_5$ does not exist.

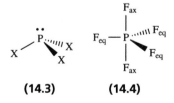

(14.3)        (14.4)

(b) CsCl has structure shown in Figure 5.5b (p. 46); $Cs^+$ and $Cl^-$ ions are spherical. In $[PCl_4]^+[PCl_6]^-$, ions are tetrahedral and octahedral respectively. Can be considered to be spherical if ions occupy lattice sites in random orientations, or if ions are freely rotating about the lattice site. $[PCl_4]^+$ and $[PCl_6]^-$ are relatively large ions, and random orientations are more likely than rotational motion in solid state.

14.13

➤

Nuclear spin and abundance data:

$^{31}P$    $I = \frac{1}{2}$    100%

$^{121}Sb$    $I = \frac{5}{2}$    57.3%

$^{123}Sb$    $I = \frac{7}{2}$    42.7%

(a) $[PF_6]^-$ is octahedral with 6 equivalent F atoms. Coupling of $^{19}F$ nuclei to one $^{31}P$ nucleus gives a doublet in the $^{19}F$ NMR spectrum.

(b) $[SbF_6]^-$ is octahedral with 6 equivalent F atoms. Coupling occurs to both $^{121}Sb$ (57.3%) and $^{123}Sb$ (42.7%) and the resulting signals are superimposed on top of each other. Coupling to $^{121}Sb$ gives a 1:1:1:1:1:1 signal; 6 spin states ($+\frac{5}{2}$, $+\frac{3}{2}$, $+\frac{1}{2}$, $-\frac{1}{2}$, $-\frac{3}{2}$, $-\frac{5}{2}$) with equal probability of coupling to $^{19}F$. Coupling to $^{123}Sb$ gives a 1:1:1:1:1:1:1:1 signal; 8 spin states with equal probability of coupling to $^{19}F$. The values of $J(^{121}Sb-^{19}F)$ and $J(^{123}Sb-^{19}F)$ are likely to be different, and the exact appearance of the signal in the $^{19}F$ NMR spectrum depends on their values.

14.14

(a) $PCl_5$ donates $Cl^-$ to $SbCl_5$ (*not* the other way around):

$$PCl_5 + SbCl_5 \rightarrow [PCl_4]^+[SbCl_6]^-$$

(b) KF is source of $F^-$, and $AsF_5$ is an $F^-$ acceptor:

$$KF + AsF_5 \rightarrow K^+[AsF_6]^-$$

(c) NOF donates $F^-$, and $SbF_5$ is a $F^-$ acceptor:

$$NOF + SbF_5 \rightarrow [NO]^+[SbF_6]^-$$

➤

See Section 8.9 in H&S

(d) HF donates $F^-$ to $SbF_5$ (a good $F^-$ acceptor) and $H^+$ combines with HF:

$$2HF + SbF_5 \rightarrow [H_2F]^+[SbF_6]^-$$

In (c) and (d), tendency for Sb–F–Sb bridge formation, for example:

$$SbF_5 + [SbF_6]^- \rightarrow [Sb_2F_{11}]^-$$

14.15

(a) See structures **14.5** and **14.6**.

$[Sb_2F_{11}]^-$ has 6 pairs of electrons in the valence shell of each Sb(V) centre: the formal scheme shown on the right can be used to confirm this, although remember that the Sb centres are equivalent. By VSEPR, 6 pairs of electrons are consistent with an octahedral environment as observed.

$[Sb_2F_7]^-$ has 5 pairs of electrons in the valence shell of each Sb(III) centre: the formal scheme shown on the right confirms this, but keep in mind that the Sb centres are actually equivalent. By VSEPR, 5 pairs of electrons are consistent with trigonal bipyramidal environments.

(b) $[\{BiX_4\}_n]^{n-}$ and $[\{BiX_5\}_n]^{2n-}$ are chains (see right) with octahedral Bi(III). Stoichiometry 'BiX_4' follows by having *two* X atoms in bridging sites and two terminal; stoichiometry 'BiX_5' requires *one* bridging X.

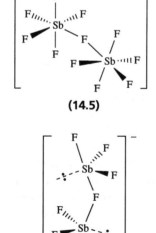

(14.5)

(14.6)

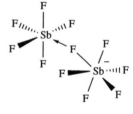

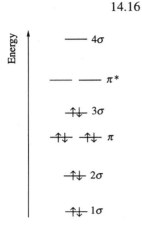

Figure 14.5 Approximate ordering of the MOs in $[NO]^+$.

14.16   $[NO]^+$ is isoelectronic with CO; an MO diagram similar to that for CO can be constructed. Starting with the ordering of the MOs in CO, one expects that the MOs in $[NO]^+$ lie in approximately the same order (Figure 14.5). NO has one more electron than $[NO]^+$ and this occupies a $\pi^*$ MO (add an electron to Figure 14.5). On going from NO to $[NO]^+$, the bond order in $[NO]^+$ increases because occupation of the $\pi^*$ level cancels out the bonding contribution of one of the $\pi$ electrons. It follows that the bond distance decreases, and the bond strengthens. Frequency of vibration (proportional to wavenumber) depends on force constant, $k$:

$$\bar{v} = \frac{1}{2\pi c}\sqrt{\frac{k}{\mu}} \qquad \text{where } \mu = \text{reduced mass}$$

Stronger bond, larger force constant, and higher value of $\bar{v}$ .

14.17   For the reaction between oxalate and permanganate:

$$2CO_2 + 2e^- \rightleftharpoons [C_2O_4]^{2-}$$
$$[MnO_4]^- + 8H^+ + 5e^- \rightleftharpoons Mn^{2+} + 4H_2O$$

Overall:
$$2[MnO_4]^- + 5[C_2O_4]^{2-} + 16H^+ \rightarrow 10CO_2 + 2Mn^{2+} + 8H_2O$$

The amount of $[C_2O_4]^{2-}$ present $= 25.0 \times 0.0500 \times 10^{-3} = 1.25 \times 10^{-3}$ moles
$\therefore$  Amount of $[MnO_4]^-$ in 24.8 cm$^3$ $= \frac{2}{5} \times 1.25 \times 10^{-3} = 0.500 \times 10^{-3}$ moles

$\therefore$  Concentration of KMnO$_4$ solution, $\mathbf{A}$ $= \dfrac{0.500 \times 10^{-3} \times 10^3}{24.8} = 0.0202$ mol dm$^{-3}$

Now consider the oxidation of $NH_2OH$.
$Fe^{3+}$ is reduced to $Fe^{2+}$, and the amount of $Fe^{2+}$ produced is equivalent to 24.65 cm$^3$ of KMnO$_4$ solution:

$$5Fe^{2+} + [MnO_4]^- + 8H^+ \rightarrow 5Fe^{3+} + Mn^{2+} + 4H_2O$$

Amount of $[MnO_4]^-$ in 24.65 cm$^3$ $= 24.65 \times 0.0202 \times 10^{-3} = 4.98 \times 10^{-4}$ moles
$\therefore$  Amount of $Fe^{2+}$ $= 5 \times 4.98 \times 10^{-4} = 2.49 \times 10^{-3}$ moles
$\therefore$  Amount of $Fe^{3+}$ reacting with $NH_2OH$ $= 2.49 \times 10^{-3}$ moles
Amount of $NH_2OH$ used $= 25.0 \times 0.0494 \times 10^{-3} = 1.24 \times 10^{-3}$ moles
$\therefore$  Ratio of moles of $Fe^{3+}$ : $NH_2OH$ $= 2.49 : 1.24 = 2 : 1$

The reduction of $Fe^{3+}$ to $Fe^{2+}$ is a 1-electron reduction, and (following from the stoichiometry above), the oxidation of $NH_2OH$ to compound $\mathbf{B}$ is a 2-electron oxidation per N, taking N from oxidation state –1 to +1. $\mathbf{B}$ may be $N_2O$, and the suggested oxidation half-equation is:

$$2NH_2OH \rightleftharpoons N_2O + 4H^+ + H_2O + 4e^-$$

Overall:
$$2NH_2OH + 4Fe^{3+} \rightarrow N_2O + 4H^+ + H_2O + 4Fe^{2+}$$

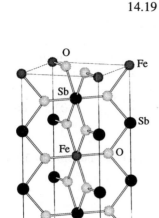

**(14.7)**

**(14.9)**

14.18    Information for this answer is in Sections 14.10 and 14.11 in H&S. Points to include:

- Structure of phosphorus(V) oxide (**14.7**); other polymorphs also contain {O=P(–O)$_3$} units;
- Action of water on P$_4$O$_{10}$ gives phosphoric acid with tetrahedral PO$_4$ unit (**14.8**), and salts retain this unit in [PO$_4$]$^{3-}$, [HPO$_4$]$^{2-}$ and [H$_2$PO$_4$]$^-$.

**(14.8)**

- Formation of condensed phosphates, e.g. H$_4$P$_2$O$_7$ and H$_5$P$_3$O$_{10}$, give a general representation to show retention of tetrahedral PO$_4$ units:

- Cyclic phosphates include [P$_3$O$_9$]$^{3-}$ (**14.9**), [P$_4$O$_{12}$]$^{4-}$ and [P$_6$O$_{18}$]$^{6-}$.
- Biological role of phosphates, e.g. ATP and ADP (see Box 14.10 in H&S) and DNA (Figure 9.9 in H&S).

14.19    (a) The unit cell of FeSb$_2$O$_6$ is shown in Figure 14.6, and that of rutile in Figure 5.5c (p. 46). Compare the two. The unit cell of FeSb$_2$O$_6$ consists of *three* stacked rutile unit cells (a triple-rutile lattice).

(b) The three rutile type subunits that make up one unit cell of FeSb$_2$O$_6$ are not identical: two subunits have 4 Sb and 4 Fe in the corner sites, while the central subunit has 8 Sb in the corner sites. Therefore, a unit cell consisting of all three subunits is needed in order that the infinite lattice of FeSb$_2$O$_6$ is unambiguously described.

(c) O$^{2-}$ is 3-coordinate (trigonal planar); Fe(II) is 6-coordinate (octahedral); Sb(V) is 6-coordinate (octahedral).

(d) Per unit cell of FeSb$_2$O$_6$:

**Figure 14.6** Unit cell of FeSb$_2$O$_6$.

| Site | Number of Fe(II) | Number of Sb(V) | Number of O$^{2-}$ |
|---|---|---|---|
| Within unit cell | 1 | 2 | 10 |
| Corner site | $8 \times \frac{1}{8} = 1$ | 0 | 0 |
| Edge site | 0 | $8 \times \frac{1}{4} = 2$ | 0 |
| Face site | 0 | 0 | $4 \times \frac{1}{2} = 2$ |
| Total | 2 | 4 | 12 |

The stoichiometry is therefore Fe(II):Sb(V):O$^{2-}$ = 2:4:12 = 1:2:6, or FeSb$_2$O$_6$.

14.20    (a) Salts contain [P$_3$O$_{10}$]$^{5-}$ and [P$_4$O$_{13}$]$^{6-}$, condensed phosphates (**14.10**) and **14.11**).

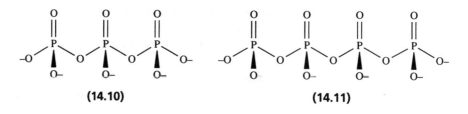

**(14.10)**                                      **(14.11)**

$[P_3O_{10}]^{5-}$ contains two P environments and gives two signals in the $^{31}P$ NMR spectrum (relative integrals 2:1); $[P_4O_{13}]^{6-}$ also contains two P environments but in a 2:2 ratio, therefore gives two signals of equal integrals.

(b) $AsF_5$ has structure **14.12**. Static structure would give rise to two signals in the $^{19}F$ NMR spectrum (relative integrals 3:2 for eq:ax). Dynamic behaviour exchanges $F_{ax}$ and $F_{eq}$ and if the rate of exchange is faster than the NMR timescale, only one signal is seen in the $^{19}F$ NMR spectrum. Recording the spectrum over a range of temperatures may allow both the high-temperature limiting spectrum (one signal) and the low-temperature limiting spectrum (two signals) to be observed (or at least, some broadening of signals) but this depends on the activation energy of the exchange process.

(c) There are three isomers:

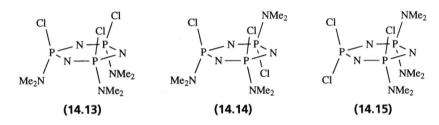

**(14.13)**          **(14.14)**          **(14.15)**

$^{31}P$ or $^{1}H$ NMR spectroscopy could be used. **14.13** has one P environment, **14.14** has two (although the signals may be near to coincident because the environments are very similar), and **14.15** has three. Assuming (on NMR spectroscopic timescale) free rotation about the P–N and C–N bonds, the $^{1}H$ NMR spectrum of **14.13** shows one signal, that of **14.14** two, and that of **14.15** three.

14.21 (a) From reaction stoichiometry, 2-electron oxidation step is:

$$2Ti(III) \rightarrow 2Ti(IV) + 2e^-$$

and this must be balanced by a 2-electron reduction. In $NH_2OH$, oxidation state of N is –1, so oxidation state in product is –3. Product is $NH_3$:

$$NH_2OH + H_2O + 2e^- \rightarrow NH_3 + 2[OH]^-$$

(b) From reaction stoichiometry, 2-electron reduction step is:

$$2Ag^+ + 2e^- \rightarrow 2Ag$$

and this must be balanced by a 2-electron oxidation. $\therefore$ P(III) oxidized to P(V); product is $[PO_4]^{3-}$:

$$[HPO_3]^{2-} + H_2O \rightarrow [PO_4]^{3-} + 2e^- + 3H^+$$

(c) The 2-electron reduction:

$$I_2 + 2e^- \rightarrow 2I^-$$

occurs twice, and so change in oxidation state for P is +1 to +3 to +5. This corresponds to $H_3PO_2 \rightarrow H_3PO_3 \rightarrow H_3PO_4$.

14.22    (a) $[NF_4]^+$

**(14.16)**

$N^+$, number of valence electrons = 4
Number of bonding pairs (4 N–F bonds) = 4
No lone pairs
'Parent' shape = molecular shape = tetrahedral, see **14.16**.
(b) $[N_2F_3]^+$, consider as $[F_2NNF]^+$, derived from $N_2F_4$ with loss of $F^-$:

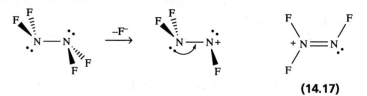

**(14.17)**

Loss of $F^-$ generates a sextet N that is stabilized by $\pi$-donation to give planar structure **14.17**.
(c) $NH_2OH$, consider N and O centres separately:
    N, number of valence electrons = 5
    Number of electron pairs = 1 lone + 3 bonding pairs
    'Parent' shape = tetrahedral; molecular shape = trigonal pyramidal.
    O, number of valence electrons = 6
    Number of electron pairs = 2 bonding + 2 lone pairs
    'Parent' shape = tetrahedral; molecular shape = bent.

See structure **14.18**.

**(14.18)**

(d) $SPCl_3$
    P, number of valence electrons = 5
    Number of bonding 'pairs' (1 P=S + 3 P–Cl bonds) = 4
    No lone pairs
    'Parent' shape = molecular shape = tetrahedral, see **14.19**

**(14.19)**

(e) $PCl_3F_2$
    P, number of valence electrons = 5
    Number of bonding pairs (5 P–X bonds) = 5
    No lone pairs
    'Parent' shape = molecular shape = trigonal bipyramidal.

**(14.20)**

Three isomers are possible depending on positions of F and Cl atoms; on steric grounds, it is most likely that Cl atoms occupy equatorial positions, see **14.20**.

14.23    The fate of the $^{15}N$ label in each step is critical; in each case, start with $[^{15}NO_3]^-$.

(a) Reduce $[NO_3]^-$ to $NH_3$ using Al or Zn; redox half-equations are:

➤

For reduction by Zn, see equation 14.15 in H&S

$$[^{15}NO_3]^- + 6H_2O + 8e^- \rightleftharpoons {}^{15}NH_3 + 9[OH]^-$$

$$Al + 4[OH]^- \rightleftharpoons [Al(OH)_4]^- + 3e^-$$

Overall:
$$3[^{15}NO_3]^- + 8Al + 5[OH]^- + 18H_2O \rightarrow 3\,{}^{15}NH_3 + 8[Al(OH)_4]^-$$

Now react $^{15}NH_3$ with sodium:

$$2\,{}^{15}NH_3 + 2Na \rightarrow 2Na[{}^{15}NH_2] + H_2$$

(b) First, prepare $^{15}NH_3$ as in part (a). Oxidize $^{15}NH_3$ with CuO or NaOCl:

$$2\,^{15}NH_3 + 3CuO \rightarrow \,^{15}N_2 + 3Cu + 3H_2O$$

or

$$2\,^{15}NH_3 + 3[OCl]^- \rightarrow \,^{15}N_2 + 3Cl^- + 3H_2O$$

> (c) Reduction of $[^{15}NO_3]^-$ with $Fe^{2+}$ or Hg, e.g.

See equations 14.76 and
14.77 in H&S

$$2[^{15}NO_3]^- + 6Hg + 8H^+ \rightarrow 2\,^{15}NO + 3[Hg_2]^{2+} + 4H_2O$$

To convert NO to $[NO]^+$ requires oxidation, but choice of oxidant and conditions affect salt formed. Use $Cl_2$ (which produces $Cl^-$) in the presence of $AlCl_3$ (Lewis acid which accepts $Cl^-$):

$$2\,^{15}NO + Cl_2 + 2AlCl_3 \rightarrow 2[^{15}NO]^+ + 2[AlCl_4]^-$$

14.24    The fate of the $^{32}P$ label in each step is critical; each reaction scheme starts with $Ca_3[^{32}PO_4]_2$ (phosphate rock).

> (a) Reduce $[^{32}PO_4]^{3-}$ to $^{32}P_4$ and treat with NaOH(aq):

See Section 14.2 and
equation 14.9 of H&S

$$2Ca_3[^{32}PO_4]_2 + 6SiO_2 + 10C \xrightarrow{1700\ K} \,^{32}P_4 + 6CaSiO_3 + 10CO$$

$$^{32}P_4 + 3NaOH + 3H_2O \rightarrow 3NaH_2PO_2 + \,^{32}PH_3$$

(b) First prepare $^{32}P_4$ as above, then convert to $PCl_3$ and treat with ice-cold water:

$$^{32}P_4 + 6Cl_2 \text{ (limited supply)} \rightarrow 4\,^{32}PCl_3$$

$$^{32}PCl_3 + 3H_2O \rightarrow H_3{}^{32}PO_3 + 3HCl$$

(c) First prepare $^{32}P_4$ as above, then convert to $^{32}P_4S_{10}$, followed by treatment with $Na_2S$:

$$^{32}P_4 \xrightarrow{\text{excess } S_8;\ >570\ K} \,^{32}P_4S_{10}$$

$$^{32}P_4S_{10} + 16Na_2S \rightarrow 4Na_3{}^{32}PS_4 + 10H_2S$$

14.25    For the reaction between oxalate and permanganate:

$$2CO_2 + 2e^- \rightleftharpoons [C_2O_4]^{2-}$$
$$[MnO_4]^- + 8H^+ + 5e^- \rightleftharpoons Mn^{2+} + 4H_2O$$

Overall:
$$2[MnO_4]^- + 5[C_2O_4]^{2-} + 16H^+ \rightarrow 10CO_2 + 2Mn^{2+} + 8H_2O$$

The amount of $[C_2O_4]^{2-}$ present $= 25.0 \times 0.0500 \times 10^{-3} = 1.25 \times 10^{-3}$ moles
$\therefore$ Amount of $[MnO_4]^-$ in 24.8 $cm^3$ = $^2/_5 \times 1.25 \times 10^{-3} = 0.500 \times 10^{-3}$ moles

$\therefore$ Concentration of $KMnO_4$ solution, $C = \dfrac{0.500 \times 10^{-3} \times 10^3}{24.7} = 0.0202$ mol $dm^{-3}$

Now consider the oxidation of $N_2H_4$.

$[Fe(CN)_6]^{3-}$ is reduced to $[Fe(CN)_6]^{4-}$, and the latter reacts with 24.80 cm$^3$ of solution **C**:

$$5[Fe(CN)_6]^{4-} + [MnO_4]^- + 8H^+ \rightarrow 5[Fe(CN)_6]^{3-} + Mn^{2+} + 4H_2O$$

Amount of $[MnO_4]^-$ in 24.80 cm$^3$ = $24.80 \times 0.0202 \times 10^{-3} = 4.96 \times 10^{-4}$ moles
$\therefore$ Amount of $[Fe(CN)_6]^{4-} = 5 \times 4.96 \times 10^{-4} = 2.48 \times 10^{-3}$ moles
$\therefore$ Amount of $[Fe(CN)_6]^{3-}$ reacting with $N_2H_4$ = $2.48 \times 10^{-3}$ moles
Amount of $N_2H_4$ used = $25 \times 0.0250 \times 10^{-3} = 6.25 \times 10^{-4}$ moles
$\therefore$ Ratio of moles of $[Fe(CN)_6]^{3-} : N_2H_4 = 4 : 1$

The reduction of Fe(III) to Fe(II) is a 1-electron reduction, and (following from the stoichiometry above), the oxidation of $N_2H_4$ to compound **D** is a 4-electron oxidation, i.e. a 2-electron oxidation per N. This takes N from oxidation state –2 to 0. **D** is $N_2$. The oxidation step is:

$$N_2H_4 \rightleftharpoons N_2 + 4H^+ + 4e^-$$

Overall:
$$N_2H_4 + 4[Fe(CN)_6]^{3-} \rightarrow N_2 + 4H^+ + 4[Fe(CN)_6]^{4-}$$

14.26   Write $AlPO_4$ as $2\{(Al,P)O_2\}$ to see relationship with $SiO_2$. Think of the periodic relationship between Al, Si and P (**14.21**): replacement of 2Si(IV) by {P(V) + Al(III)} retains electrical neutrality. Sizes of Si(IV), P(V) and Al(III) centres comparable and $SiO_2$ macromolecular structure not perturbed significantly on going from $SiO_2$ to $(Al,P)O_2$.

| 13 | 14 | 15 |
|----|----|----|
| B  | C  | N  |
| Al | Si | P  |
| Ga | Ge | As |

**(14.21)**

# 15 The group 16 elements

15.1 (a)

| | |
|---|---|
| O | oxygen |
| S | sulfur |
| Se | selenium |
| Te | tellurium |
| Po | polonium |

(b) $ns^2np^4$

15.2 Start with $^{209}Bi$ ($Z = 83$). In the (n,γ) reaction, a neutron reacts with the substrate and γ-radiation is emitted. Addition of the neutron converts $^{209}Bi$ to $^{210}Bi$. Loss of a β-particle leads to an increase in $Z$ by 1, but mass number stays the same:

$$^{209}_{83}Bi(n, \gamma)^{210}_{83}Bi \xrightarrow{\beta^-} {}^{210}_{84}Po$$

15.3 Aqueous alkali = aqueous solution of $[OH]^-$, e.g. aqueous NaOH:

At the anode: $4[OH]^-(aq) \rightarrow O_2(g) + 2H_2O(l) + 4e^-$

At the cathode: $2H^+(aq) + 2e^- \rightarrow H_2(g)$

15.4 For E = O:

$$8O(g) \rightarrow 4O_2(g)$$

$O_2$ contains O=O bond; enthalpy change (exothermic since bonds are formed) is:

$$\Delta_rH^o = -4D(O=O) = -4(498) = -1992 \text{ kJ mol}^{-1}$$

$$8O(g) \rightarrow O_8(g)$$

(15.1)

Assume $O_8$ is a structural analogue of $S_8$ (15.1) with 8 O–O bonds; for this reaction:

$$\Delta_rH^o = -8D(O-O) = -8(146) = -1168 \text{ kJ mol}^{-1}$$

∴ Formation of diatomic $O_2$ is thermodynamically favoured over $O_8$ ring formation. For E = S:

$$8S(g) \rightarrow 4S_2(g)$$

$$\Delta_rH^o = -4D(S=S) = -4(427) = -1708 \text{ kJ mol}^{-1}$$

$$8S(g) \rightarrow S_8(g)$$

$$\Delta_rH^o = -8D(S-S) = -8(266) = -2128 \text{ kJ mol}^{-1}$$

∴ Formation of cyclic $S_8$ is thermodynamically favoured over $S_2$ formation.

15.5    (a) Reaction to consider:

$$2H_2O_2 \rightarrow 2H_2O + O_2$$

which is the combination of half-equations:

➤

See Chapter 7 for further
details of the method used in
this answer

$$O_2 + 2H^+ + 2e^- \rightleftharpoons H_2O_2 \qquad\qquad E^o = +0.70 \text{ V}$$
$$H_2O_2 + 2H^+ + 2e^- \rightleftharpoons 2H_2O \qquad\qquad E^o = +1.78 \text{ V}$$

For the overall reaction:    $E^o_{cell} = +1.78 - (+0.70) = 1.08$ V
$\Delta G^o = -(2)(96\ 485)(1.08) \times 10^{-3} = -208$ kJ mol$^{-1}$

The large, negative value of $\Delta G^o$ indicates that $H_2O_2$ is thermodynamically unstable with respect to $H_2O$ and $O_2$.

(b) '20 volume' $H_2O_2$ : 1 volume of solution liberates 20 volumes of $O_2$.
Assume an ideal gas: 273 K and 1 bar pressure are standard conditions.
    Let 1 vol. of solution = 1 dm$^3$
        $\therefore$    20 vols. $O_2 = 20$ dm$^3$
For an ideal gas, 1 mole occupies a volume of 22.7 dm$^3$ at 273 K and 1 bar pressure.

$$\therefore \quad 20 \text{ dm}^3 \text{ } O_2 = \frac{20}{22.7} = 0.88 \text{ moles}$$

From the equation in part (a), 2 moles of $H_2O_2$ produce 1 mole of $O_2$
        $\therefore$    0.88 moles of $O_2$ are produced by 1.76 moles $H_2O_2$

To find the number of grams per dm$^3$ :

$$M_r(H_2O_2) = 2(1.008) + 2(16.00) = 34.016$$

Mass of $H_2O_2$ present in 1 dm$^3$ = 34.016 $\times$ 1.76 = 59.9 g

15.6    Half-equations from Appendix 11 in H&S are needed for this problem. For $H_2O_2$:

$$H_2O_2 + 2H^+ + 2e^- \rightleftharpoons 2H_2O \qquad\qquad E^o = +1.78 \text{ V}$$
$$O_2 + 2H^+ + 2e^- \rightleftharpoons H_2O_2 \qquad\qquad E^o = +0.70 \text{ V}$$

(a) For the reaction of $H_2O_2$ with $Ce^{4+}$, the relevant half-equation is:

$$Ce^{4+} + e^- \rightleftharpoons Ce^{3+} \qquad\qquad E^o = +1.72 \text{ V}$$

The overall spontaneous reaction is:

$$2Ce^{4+} + H_2O_2 \rightarrow 2Ce^{3+} + O_2 + 2H^+$$

(b) For the reaction of $H_2O_2$ with $I^-$, the relevant half-equation is:

$$I_2 + 2e^- \rightleftharpoons 2I^- \qquad\qquad E^o = +0.54 \text{ V}$$

The overall spontaneous reaction is:

$$2I^- + H_2O_2 + 2H^+ \rightarrow I_2 + 2H_2O$$

15.7

Potential diagrams for Mn:
see p. 72
Half-equations for $H_2O_2$:
see p. 130

The half-equation for the oxidation is:

$$MnO_2 + 2H_2O + 2e^- \rightleftharpoons Mn(OH)_2 + 2[OH]^- \qquad E^\circ_{[OH^-]=1} = -0.04 \text{ V}$$

(a)    $Mn(OH)_2 + H_2O_2 \rightarrow MnO_2 + 2H_2O$

(b) The $MnO_2$ produced will act as a catalyst for the decomposition of $H_2O_2$, so secondary reaction is:

$$2H_2O_2 \rightarrow 2H_2O + O_2$$

15.8

VSEPR theory:
see answer 1.29, p. 11

(a) $H_2Se$
Se, number of valence electrons = 6
Number of bonding pairs (2 Se–H bonds) = 2
2 lone pairs
'Parent' shape = tetrahedral
Molecular shape = bent, see **15.2**.

**(15.2)**

(b) $[H_3S]^+$
$S^+$, number of valence electrons = 5
Number of bonding pairs (3 S–H bonds) = 3
1 lone pair
'Parent' shape = tetrahedral
Molecular shape = trigonal pyramidal, see **15.3**.

**(15.3)**

(c) $SO_2$
S, number of valence electrons = 6
Number of bonding 'pairs' (2 S=O bonds) = 2
1 lone pair
'Parent' shape = trigonal planar
Molecular shape = bent, see **15.4**.

**(15.4)**

(d) $SF_4$
S, number of valence electrons = 6
Number of bonding pairs (4 S–F bonds) = 4
1 lone pair
'Parent' shape = trigonal bipyramidal
Molecular shape = disphenoidal, see **15.5**.

**(15.5)**

(e) $SF_6$
S, number of valence electrons = 6
Number of bonding pairs (6 S–F bonds) = 6
No lone pairs
'Parent' shape = molecular shape = octahedral, see **15.6**.

**(15.6)**

(f) $S_2F_2$ – for each S centre, assuming each S is in a similar environment:
S, number of valence electrons = 6
Number of bonding pairs (1 S–S + 1 S–F bond) = 2
2 lone pairs
'Parent' shape = tetrahedral at each S
Molecular shape = bent (at each S), see **15.7**.

There is, however, a second isomer (see Section 15.7 in H&S) which possesses structure **15.8**.

**(15.7)**

**(15.8)**

15.9    (a) SF$_4$ (**15.5**) can act both as an F$^-$ donor or acceptor; behaviour depends on its reaction partner. BF$_3$ is an F$^-$ acceptor:

*Exercise*:
Suggest shapes for [SF$_3$]$^+$ and [SF$_5$]$^-$. See Figure 15.10 in H&S for the answer

$$SF_4 + BF_3 \rightarrow [SF_3]^+ + [BF_4]^-$$

CsF is ionic and an F$^-$ donor:

$$SF_4 + CsF \rightarrow Cs^+ + [SF_5]^-$$

(b) SF$_4$ is a selective fluorinating agent; converts C=O into CF$_2$ group, and will fluorinate C–OH:

$$RCO_2H \xrightarrow{\phantom{x}SF_4\phantom{x}} RCF_3$$

15.10    (a) Approach the problem by first considering an appropriate MO diagram for O$_2$, [O$_2$]$^+$ and [O$_2$]$^{2-}$, see answer 1.26, p. 10. From the MO diagram, the bond orders are O$_2$, 2; [O$_2$]$^+$, $2\frac{1}{2}$; [O$_2$]$^{2-}$, 1, giving a trend of:

$$[O_2]^+ > O_2 > [O_2]^{2-}$$

consistent with the trend in O–O bond distances:

$$[O_2]^+ < O_2 < [O_2]^{2-}$$

Now consider H$_2$O$_2$ and O$_2$F$_2$. Both have structure **15.9**, and might be expected to possess similar O–O bond lengths consistent with single bonds. In H$_2$O$_2$, distance of 147.5 pm is similar to that in [O$_2$]$^{2-}$ (149 pm) – single bond. In O$_2$F$_2$, O–O bond length of 122 pm is almost the same as that in O$_2$, i.e. indicates double bond character; rationalize this in terms of the resonance forms **15.10**.

X = H or F

**(15.9)**

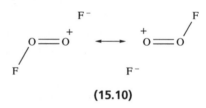

**(15.10)**

(b) For S, $r_{cov}$ = 103 pm, therefore expect a single S–S bond distance to be ≈ 206 pm. This is observed in S$_8$ (**15.1**) and H$_2$S$_2$ (isostructural with H$_2$O$_2$, **15.9**). Consider remaining species:

S$_2$: 189 pm is consistent with a diatomic with S=S double bond.

[S$_4$]$^{2+}$ : the cation is a planar square with equal bond lengths; 198 pm suggests an S–S bond order in between 1 and 2; **15.11** is one of a set of resonance structures; the overall delocalized bonding is represented in **15.12**.

S$_2$F$_2$ : 189 pm, i.e. the same as in S$_2$, so indicates double bond character despite the fact structure **15.7** shows a single bond. Rationalized in terms of contributions by ionic resonance structures as for O$_2$F$_2$ in **15.10**.

S$_2$Cl$_2$: 193 pm, not very different from that in S$_2$F$_2$, but less than 2$r_{cov}$.

S$_2$F$_{10}$ : 221 pm (significantly > 2$r_{cov}$) indicates a weak S–S bond; compound readily disproportionates with S–S cleavage (see Section 15.7 in H&S).

**(15.11)**    **(15.12)**

15.11    Structures of the compounds are given in **15.13-15.18** with the resultant dipole moment indicated where appropriate. The Pauling electronegativities that need to be considered are: $\chi^P$(Se) = 2.6, $\chi^P$(S) = 2.6, $\chi^P$(F) = 4.0, $\chi^P$(Cl) = 3.2, $\chi^P$(O) = 3.4.

**(15.13)    (15.14)    (15.15)    (15.16)    (15.17)    (15.18)**

From the values given for the molecular dipole moments, you can comment to some extent on correlations between electronegativity differences, molecular shape and effects of lone pairs. $SeF_6$ (**15.13**) is non-polar. $SeF_4$ and $SF_4$ are isostructural, although bond lengths and angles will be different; values of $\chi^P(S)$ and $\chi^P(Se)$ are the same. The directions of the resultant dipole moments are shown in **15.14** and **15.15**. In $SCl_2$ (**15.16**), each S–Cl bond is polar in the sense $S^{\delta+}$–$Cl^{\delta-}$, and the resultant dipole of the two bond moments is larger than that due to the lone pairs. Structure **15.17** shows the approximate direction in which the resultant dipole moment in $SOCl_2$ acts; the resultant of the S=O and S–Cl bond moments compete with the effects of the lone pair. In $SO_2Cl_2$ (**15.18**), the resultant of the two S=O bond moments is greater than that of the two S–Cl moments.

15.12    First, consider the nuclear spin data:
$^{19}F$     100%    $I = \frac{1}{2}$
$^{125}Te$   7%      $I = \frac{1}{2}$
$^{123}Te$   0.9%    $I = \frac{1}{2}$

In the $^{125}Te$ NMR spectrum, the binomial octet is due to coupling to 7 equivalent $^{19}F$ nuclei. $[TeF_7]^-$ is pentagonal bipyramidal (**15.19**). NMR spectroscopic data indicate that in solution at 298 K, the anion is stereochemically non-rigid (fluxional). In the $^{19}F$ NMR spectrum, the singlet is due to the 92.1% $^{19}F$ nuclei that are *not* attached to *spin-active* Te. $^{19}F$ bonded to $^{125}Te$ ($I = \frac{1}{2}$) gives rise to a doublet, and $^{19}F$ bonded to $^{123}Te$ ($I = \frac{1}{2}$) also gives rise to a doublet. Coupling constants of 2876 and 2385 Hz are assigned to $J(^{125}Te\text{-}^{19}F)$ and $J(^{123}Te\text{-}^{19}F)$, but it is not possible from the data available here to say which is which.

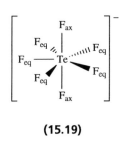

**(15.19)**

15.13    Use the periodic table (**15.20**) to help you with identifying isoelectronic species. Note that 'isoelectronic' is being used in respect only of the *valence electrons*.

(a) $[SiO_4]^{4-}$, $[PO_4]^{3-}$ and $[SO_4]^{2-}$ are all isoelectronic and isostructural (tetrahedral).
(b) $CO_2$, $SiO_2$ and $[NO_2]^+$ are isoelectronic; $CO_2$ and $[NO_2]^+$ are isostructural (linear) but $SiO_2$ has a macromolecular lattice with tetrahedral Si.
$SO_2$ and $TeO_2$ are isoelectronic but not isostructural; $SO_2$ is molecular (**15.22**) but $TeO_2$ adopts an infinite lattice with Te–O–Te bridges connecting $TeO_4$ units (**15.23**).

| 14 | 15 | 16 |
|----|----|----|
| C  | N  | O  |
| Si | P  | S  |
| Ge | As | Se |
| Sn | Sb | Te |

**(15.20)**

**(15.21)**

$(TeO_2)_n$

**(15.22)**

(c) $SO_3$, $[PO_3]^-$ and $SeO_3$ are all isoelectronic; $SO_3$ and $[PO_3]^-$ are isostructural (trigonal planar) but $SeO_3$ is tetrameric (**15.23**).
(d) $[P_4O_{12}]^{4-}$, $Se_4O_{12}$ and $[Si_4O_{12}]^{8-}$ are isoelectronic and isostructural (cyclic as in **15.23**).

**(15.23)**

15.14　(a) $SO_3$, trigonal planar; $[SO_3]^{2-}$, trigonal pyramidal.
Rationalized in terms of VSEPR: for $SO_3$, all 6 valence electrons of S used in bonding, but in the dianion, there is a lone pair (**15.24** shows one resonance form).
(b) Dissolution of $SO_2$ in water gives sulfurous acid, $H_2SO_3$, although the equilibrium constant shows that the solution contains mainly dissolved $SO_2$:

**(15.24)**

➤

$SO_2$ in aqueous solution: see Section 6.4 in H&S

$$SO_2(aq) + H_2O(l) \rightleftharpoons H_2SO_3(aq) \qquad K < 10^{-9}$$

$H_2SO_3$ (cannot be isolated) dissociates in solution according to the equilibria:

$$H_2SO_3(aq) + H_2O(l) \rightleftharpoons [H_3O]^+(aq) + [HSO_3]^-(aq) \qquad pK_a = 1.82$$

$$[HSO_3]^-(aq) + H_2O(l) \rightleftharpoons [H_3O]^+(aq) + [SO_3]^{2-}(aq) \qquad pK_a = 6.92$$

Sulfite, $[SO_3]^{2-}$, is a reducing agent in alkaline solution:

$$[SO_4]^{2-}(aq) + H_2O(l) + 2e^- \rightleftharpoons [SO_3]^{2-}(aq) + 2[OH]^- \qquad E^\circ_{[OH^-]=1} = -0.93\,V$$

and in acidic solution, aqueous $SO_2$ acts as a weak reducing agent:

$$[SO_4]^{2-}(aq) + 4H^+(aq) + 2e^- \rightleftharpoons H_2SO_3(aq) + H_2O(l) \qquad E^\circ_{[H^+]=1} = +0.17\,V$$

but value of $E$ depends on $[H^+]$, and $[SO_4]^{2-}$ can be reduced to $SO_2$ in the presence of high $[H^+]$:

$$Cu + 2H_2SO_4(conc) \rightarrow CuSO_4 + SO_2 + 2H_2O$$

Examples of aqueous $SO_2$ operating as a reducing agent:

$$H_2SO_3(aq) + H_2O(l) + 2Fe^{3+} \rightarrow [SO_4]^{2-}(aq) + 4H^+(aq) + 2Fe^{2+}(aq)$$

$$H_2SO_3(aq) + H_2O(l) + I_2 \rightarrow [SO_4]^{2-}(aq) + 4H^+(aq) + 2I^-(aq)$$

A commercial application of aqueous solutions of $SO_2$ is in the wine industry where it is used to kill microorganisms and to stabilize the wine against oxidation – see Box 15.7 in H&S.

15.15　(a) All the structures are based on $S_8$ (**15.1**) with NH replacing S; no N–N bonds (Figure 15.1).

**Figure 15.1** Diagrams and isomers of $S_7NH$, $S_6N_2H_2$, $S_5N_3H_3$ and $S_4N_4H_4$. The rings are non-planar.

**(b) Preparation of $S_4N_4$ (15.25):**

$$6S_2Cl_2 + 16NH_3 \xrightarrow{CCl_4, \ 320 \ K} S_4N_4 + 12NH_4Cl + S_8$$

Shock sensitivity of $S_4N_4$ means it must be handled with care; pure samples tend to explode when struck. For a summary of reactions, see Figure 15.14 in H&S. Examples chosen should illustrate:

- reduction to $S_4N_4H_4$
- halogenation
- ring degradation to NSF and $NSF_3$
- oxidation to $[S_4N_4]^{2+}$
- $(SN)_x$ polymer formation

and should show changes in ring conformations that accompany reactions, e.g. unfolding of the cage in $S_4N_4$ to the crown ring of $S_4N_4H_4$.

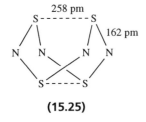

(15.25)

15.16   (a) The primary reaction is reduction of $[SO_4]^{2-}$ to $SO_2$ (see opposite page). For the formation of CuS, the reduction half-equation is:

$$[SO_4]^{2-} + 8H^+ + 8e^- \rightleftharpoons S^{2-} + 2H_2O$$

Two factors to consider: (i) $S^{2-}$ combines with $Cu^{2+}$ to give CuS which is sparingly soluble. Very low solubility of CuS ($K_{sp} \approx 10^{-44}$) facilitates reduction of sulfate to $S^{2-}$ (see Section 7.3 in H&S). (ii) Reduction potential is affected by concentration of $H^+$:

$$E = E^\circ - \left\{ \frac{RT}{zF} \times \left( \ln \frac{[\text{reduced form}]}{[\text{oxidized form}]} \right) \right\} = E^\circ - \left\{ \frac{RT}{zF} \times \left( \ln \frac{[S^{2-}]}{[SO_4^{2-}][H^+]^8} \right) \right\}$$

Very high $[H^+]$ makes $E^\circ$ more positive, making $[SO_4]^{2-}$ a stronger oxidizing agent.

(b) Apply VSEPR model to rationalize the structure of $[TeF_5]^-$ :

   $Te^-$, number of valence electrons = 7
   Number of bonding pairs (5 Te–F bonds) = 5
   1 lone pair
   'Parent' shape = octahedral
   Molecular shape = square-based pyramidal, see **15.26**.

(15.26)

(c) Initial precipitate is $Ag_2S_2O_3$, and this dissolves in excess $[S_2O_3]^{2-}$:

$$Na_2S_2O_3(aq) + 2AgNO_3(aq) \rightarrow Ag_2S_2O_3(s) + 2NaNO_3(aq)$$

$$Ag_2S_2O_3(s) + 5[S_2O_3]^{2-}(aq) \rightarrow 2[Ag(S_2O_3)_3]^{5-}(aq)$$

Black colour is $Ag_2S$, formed by the disproportionation of $[S_2O_3]^{2-}$:

$$[S_2O_3]^{2-} + H_2O \rightarrow S^{2-} + [SO_4]^{2-} + 2H^+$$

The sulfate gives a white precipitate with acidified aqueous $Ba(NO_3)_2$:

$$[SO_4]^{2-}(aq) + Ba(NO_3)_2(aq) \rightarrow BaSO_4(s) + 2[NO_3]^-(aq)$$

15.17    (a) $0.0261$ g $Na_2S_2O_4 = \dfrac{0.0261}{174.1} = 1.50 \times 10^{-4}$ moles

Values of $A_r$:

| Na | 22.99 |
|----|-------|
| S  | 32.06 |
| O  | 16.00 |

Ag is oxidized in $HNO_3$ to $Ag^+$ (i.e. a 1-electron oxidation) and this is equivalent to 30.0 cm$^3$ 0.1 M $[NCS]^-$:

Amount of $[NCS]^- = 30.0 \times 0.1 \times 10^{-3} = 3.0 \times 10^{-3}$ moles

Amount of $Ag^+ = 3.0 \times 10^{-3}$ moles

$\therefore$ Ratio of $[S_2O_4]^{2-} : Ag^+ = 1 : 2$

Therefore, the $[S_2O_4]^{2-}$ undergoes a 2-electron reduction. Reaction is:

$$[S_2O_4]^{2-} + 2Ag^+ + H_2O \rightarrow [S_2O_5]^{2-} + 2Ag + 2H^+$$

(b) $0.0725$ g $Na_2S_2O_4 = \dfrac{0.0725}{174.1} = 4.16 \times 10^{-4}$ moles

*Residual* $I_2$ is reduced by 23.75 cm$^3$ 0.1050 M $[S_2O_3]^{2-}$:

$$2[S_2O_3]^{2-} + I_2 \rightarrow [S_4O_6]^{2-} + 2I^-$$

Amount of $[S_2O_3]^{2-} = 23.75 \times 0.1050 \times 10^{-3} = 2.49 \times 10^{-3}$ moles

Amount of residual $I_2 = 1.25 \times 10^{-3}$ moles

Initial amount $I_2 = 50.0 \times 0.0500 \times 10^{-3}$ moles $= 2.50 \times 10^{-3}$ moles

Amount $I_2$ reacted with $Na_2S_2O_4 = (2.50 - 1.25) \times 10^{-3} = 1.25 \times 10^{-3}$ moles

$\therefore$ Ratio $[S_2O_4]^{2-} : I_2 = (4.16 \times 10^{-4}) : (1.25 \times 10^{-3}) = 1 : 3$

Therefore, the reaction is:

$$[S_2O_4]^{2-} + 3I_2 + 4H_2O \rightarrow 2[SO_4]^{2-} + 6I^- + 8H^+$$

**(15.27)**

15.18    Urea has structure **15.27**. After reaction of urea with $H_2SO_4$, the product **X** reacts with $NaNO_2$ and $H^+$ to give $N_2$, and adding $BaCl_2$ yields $BaSO_4$. For the latter:

$$BaCl_2(aq) + [SO_4]^{2-}(aq) \rightarrow BaSO_4(s) + 2Cl^-(aq)$$

**(15.28)**

The formula of **X** is $H_3NO_3S$. The fact that this is formed from $OC(NH_2)_2$ and $H_2SO_4$ suggests that $H_3NO_3S$ is $HOSO_2NH_2$ (**15.28**, sulfamic acid). The reaction liberating $N_2$ is:

$$HOSO_2NH_2 + NaNO_2 \rightarrow N_2 + NaHSO_4 + H_2O$$

15.19   Oxoacids of sulfur are described in Section 15.9 in H&S. Points to include:
  - Acids that are *not* isolable but for which salts can be isolated are $H_2S_2O_4$ (dithionous acid, **15.29**), $H_2SO_3$ (sulfurous acid, **15.30**), $H_2S_2O_5$ (disulfurous acid) and $H_2S_2O_6$ (dithionic acid).
  - Acids that can be isolated are $H_2SO_4$ (sulfuric acid, **15.31**), $HSO_3F$ and $HSO_3Cl$ (fluoro- and chlorosulfonic acids, **15.32**), $H_2SO_5$ (peroxomonosulfuric acid, **15.33**) and $H_2S_2O_8$ (peroxodisulfuric acid, **15.34**).
  - Thiosulfuric acid ($H_2S_2O_3$, **15.35**) can be isolated but is very unstable.
  - Most important acid commercially is $H_2SO_4$; scale of manufacture is huge.
  - Structures: trigonal pyramidal or tetrahedral S in each oxoacid or oxoanion:

(15.29)          (15.30)          (15.31)          (15.32)

(15.33)          (15.34)          (15.35)

  - Show how each acid ionizes in aqueous solution, indicting whether strong or weak acids; give examples of salts formed, noting series of salts for polybasic acids.
  - Dithionate and dithionite anions possess weak S–S bonds.
  - Concentrated $H_2SO_4$ as an oxidizing agent.
  - Use of liquid $H_2SO_4$ as a non-aqueous solvent (see Section 8.8 in H&S); self-ionization is:

$$2H_2SO_4 \rightleftharpoons [H_3SO_4]^+ + [HSO_4]^-$$

  - 'Superacid' behaviour of $HSO_3F$ (see Section 8.9 in H&S).

15.20   $S_2O$, **15.36**. Structure analogous to that of $SO_2$ (**15.21**) with S replacing O; bent rather than linear because of presence of lone pair on central S atom.

(15.36)

$[S_2O_3]^{2-}$ – **15.37** shows one resonance structure. Analogue of $[SO_4]^{2-}$ with tetrahedral S (6 valence electrons) in centre.

(15.37)

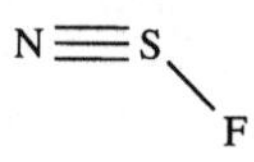

N≡S with F below (15.38)

N≡S with F, F, F (15.39)

**(15.38)** 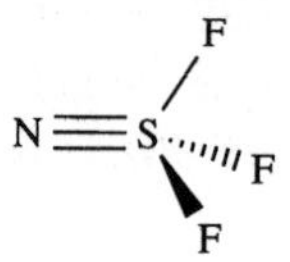       **(15.39)**

S=N⁺=S

**(15.40)**

NSF, **15.38**, possesses an N≡S triple bond and is bent because the S (6 valence electrons) has a lone pair. On going from NSF to $NSF_3$, S is oxidized. In $NSF_3$, **15.39**, all 6 valence electrons of S are used for bonding. Tetrahedral S, with N≡S triple bond.

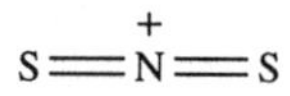

$[NS_2]^+$, **15.40**, is linear. This is consistent with VSEPR theory; note that $[NS_2]^+$ is isoelectronic (with respect to its valence electrons) with $[NO_2]^+$ and $CO_2$ (both linear).

$S_2N_2$ is cyclic with equal S–N bond lengths; equality of bond lengths can be explained in terms of a series of resonance structures, e.g. **15.41**:

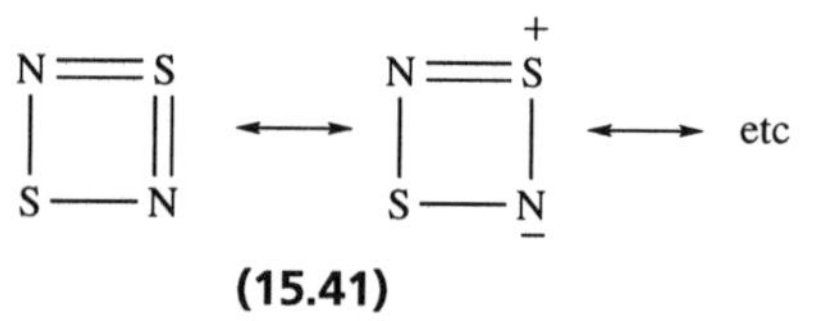

N=S / S—N  ⟷  N=S⁺ / S—N⁻  ⟷  etc

**(15.41)**

or in terms of $\pi$-delocalization using N $2p$ and S $3p$ atomic orbitals. $S_2N_2$ has 22 valence electrons; allocate 8 electrons for 4 $\sigma$-bonds, and 1 lone pair per atom, leaving 6 electrons for $\pi$-bonding; these occupy the $\pi$-MOs shown in Figure 15.2.

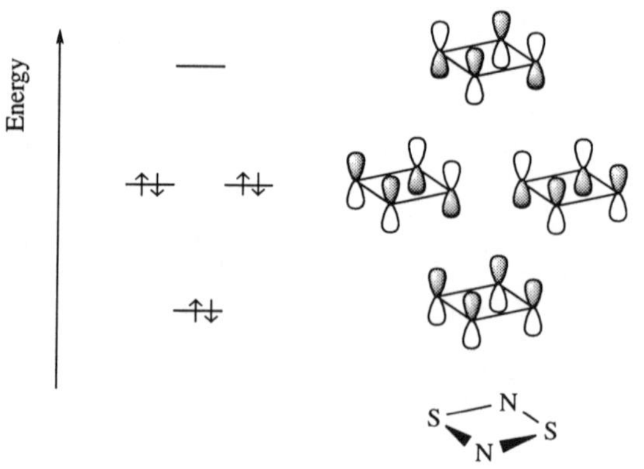

**Figure 15.2**  $\pi$-Molecular orbitals in $S_2N_2$, illustrating that there is delocalization of $\pi$-electrons around the ring.

# 16 The group 17 elements

16.1 (a) Halogens

(b)
| | |
|---|---|
| F | fluorine |
| Cl | chlorine |
| Br | bromine |
| I | iodine |
| At | astatine |

(c) $ns^2np^5$

16.2 (a) Displacement reactions (look at appropriate $E^{\circ}$ values in Appendix 11 in H&S):

$$2X^- + Cl_2 \rightarrow X_2 + 2Cl^- \qquad X = Br\ or\ I$$

(b) Downs process is electrolysis of molten NaCl; overall:

$$2Na^+(l) + 2Cl^-(l) \rightarrow 2Na(l) + Cl_2(g)$$

Products must be kept apart to prevent recombination and formation of NaCl.
(c) Recombination of $H_2$ and $F_2$ in an explosive radical chain reaction:

$$F_2 + H_2 \rightarrow 2HF$$

16.3 Lone pair-lone pair repulsions between lone pairs on O and F weaken bond. For $O_2F_2$, see resonance forms **15.10** (p. 132) for ionic contributions made in this case.

16.4 Hydrogen bonding for HF gives anomalously high values (see answer 9.9, p. 83); HCl, HBr and HI follow trend expected: increase with increasing molecular weight.

| $r_{cov}$ / pm | |
|---|---|
| F | 71 |
| Cl | 99 |
| Br | 114 |
| I | 133 |

16.5 Estimate bond length from sum of covalent radii (all single bonds): ClF, 170; BrF, 185; BrCl, 213; ICl, 232; IBr, 247 pm. The agreement with tabulated data for XY is good where the difference in electronegativities $\{\chi^P(Y) - \chi^P(X)\}$ is small. In the homonuclear diatomic molecule $X_2$, contribution made to the bonding by covalent resonance structure X–X dominates, and there is negligible contribution from $X^+X^-$; similarly for $Y_2$. This is also true for XY if $\chi^P(Y) \approx \chi^P(X)$, but if $\chi^P(Y) > \chi^P(X)$ (e.g. BrF), then the ionic form contributes significantly (e.g. **16.1**) and the bond is shorter than predicted from covalent radii.

$$Br \text{——} F \quad \longleftrightarrow \quad Br^+ \quad F^- \quad \longleftrightarrow \quad Br^- \quad F^+$$

**(16.1)** Negligible contribution because $\chi^P(F) > \chi^P(Br)$

16.6 (a) $ClF_3$ is a very strong oxidizing agent, and therefore $Ag^+$ is oxidized, rather than AgCl simply being fluorinated to AgF:

$$2AgCl + 2ClF_3 \rightarrow 2AgF_2 + Cl_2 + 2ClF$$

(b) $BF_3$ is an $F^-$ acceptor, and ClF donates $F^-$ but will not form 'naked' $Cl^+$ :

$$2ClF + BF_3 \rightarrow [Cl_2F]^+[BF_4]^-$$

(c) CsF is ionic and an F⁻ donor; IF$_5$ acts as an F⁻ acceptor:

$$CsF + IF_5 \rightarrow Cs^+[IF_6]^-$$

(d) SbF$_5$ is an F⁻ acceptor (will not act as a donor); the [SbF$_6$]⁻ ion formed can form an adduct with SbF$_5$:

$$SbF_5 + ClF_5 \rightarrow [ClF_4]^+[SbF_6]^-$$
or
$$2SbF_5 + ClF_5 \rightarrow [ClF_4]^+[Sb_2F_{11}]^-$$

(e) [Me$_4$N]F is ionic and an F⁻ donor, and IF$_7$ acts as an F⁻ acceptor:

$$[Me_4N]F + IF_7 \rightarrow [Me_4N]^+[IF_8]^-$$

(f)    $K[BrF_4] \xrightarrow{\Delta} KF + BrF_3$

16.7    Section 16.7 in H&S contains information for this problem. Points to include:
- Self-ionization of BrF$_3$ gives [BrF$_2$]⁺ and [BrF$_4$]⁻.
- Interhalogens such as ClF and ClF$_3$ act as F⁻ donors or acceptors depending on their reaction partner, e.g. with ionic fluoride, the interhalogen accepts F⁻ but with a potent fluoride acceptor (e.g. BF$_3$, AsF$_5$), the interhalogen donates F⁻.
- Higher interhalogens are mainly fluorides, and hence F⁻ donor/acceptor chemistry (as opposed to Cl⁻, Br⁻ or I⁻) dominates.
- Large cation (e.g. Cs⁺) needed to stabilize [XY$_n$]⁻ anion.

16.8    Apply VSEPR model; localize the overall charge on central atom for purposes of electron counting.
   (a) [ICl$_4$]⁻
      I⁻, number of valence electrons = 8
      Number of bonding pairs (4 I–Cl bonds) = 4
      2 lone pairs
      Molecular shape = square planar, see **16.2**.
   (b) [BrF$_2$]⁺
      Br⁺, number of valence electrons = 6
      Number of bonding pairs (2 Br–F bonds) = 2
      2 lone pairs
      Molecular shape = bent, see **16.3**.
   (c) [ClF$_4$]⁺
      Cl⁺, number of valence electrons = 6
      Number of bonding pairs (4 Cl–F bonds) = 4
      1 lone pair
      Molecular shape = disphenoidal, see **16.4**.
   (d) IF$_7$
      I, number of valence electrons = 7
      Number of bonding pairs (7 I–F bonds) = 7
      No lone pairs
      Molecular shape = pentagonal bipyramidal, see **16.5**.

**(16.2)**

**(16.3)**

**(16.4)**

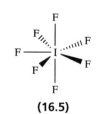

**(16.5)**

(e) $I_2Cl_6$; assume both I atoms are in the same environment and are connected by 2 bridging Cl atoms (must be 2, not 1, for symmetry).

μ- means 'bridging'

I, number of valence electrons = 7

Number of bonding pairs (3 I–Cl bonds) = 3

2 lone pairs

One extra pair of electrons from the coordinate bond of μ-Cl

Molecular shape = square planar at each I, see **16.6**.

**(16.6)**

(f) $[IF_6]^+$

$I^+$, number of valence electrons = 6

Number of bonding pairs (6 I–F bonds) = 6

No lone pairs

Molecular shape = octahedral.

(g) $BrF_5$

Br, number of valence electrons = 7

Number of bonding pairs (5 Br–F bonds) = 5

1 lone pair

Molecular shape = square-based pyramidal, see **16.7**.

**(16.7)**

16.9    (a) Static structure of $BrF_5$ (**16.7**) has two F environments – $F_{apical} : F_{basal} = 1 : 4$. Assume only $^{19}F$-$^{19}F$ spin-spin coupling is important. The $^{19}F$ NMR spectrum contains 2 signals. Signal assigned to $F_{apical}$ is a binomial quintet (coupling to 4 equivalent $^{19}F_{basal}$), and signal due to $F_{basal}$ is a doublet (coupling to 1 $^{19}F_{apical}$). Relative integrals of quintet : doublet = 1 : 4.

$[IF_6]^+$ is octahedral (see answer 16.8f). In the $^{19}F$ NMR spectrum of $[IF_6]^+$, a singlet is observed because all $^{19}F$ nuclei are equivalent.

(b) The $^{19}F$ NMR spectrum of $BrF_5$ is likely to be temperature dependent. Exchange of $F_{apical}$ and $F_{basal}$ sites may occur. The high temperature limiting spectrum is a singlet; the low temperature limiting spectrum is as described in part (a).

The $^{19}F$ NMR spectrum of $[IF_6]^+$ is not temperature dependent – a singlet at all NMR-observable temperatures.

16.10    (a) $I_2Cl_6$ is planar, with I in square planar environments; the planarity arises because of stereochemically active lone pairs (see answer 16.8e). Valence shell of each Al in $Al_2Cl_6$ has 8 electrons, all involved in bonding; each Al is tetrahedral (**16.8**).

**(16.8)**

(b) Thermal decomposition of $[Me_4N]^+[ClHI]^-$ could lead to $[Me_4N]^+I^-$ and HCl, $[Me_4N]^+Cl^-$ and HI, or $[Me_4N]^+H^-$ and ICl. $[Me_4N]^+$ is a large cation, and therefore all three salts are expected to have about the same lattice energy. The bond strength of the second product is therefore the decisive factor. The bond enthalpies of HCl, HI and ICl are 432, 298 and 211 kJ mol$^{-1}$ (value for ICl can be found from standard tables of data or estimated from Pauling electronegativity values of Cl and I, bond enthalpies of $Cl_2$ and $I_2$, see Section 1.15 in H&S).

(c) The purple solution of $I_2$ in *n*-hexane contains uncomplexed $I_2$. In benzene, ethanol and pyridine, charge transfer complexes **16.9** form.

**(16.9)**

Most stable complex is that with pyridine; pyridine preferentially complexes $I_2$ in each of the solutions to which it is added, making each solution the same colour.

**(16.10)**

**(16.11)**

**(16.13)**

16.11    In each oxohalide: X=O double bonds and X–Y single bonds (X and Y = halogens). The halogen in the higher oxidation state is the central atom; F is always terminal (ox. state –1).

(a) $[F_2ClO_2]^-$

    $Cl^-$, number of valence electrons = 8

    Number of bonding 'pairs' (2 Cl–F + 2 Cl=O) = 4

    1 lone pair

    Molecular shape = disphenoidal, probably with equatorial O, see **16.10**.

(b) $FBrO_3$

    Br, number of valence electrons = 7

    Number of bonding 'pairs' (1 Br–F + 3 Br=O) = 4

    No lone pairs

    Molecular shape = tetrahedral, see **16.11**.

(c) $[ClO_2]^+$ (note: isoelectronic with $SO_2$)

    $Cl^+$, number of valence electrons = 6

    Number of bonding 'pairs' (2 Cl=O) = 2

    1 lone pair

    Molecular shape = bent, see **16.12**.

**(16.12)**

(d) $[F_4ClO]^-$

    $Cl^-$, number of valence electrons = 8

    Number of bonding 'pairs' (4 Cl–F + 1 Cl=O) = 5

    1 lone pair

    Molecular shape = square-based pyramidal with axial O, see **16.13**.

16.12    (a) Section 16.9 in H&S has relevant information. In cold alkali:

$$Cl_2 + 2NaOH \rightarrow NaCl + NaOCl + H_2O$$

but on heating, $[OCl]^-$ disproportionates and overall reaction in hot alkali is:

$$3Cl_2 + 6NaOH \rightarrow NaClO_3 + 5NaCl + 3H_2O$$

(b) In the presence of excess $I^-$, only 1 mole of $I_2$ is produced per mole of $[IO_4]^-$. Since the oxidation step is:         $2I^- \rightarrow I_2 + 2e^-$

reduction must be a 2-electron step:     $[IO_4]^- + 2e^- + H_2O \rightarrow [IO_3]^- + 2[OH]^-$

Overall:

    $[IO_4]^- + 2I^- + H_2O \rightarrow [IO_3]^- + I_2 + 2[OH]^-$

In the presence of the excess $I^-$, final step is:

    $[IO_3]^- + 5I^- + 6H^+ \rightarrow 3I_2 + 3H_2O$

(c) $BaMnO_4$ contains $[MnO_4]^{2-}$, so $[MnO_4]^-$ undergoes a 1-electron reduction.

    Amount of $[MnO_4]^-$ used = $10 \times 0.1 \times 10^{-3} = 1 \times 10^{-3}$ moles

    Amount of $[I]^-$ used = $\dfrac{0.01587}{126.90} = 1.25 \times 10^{-4}$ moles

∴    Ratio $[MnO_4]^- : I^- = 1 \times 10^{-3} : 1.25 \times 10^{-4} = 8 : 1$

$I^-$ must undergo an 8-electron oxidation and the product is $[IO_4]^-$ (ox. state +7).

16.13    (a) $ClO_2$ and $[ClO_2]^-$ are bent, each with equivalent Cl–O bond lengths: 147 pm in $ClO_2$ and 157 pm in $[ClO_2]^-$. Resonance structures can be drawn to rationalize equivalence of bonds:

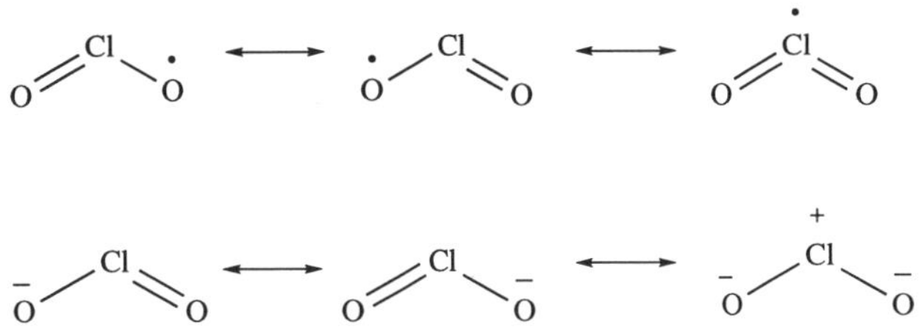

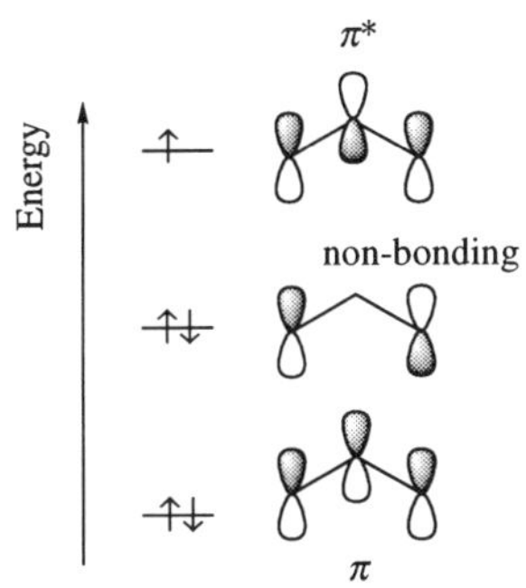

Or, use MO theory to show delocalized bonding schemes analogous to $\pi$-allyl-type bonding. There are 12 atomic orbitals in the basis set of $ClO_2$ (4 valence orbitals per atom) and therefore 12 MOs; these are occupied by 19 valence electrons. In $[ClO_2]^-$, 10 MOs are fully occupied. The $\pi$-MOs are shown in Figure 16.1. In $ClO_2$, the HOMO is the $\pi^*$ MO and is singly occupied. In $[ClO_2]^-$, this MO is fully occupied, giving more antibonding character to the Cl–O interactions. The MO scheme illustrates delocalization of electrons over the O–Cl–O framework, and rationalizes the lengthening of bonds on going from $ClO_2$ to $[ClO_2]^-$.

(b) Compare the ionic salt formulae: $K^+[ClO_4]^-$ and $Ba^{2+}[SO_4]^{2-}$. $[ClO_4]^-$ is isostructural with $[SO_4]^{2-}$, similar sized ions which can be regarded as being spherical. Ionic radii of $K^+$ (138 pm) and $Ba^{2+}$ (142 pm) are almost the same. The two salts can crystallize with the same lattice structure, i.e. they are isomorphous.

16.14    Use reduction half-equations and $E^o$ values from H&S Appendix 11 to help you.
(a) $[ClO_3]^-$ is a strong oxidizing agent:

$$[ClO_3]^- + 6Fe^{2+} + 6H^+ \rightarrow Cl^- + 6Fe^{3+} + 3H_2O$$

(b) Reduction of $[IO_3]^-$ occurs in acidic solution, therefore the half-equation for $[SO_3]^{2-}$ oxidation must also be in acidic solution (see Appendix 11 in H&S). Possible reaction is:

$$[IO_3]^- + 3[SO_3]^{2-} \rightarrow I^- + 3[SO_4]^{2-}$$

but partial reduction also possible:

$$2[IO_3]^- + 2H^+ + 5[SO_3]^{2-} \rightarrow I_2 + 5[SO_4]^{2-} + H_2O$$

(c) A good starting point is reaction 16.65 in H&S in which $[IO_3]^-$ and $I^-$ react to give $I_2$. The reaction with $Br^-$ gives:

$$[IO_3]^- + 5Br^- + 6H^+ \rightarrow 2Br_2 + IBr + 3H_2O$$

16.15    (a) Note that it is an *equilibrium* being considered. Determine the total chlorine by addition of excess of $I^-$ and titration with thiosulfate:

$$Cl_2 + 2I^- \rightarrow I_2 + 2Cl^-$$

$$I_2 + 2[S_2O_3]^{2-} \rightarrow 2I^- + [S_4O_6]^{2-}$$

**Figure 16.1** $\pi$-Molecular orbitals in $ClO_2$.

Of the products, only HCl is a strong acid, and so its concentration can be determined by pH measurement (not by titration which would upset the equilibrium). From these data, the equilibrium constant $K$ at a given temperature can be determined. Determine $K$ at different temperatures, and from these data, $\Delta_r H^\circ$ can be found from:

$$\frac{\mathrm{d}\ln K}{\mathrm{d}T} = \frac{\Delta H^\circ}{RT^2}$$

(b) Take a weighed amount of $I_4O_9$ and dissolve in water. Neutralize the solution with $NaHCO_3$ and titrate the $I_2$ against thiosulfate (see part (a)). This gives the amount of $I_2$ from the initial hydrolysis of $I_4O_9$. The reaction with thiosulfate produces $I^-$ which, under acidic conditions (created by adding, for example, excess dilute HCl), converts $[IO_3]^-$ to $I_2$:

$$[IO_3]^- + 5I^- + 6H^+ \rightarrow 3I_2 + 3H_2O$$

The $I_2$ formed in this step is then titrated against thiosulfate.

(c) In the solidified noble gas matrix, species can be studied by vibrational spectroscopy. Raman (not IR) spectroscopy must be used to study a *homonuclear* diatomic since stretching mode is IR inactive. Going from $Cl_2$ to $[Cl_2]^-$ weakens the Cl–Cl bond because the extra electron occupies an antibonding MO (see Figure 1.21 in H&S for related $F_2$ MO diagram). The formation of $[Cl_2]^-$ from $Cl_2$ should be accompanied by a shift to lower frequency for the absorption corresponding to the Cl–Cl stretch.

16.16    (a) HF remains strongly hydrogen-bonded (**16.14**) in the vapour state and few hydrogen bonds are broken upon vaporization. Liquid $H_2O$ is hydrogen-bonded but on vaporization, these intermolecular interactions are broken. Greater amount of energy is therefore needed to vaporize $H_2O$ than HF.
(b) The fact that AgCl and AgI are both soluble in saturated aqueous KI (in which $I^-$ is present), but insoluble in saturated aqueous KCl (in which $Cl^-$ is present) suggests that the complex formation between $Ag^+$ and $I^-$ gives a more stable species than is formed between $Ag^+$ and $Cl^-$. Species that may be formed are $[AgX_2]^-$ and $[AgX_3]^{2-}$.

(**16.14**)

16.17    (a) First note that the structural descriptions likening the ammonium salts to NaCl, CsCl or ZnS arise from assumption that $[NH_4]^+$ can be treated as a spherical ion. In $[NH_4]^+X^-$ (X = Cl, Br, I), the ions are arranged in either an NaCl or CsCl lattice. The preference for the wurtzite structure for $[NH_4]F$ arises from N–H·····F hydrogen bond formation which gives a structure similar to that of ice.
(b) The decomposition reaction is:

$$[PH_4]^+X^- \rightarrow PH_3 + HX$$

For the product HX where X = F, Cl, Br or I, the weakest bond is H–I; in the absence of numerical data, you can only infer that this factor is the most important. For a detailed answer in terms of a thermochemical cycle, see answer 6.24, p. 59.

# 17 The group 18 elements

17.1  (a) Noble gases
(b)  He    helium
     Ne    neon
     Ar    argon
     Kr    krypton
     Xe    xenon
     Rn    radon
(c) A fully occupied quantum shell, $2s^2$ for He, and $ns^2np^6$ for the later elements.

17.2  See answer 1.25, p. 9.

17.3  $XeF_2$
      Xe, number of valence electrons = 8
      Number of bonding pairs (2 Xe–F) = 2
      3 lone pairs
      Molecular shape = linear, see **17.1**.

**(17.1)**

$XeF_4$
      Xe, number of valence electrons = 8
      Number of bonding pairs (4 Xe–F) = 4
      2 lone pairs
      Molecular shape = square planar, see **17.2**.

**(17.2)**

$XeF_6$
      Xe, number of valence electrons = 8
      Number of bonding pairs (6 Xe–F) = 6
      1 lone pair
      Molecular shape = distorted octahedral, see **17.3**.

**(17.3)**

17.4  $[XeF_8]^{2-}$
      $Xe^{2-}$, number of valence electrons = 10
      Number of bonding pairs (8 Xe–F) = 8
      1 lone pair
      The observation that the anion is square antiprismatic (**17.4**) indicates a stereo-chemically *inactive* lone pair.

**(17.4)**

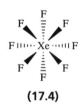

17.5  (a) To determine a value of $\Delta_fH^o(XeF_2, 298\ K)$, measure the enthalpy change for the hydrolysis of $XeF_2$:

$$2XeF_2(s) + 2H_2O(l) \xrightarrow{\Delta_rH^o(298\ K)} 2Xe(g) + 4HF(g) + O_2(g)$$

$2\Delta_fH^o(XeF_2, 298\ K) +$        $4\Delta_fH^o(HF, 298\ K)$ (available
$2\Delta_fH^o(H_2O, 298\ K)$            from tables of standard data)

$$2Xe(g) + 2H_2(g) + 2F_2(g) + O_2(g)$$

$$2\Delta_fH^o(XeF_2, 298\ K) = 4\Delta_fH^o(HF, 298\ K) - \Delta_rH^o(298\ K) - 2\Delta_fH^o(H_2O, 298\ K)$$
            ↑                              ↑                      ↑
      available data                  measured            available data

(b) Set up the following thermochemical cycle (sub = sublimation), all data at 298 K:

$$XeF_2(s) \xrightarrow{\Delta_{sub}H^o(XeF_2)} XeF_2(g)$$

$\Delta_f H^o(XeF_2,s)$        $\Delta_a H^o(XeF_2) = 2D(Xe-F)$

$$Xe(g) + F_2(g) \longrightarrow Xe(g) + 2F(g)$$
$$2\Delta_a H^o(F_2) = D(F-F)$$

The value of $\Delta_f H^o(XeF_2,s)$ comes from part (a) of this problem; $\Delta_{sub}H^o(XeF_2)$ can be measured; $D(F-F) = 158$ kJ mol$^{-1}$. The Xe–F bond energy can be found from:

$$2D(Xe-F) = D(F-F) - \Delta_f H^o(XeF_2,s) - \Delta_{sub}H^o(XeF_2)$$

17.6    Consider the reactions:

$$Xe + Cl_2 \rightarrow XeCl_2$$

$$Xe + F_2 \rightarrow XeF_2$$

From the thermochemical cycle:

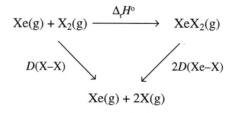

enthalpy change for each reaction is:

$$\Delta_r H^o = D(X-X) - 2D(Xe-X) \qquad\qquad X = Cl \text{ or } F$$

The bond enthalpies (from Tables 16.1 and 17.2 in H&S) are:
    $D(F-F) = 158$ kJ mol$^{-1}$         $D(Cl-Cl) = 242$ kJ mol$^{-1}$
    $D(Xe-F) = 133$ kJ mol$^{-1}$
It is expected that $D(Xe-Cl) < D(Xe-F)$ (e.g. see Tables 13.2 or 13.2 in H&S for trends in other halides).
For X = F:
    $\Delta_r H^o = (158) - 2(133) = -108$ kJ mol$^{-1}$

For X = Cl, $D(Xe-Cl)$ the reaction cannot be more exothermic than for X = F unless the Xe–Cl bond is stronger than the Xe–F bond which will not be the case.

17.7    Xe is next to Cs in the periodic table. Assume that the lattice energy of Xe$^+$F$^- \approx$ lattice energy of Cs$^+$F$^-$. Set up a Born-Haber cycle (e.g. see answer 5.15, p. 48) and estimate $\Delta_f H^o(XeF, 298$ K$)$. Other data required including $IE_1$ of Xe are obtained from tables of standard data.

17.8    $[XeO_6]^{4-}$

$Xe^{4-}$, number of valence electrons = 12
Number of bonding 'pairs' (6 Xe=O) = 6
No lone pairs
Molecular shape = octahedral.

XeOF$_2$

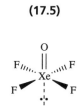

**(17.5)**

Xe, number of valence electrons = 8
Number of bonding 'pairs' (1 Xe=O + 2 Xe–F) = 3
2 lone pairs
Molecular shape = T-shaped, see **17.5**; O atom is in equatorial site.

XeOF$_4$

**(17.6)**

Xe, number of valence electrons = 8
Number of bonding 'pairs' (1 Xe=O + 4 Xe–F) = 5
1 lone pair
Molecular shape = square pyramidal, see **17.6**; O atom is in apical site.

$XeO_2F_2$

**(17.7)**

Xe, number of valence electrons = 8
Number of bonding 'pairs' (2 Xe=O + 2 Xe–F) = 4
1 lone pair
Molecular shape = disphenoidal, see **17.7**; O atoms are in equatorial sites.

$XeO_2F_4$

Xe, number of valence electrons = 8
Number of bonding 'pairs' (2 Xe=O + 4 Xe–F) = 6
No lone pairs
Molecular shape = octahedral; O atoms could be *trans* or *cis*.

$XeO_3F_2$

**(17.8)**

Xe, number of valence electrons = 8
Number of bonding 'pairs' (3 Xe=O + 2 Xe–F) = 5
No lone pairs
Molecular shape = trigonal bipyramidal, **17.8**; F atoms are axial.

17.9    (a) CsF is ionic, and a source of F⁻. XeF$_4$ acts as F⁻ acceptor:

$$CsF + XeF_4 \rightarrow Cs[XeF_5] \qquad \text{i.e. } Cs^+[XeF_5]^-$$

(b) F for O exchange; reaction driven by formation of SiF$_4$ (formation enthalpy = –1615 kJ mol⁻¹):

$$SiO_2 + 2XeOF_4 \rightarrow SiF_4 + 2XeO_2F_2$$

or

$$SiO_2 + XeOF_4 \rightarrow SiF_4 + XeO_3$$

(c) SbF$_5$ is potent F⁻ acceptor; $[SbF_6]^-$ forms adducts with SbF$_5$. The $[XeF]^+$ ion may associate with XeF$_2$. Possible reactions are:

$$XeF_2 + SbF_5 \rightarrow [XeF]^+[SbF_6]^-$$

or

$$2XeF_2 + SbF_5 \rightarrow [Xe_2F_3]^+[SbF_6]^-$$

or

$$XeF_2 + 2SbF_5 \rightarrow [XeF]^+[Sb_2F_{11}]^-$$

(d) Hydrolysis of $XeF_6$ by water gives $XeO_3$, but in alkaline solution, this dispropotionates to Xe and perxenate ion, $[XeO_6]^{4-}$. Overall:

$$2XeF_6 + 16[OH]^- \rightarrow [XeO_6]^{4-} + Xe + O_2 + 8H_2O + 12F^-$$

(e) Hydrolysis by water of $KrF_2$ is analogous to hydrolysis of $XeF_2$:

$$2KrF_2 + 2H_2O \rightarrow 2Kr + O_2 + 4HF$$

17.10    Section 17.4 in H&S gives information for the answer. Points to include:
- preparations, and difficulties of obtaining pure samples;
- structures in gas phase and solid state;
- hydrolysis by $H_2O$;
- reactions showing $XeF_n$ acting as oxidizing agents;
- reactions with silica, and the unsuitability of silica glassware for studying reactions of Xe fluorides;
- reactions illustrating $F^-$ donor and acceptor properties.

# 18 Organometallic compounds of s- and p-elements

18.1 (a) $MeBr + 2Li \xrightarrow{Et_2O} MeLi + LiBr$    (lithiation; $\Delta_fH°$ of LiBr drives reaction)

(b) $Na + (C_6H_5)_2 \xrightarrow{THF} Na^+[(C_6H_5)_2]^-$    (Na reduction of biphenyl)

(c) $^nBuLi + H_2O \rightarrow {}^nBuH + LiOH$    (hydrolysis of a lithium alkyl gives corresponding alkane)

(d) $Na + C_5H_6 \rightarrow Na^+[C_5H_5]^- + {}^1/_2H_2$    (route to prepare synthetically important Na[Cp])

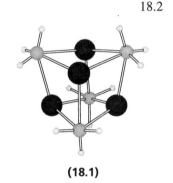

**(18.1)**

18.2 In the tetramer $(MeLi)_4$ (**18.1**), each C atom appears to be 6-coordinate. A localized bonding scheme within the $C_4Li_4$ unit is inappropriate. It is convenient to consider each C atom as $sp^3$ hybridized; allocate localized 2c-2e C–H bonds. Each Me group provides one $sp^3$ hybrid and one electron to the bonding in the $C_4Li_4$ unit. Each Li atom has one valence 2s electron, but can make use of empty 2p atomic orbitals to form $sp^3$ hybrids, three of which are used to overlap with C $sp^3$ hybrids. Total number of valence electrons available = 4 (from Li) + 4 (from C) = 8. In the diagrams below, first note the positions of the C and Li atoms at the corners of an approximate cube; the Li atoms define a tetrahedron, and so do the C atoms. Each C atom caps one $Li_3$ face. Each C atom provides one $sp^3$ hybrid pointing into an $Li_3$ face. Each Li atom provides three $sp^3$ hybrids, each pointing into a different $Li_3$ face. The right-hand diagram below shows the result *for one $Li_3$ face.*

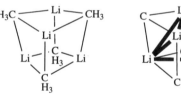  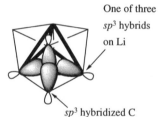

Overlap of the hybrid orbitals gives rise to a 4c-2e interaction over each $Li_3$ face. This model allows for a degree of Li–Li bonding as well as Li-C interactions.

18.3 Gas phase: $Me_2Be$ consists of monomeric, linear molecules (**18.2**). Bonding described in terms of $sp$ hybridized Be with 2 valence electrons, and $sp^3$ hybridized C. Three localized 2c–2e C–H bonds in each Me group, leaving one $sp^3$ hybrid and one valence electron for localized $\sigma$-bond formation with Be. This is analogous to descriptions of the bonding in monomeric $BeH_2$ and $BeCl_2$.

$H_3C — Be — CH_3$

**(18.2)**

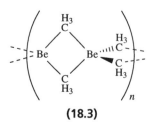

**(18.3)**

Solid state: polymeric chain structure (**18.3**). The bonding can be described in the same way as in polymeric $BeH_2$ (see answer 9.15, p. 85) using a C $sp^3$ hybrid instead of an H 1s atomic orbital. In polymeric $BeCl_2$, lone pairs of electrons from Cl supply enough electrons so that all Be–Cl interactions are localized 2c-2e (see answer 11.9b, p. 96).

18.4  (a)  $Mg + 2C_5H_6 \rightarrow (\eta^5\text{-}C_5H_5)_2Mg + H_2$     (reduction of cyclopentadiene and oxidation of Mg)

(b)  $MgCl_2 + LiR \rightarrow RMgCl + LiCl$     (preparation of Grignard reagent, and magnesium dialkyl)

 or

$MgCl_2 + 2LiR \rightarrow R_2Mg + 2LiCl$

(c)  $RBeCl \xrightarrow{\ Li[AlH_4]\ } RBeH$     (reduction using $[AlH_4]^-$ )

18.5  The coordination number of Mg is likely to be 4; shape: tetrahedral. THF is a monodentate ligand, and *O*-donor. The part of the formula '$(Me_3Si)_2C(MgBr)_2$' shows that C is bonded to 2 MgBr units. This leaves Mg with 2 vacant coordination sites.
∴ $n = 4$, see structure **18.4**.

**(18.4)**

18.6  (a) For the equilibrium:     $Al_2R_6 \rightleftharpoons 2AlR_3$

**(18.5)**

**(18.6)**

the smaller value of *K* corresponds to the smaller steric demands of R. For small Me group, a dimer is stongly favoured. For bulky $Me_2CHCH_2$ group, the dimer is less favoured.
(b) The terminal Me groups in $Al_2Me_6$ (**18.5**) involve localized 2c-2e Al–C bonds. Each Al is $sp^3$ hybridized with 3 valence electrons; after terminal Al–C bond formation, 1 electron is left for bridge bonding. Each Al–C–Al interaction is a 3c-2e interaction. In $Al_2Cl_6$, each bridging Cl provides 3 valence electrons to give localized 2c-2e interactions (**18.6**). In $Al_2Me_4(\mu\text{-}Cl)_2$, the bonding can be described in terms of a combination of the two bonding models for $Al_2Me_6$ and $Al_2Cl_6$. All bonds can be considered to be localized 2c-2e.

18.7  (a)  $Al_2Me_6 + 6H_2O \rightarrow 2Al(OH)_3 + 6CH_4$     (hydrolysis of aluminium alkyl releases alkane)

(b)  $nAlR_3 + nR'NH_2 \rightarrow (RAlNR')_n + 2nRH$     (RH elimination to give AlN ring or cage)

(c)  $Me_3SiCl + Na[C_5H_5] \rightarrow Me_3Si(\eta^1\text{-}C_5H_5) + NaCl$
(elimination of NaCl with formation of Si–C $\sigma$-bond)

(d)  $2Me_2SiCl_2 + Li[AlH_4] \rightarrow 2Me_2SiH_2 + LiCl + AlCl_3$
(reduction by $[AlH_4]^-$ )

18.8    (a) Section 18.4 in H&S contains relevant information. Points to include:

- monomeric $R_3B$ with trigonal planar B, and 2c-2e B–C bonds;
- dimeric $Al_2R_6$ with tetrahedral Al, and 3c-2e bridging interactions;
- monomeric $GaR_3$, $InR_3$ and $TlR_3$ with trigonal planar group 13 element, but tendency for interactions between molecules in solid state.

(b) $[Me_2(PhC_2)Ga]_2$ (**18.7**) exhibits $\sigma,\pi$-mode of bonding for the bridging $PhC_2$-units. In $[Ph_3Al]_2$ (**18.8**), two Ph groups bridge between Al atoms with the *ipso*-C atoms ≈tetrahedrally sited.

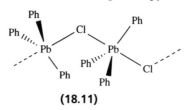

(**18.7**)

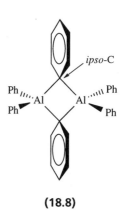

(**18.8**)

18.9    Going from $(\eta^1\text{-}C_5Me_5)_2SiBr_2$ to $(\eta^5\text{-}C_5Me_5)_2Si$ reduces Si(IV) to Si(II) – the reducing agent is prepared by treating anthracene (**18.9**) with K. Anthracene undergoes a 1-electron reduction to give the salt $K^+[(\mathbf{18.9})]^-$. The radical anion $[(\mathbf{18.9})]^-$ acts as a reducing agent, regenerating anthracene as it is oxidized. The second product of the reaction is KBr.

(**18.9**)

18.10   This question is tackled by: (i) considering the availability of halo-substituents to act as bridging groups between group 14 elements, (ii) considering the likely coordination number of the group 14 element, and (iii) checking that the proposed structure retains the correct stoichiometry.
(a) General formula $R_2EX_2$ suggests a chain with double halo-bridges, i.e. **18.10** with octahedral Pb.
(b) General formula $R_3EX$ with an R group which is not too bulky suggests a chain with single halo-bridges, i.e. **18.11** with trigonal bipyramidal Pb.

(**18.10**)

(**18.11**)

(c) In $(2,4,6\text{-}Me_3C_6H_2)_3PbCl$, the aryl substituents are very sterically demanding. Therefore, $(2,4,6\text{-}Me_3C_6H_2)_3PbCl$ is likely to be a monomer with tetrahedral Pb.
(d) $[PhPbCl_5]^{2-}$ is 6-coordinate and unlikely to increase its coordination number; therefore monomeric, with octahedral Pb.

18.11   (a)    $Et_3SnCl + H_2O \rightarrow Et_3SnOH + HCl$    (hydrolysis of Sn–Cl bond)
        or
        $2Et_3SnCl + H_2O \rightarrow (Et_3Sn)_2O + 2HCl$

(b)    $Et_3SnCl + Na[Cp] \rightarrow (\eta^1\text{-}Cp)Et_3Sn + NaCl$    (elimination of NaCl; formation of Sn–C $\sigma$-bond)

(c)    $2Et_3SnCl + Na_2S \rightarrow (Et_3Sn)_2S + 2NaCl$    (elimination of NaCl; formation of Sn–S bonds)

(d)    $Et_3SnCl + PhLi \rightarrow Et_3PhSn + LiCl$    (elimination of LiCl; formation of Sn–C $\sigma$-bond)

(e)     $2Et_3SnCl + 2Na \rightarrow Et_3SnSnEt_3 + 2NaCl$     (oxidation of Na; Sn–Sn bond formation)

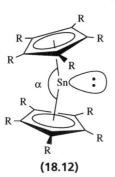

**(18.12)**

18.12    (a) Structure **18.12** shows the general structure of $(\eta^5\text{-}C_5R_5)_2Sn$. Tilt angle $\alpha$ increases as the steric demands of R increase: $\alpha = 125°$ for R = H, $144°$ for R = Me, $180°$ for R = Ph; tilting is rationalized in terms of the presence of a lone pair of electrons, but for R = Ph, the lone pair becomes stereochemically inactive as steric demands of R are large.
(b) Structure of $(\eta^5\text{-}C_5Me_5)_2Mg$ is as expected; no lone pair of electrons is present (contrast with structure **18.12**), and the $C_5$ rings are parallel. It is difficult to rationalize why the heavier group 2 metals show tilted structures – no simple explanation can be put forward but it may be solid state (crystal packing) effect.

18.13    InBr contains In(I) and oxidation will give In(III), so **A** is an In(III) compound. Oxidative addition of C–Br bond occurs. In the $^1H$ NMR spectrum, the multiplet at $\delta\,3.6$ (rel. integral 8) is assigned to 1,4-dioxane, singlet at $\delta\,5.36$ (rel. integral 1) is a CH unit. Suggests **A** is the adduct $Br_2InCHBr_2 \cdot C_4H_8O_2$. **A** contains two more C–Br bonds which react with 2 equivalents of InBr in the same manner as the first reaction. Each In(III) centre can behave as a Lewis acid, reacting with the $Br^-$ that is added:

See: J.A. Nobrega *et al.* (1998) *Chemical Communications*, p. 381 for full information

$$\underset{Br_2In}{\overset{H}{\big|}}\!\!C\!\!\underset{InBr_2}{\overset{InBr_2}{\diagup}} + 3Br^- \rightarrow \left[ \underset{Br_3In}{\overset{H}{\big|}}\!\!C\!\!\underset{InBr_3}{\overset{InBr_3}{\diagup}} \right]^{3-}$$

Product is stabilized by a large counterion, $[Ph_4P]^+$. Check the analysis of $[Ph_4P]_3[HC(InBr_3)_3]$ $(M_r = 2093.56)$:

$$In = \frac{344.46}{2093.56} \times 100 = 16.45\%$$

$$Br = \frac{719.1}{2093.56} \times 100 = 34.35\%$$

Signals in $^1H$ NMR spectrum of **B** at $\delta\,8.01\text{-}7.71$ (rel. integral 60) are assigned to Ph protons; singlet at $\delta\,0.20$ (rel. integral 1) assigned to CH proton. **B** is therefore $[Ph_4P]_3[HC(InBr_3)_3]$.

18.14    (a) B is in group 13; $[R_2BBR_2]^{2-}$ is isoelectronic with an alkene $R_2C=CR_2$, and contains a B=B double bond (**18.13**). The $B_2C_4$-framework is planar; bulky R groups needed to stabilize compound. Bonding: $sp^2$ hybridized B, overlap giving B–B $\sigma$-bond; $\pi$-bond from overlap of remaining $2p$ atomic orbitals; 4 valence electrons per $B^-$.
(b) Ga is in group 13; $[R_2GaGaR_2]^-$ is related to $[R_2BBR_2]^{2-}$, but has one electron less for bonding. Ga–Ga $\sigma$-bonding MO is fully occupied, but $\pi$-bonding MO is half-occupied giving a Ga–Ga bond order of 1.5.
(c) Sn is in group 14; by analogy with an alkene, $R_2SnSnR_2$ might be thought to have an Sn=Sn double bond, but distannenes (and digermenes) have non-planar frameworks. Bonding described in terms of $sp^2$ hybridized Sn(II), localized Sn–$C_R$ bonds and a lone pair occupying the remaining $sp^2$ hybrid orbital. Donation of lone pair into vacant $5p$ atomic orbital on adjacent Sn atom (**18.14**).

$$\left[ \underset{R}{\overset{R}{\diagdown}}B\!=\!B\underset{R}{\overset{R}{\diagup}} \right]^{2-}$$

**(18.13)**

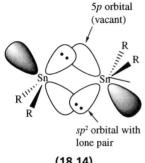

5*p* orbital (vacant)

*sp*² orbital with lone pair

**(18.14)**

(d) Ge is in group 14; $R_3GeGeR_3$ is an analogue of an alkane $R_3CCR_3$. Ge is *sp³* hybridized; Ge–Ge single bond; staggered conformation.

(e) As is in group 15; RAsAsR is an analogue of RN=NR; As is *sp²* hybridized, giving localized As–C$_R$ bonds and a lone pair occupying one hybrid orbital; overlap of *sp²* hybrid orbitals gives the As–As $\sigma$-bond, and overlap of singly occupied 4*p* atomic orbitals gives the As–As $\pi$-bond. A *trans*-arrangement of R groups is sterically favoured.

**(18.15)**

18.15  (a)  $2Me_3Sb + B_2H_6 \rightarrow 2Me_3Sb\cdot BH_3$    (Me$_3$Sb acting as a Lewis base)

(b)  $Me_3Sb + H_2O_2 \rightarrow Me_3SbO + H_2O$    (oxidation of Sb(III) to Sb(V) with accompanying O transfer)

(c)  $Me_3Sb + Br_2 \rightarrow Me_3SbBr_2$    (oxidative addition of Br$_2$)

(d)  $Me_3Sb + Cl_2 \rightarrow Me_3SbCl_2$    (oxidative addition of Cl$_2$)

$Me_3SbCl_2 + 3MeLi \rightarrow Li^+[Me_6Sb]^- + 2LiCl$
(methylation; Li$^+$ salt converted to a salt with a large cation for isolation)

(e)  $Me_3Sb + MeI \rightarrow Me_4SbI$    (oxidative addition of MeI)

(f)  $Me_3Sb + Br_2 \rightarrow Me_3SbBr_2$    (oxidative addition of Br$_2$)

$Me_3SbBr_2 + 2Na[OEt] \rightarrow Me_3Sb(OEt)_2 + 2NaBr$
(elimination of NaBr and formation of an ethoxy derivative)

18.16  Sections 18.4-18.6 in H&S contain relevant information. Points to include:
- group 13, oxidation state +3, with +1 becoming stable for the heaviest elements, exemplified by cyclopentadienyl derivatives: $R_3B$, $R_3Al$ but forms dimers $R_2Al(\mu\text{-}R)_2AlR_2$, $R_3Ga$, $R_3In$, $R_3Tl$, CpGa, CpIn, CpTl;
- group 14 (ignoring C), oxidation state +4 with +2 becoming stable for the later elements: $R_4Si$, $R_4Ge$, $R_4Sn$, $R_4Pb$, $R_2Si$ (highly reactive), $R_2Ge$ and $R_2Sn$, $R_2Pb$ (stabilized or stability increased by bulky R groups);
- group 15 (ignoring N and P), oxidation states +3 and +5, $R_3E$ and $R_5E$ with $R_3E$ being sensitive to oxidation;
- comment on relative stabilities among groups of similar compounds.

18.17  (a)  $GeCl_4 + 4RMgCl \xrightarrow{Et_2O} R_4Ge + 4MgCl_2$    (Grignard reagent)

$GeCl_4 + 4RLi \xrightarrow{Et_2O} R_4Ge + 4LiCl$    (organolithium reagent)

Range of derivatives can be made by these methods.

(b)  $^1/_2B_2H_6 + 3RCH_2=CH_2 \rightarrow B(CH_2CH_2R)_3$    (hydroboration; R = alkyl, aryl)

$Et_2O\cdot BF_3 + 3RMgCl \xrightarrow{Et_2O} R_3B + 3MgClF + Et_2O$    (R = alkyl, aryl)

(c)     $GaCl_3 + 3Li[C_5R_5] \rightarrow (C_5R_5)_3Ga + 3LiCl$

(d)     $nR_2SiCl_2 + 2nNa \rightarrow cyclo\text{-}(R_2Si)_n + 2nNaCl$     (small R favours large ring; bulky R favours small ring)

(e) Requires oxidation of $R_3As$; cannot be formed from $AsX_5$, X = halogen:

$$R_3As \xrightarrow{\ Cl_2\ } R_3AsCl_2 \xrightarrow{\ 2RLi\ } R_5As$$

(f)     $2R_2AlCl + 2K \rightarrow R_2AlAlR_2 + 2KCl$     (only works for *very* bulky R group, e.g. $(Me_3Si)_2CH$)

(g)     $SbCl_3 + 3RMgCl \xrightarrow{\ Et_2O\ } R_3Ge + 3MgCl_2$     (Grignard reagent)

18.18   Relevant information is found in Sections 18.4-18.6 in H&S. Points on which to base an answer:
- $\eta^1$, $\eta^3$ and $\eta^5$-modes of bonding;
- differences between gas phase and solid state structures, e.g. CpTl is a monomer in gas phase, but polymeric in solid, and similarly $Cp_2Pb$.
- monomer versus polymer formation for related species in solid, e.g. $Cp_3Ga$ is monomeric, $Cp_3In$ forms chains;
- in group 14, tilted versus parallel $C_5R_5$ rings in $(\eta^5\text{-}C_5R_5)_2E$;
- fluxional properties in solution observed by using (typically) $^1H$ NMR spectroscopy.

18.19   Relevant information is found in Sections 18.4-18.6 in H&S. Points to include:
- in group 13, compound types are $R_2E\text{–}ER_2$ and reduced derivatives;
- in group 14, compound types are $R_3E\text{–}ER_3$ (ethane analogue), $R_2E=ER_2$ (ethene analogue) and $R_2E\text{–}ER_2$ (non-planar $E_2C_4$ framework, see **18.14**).
- in group 15, compound types are $R_2E\text{–}ER_2$ and RE=ER (see **18.15**);
- sterically demanding substituents vital to stabilization of these compounds except for $R_2E\text{–}ER_2$ where E = group 15 element;
- common bulky substituents include **18.16** and **18.17**;
- examples: $B_2R_4$ and $[B_2R_4]^{2-}$ (ethene analogue); $Al_2R_4$ and $[Al_2R_4]^-$ (radical anion); $Ga_2R_4$, $[Ga_2R_4]^-$ and $[Ga_2R_4]^{2-}$; $R_2Si=SiR_2$, $R_2Ge=GeR_2$, but $R_2Sn\text{–}SnR_2$ (see **18.14**); $R_2E\text{–}ER_2$ for E = As, Sb, Bi and R = Ph; RE=ER for E = As, Sb and R is bulky;
- selected examples of preparations of these compounds.

R = Me, $^iPr$

**(18.16)**

$Me_3Si\!-\!\overset{\displaystyle |}{\underset{\displaystyle H}{C}}\!-\!SiMe_3$

**(18.17)**

18.20   Information for this answer is spread through Chapter 18 in H&S. Main points to include (give examples of specific reactions in each case):
- general use of organolithium reagents;
- general use of Grignard reagents;
- transmetallation using organomercury compounds, e.g. for group 1 and 2 metal M–C bonds;
- hydroboration for B–C bonds;
- for cyclopentadienyl compounds, use of $Na[C_5R_5]$ or $Li[C_5R_5]$.

# 19 *d*-Block chemistry: some general considerations

19.1 (a) Detailed information about the *s*- and *p*-block elements is found by referring to Chapters 10-17 of H&S. Ignore oxidation state of 0 for the element itself. Points to include:
- an *s*-block (group 1 or 2) element exhibits one oxidation state (+1 for group 1, +2 for group 2); exceptions are in group 1 metal alkalides containing $M^-$;
- elements in groups 13-16 exhibit either one or two states, e.g. +3 or +1 in group 13 with +1 being available to the later elements owing to the 6*s* thermodynamic inert pair effect (see Box 12.3 in H&S);
- in group 17, F shows oxidation state of –1, but variable oxidation states are observed for the remaining halogens with iodine showing the largest range (–1 to +7);
- in group 18, only Xe has an extensive chemistry and exhibits oxidation states from +2 to +8.

(b) In group 3, only the +3 state is stable, e.g. Sc(III); in group 12, Zn and Cd show only +2 while Hg shows +1 and +2; for the rest of the *d*-block, variable oxidation states are characteristic with the widest range observed in the middle of the block, e.g. Mn and Fe.

19.2 (a)
| | |
|---|---|
| Sc | $[Ar]4s^23d^1$ |
| Ti | $[Ar]4s^23d^2$ |
| V | $[Ar]4s^23d^3$ |
| Cr | $[Ar]4s^13d^5$ |
| Mn | $[Ar]4s^23d^5$ |
| Fe | $[Ar]4s^23d^6$ |
| Co | $[Ar]4s^23d^7$ |
| Ni | $[Ar]4s^23d^8$ |
| Cu | $[Ar]4s^13d^{10}$ |
| Zn | $[Ar]4s^23d^{10}$ |

**Group number**

| 4 | 8 | 11 |
|---|---|---|
| Ti | Fe | Cu |
| Zr | Ru | Ag |
| Hf | Os | Au |

**(19.1)**

**Group number**

| 8 | 9 | 10 |
|---|---|---|
| Ru | Rh | Pd |
| Os | Ir | Pt |

**(19.2)**

(b) Groups 4, 8 and 11: see **19.1**.
(c) Platinum group metals: see **19.2**.

19.3 The trend in $E^o$ values is *irregular* across the period. Factors that contribute to $E^o$ are summarized in a thermochemical cycle analogous to that in answer 7.21, p. 74. Use this answer for the way in which to tackle problem 19.3, and also refer to the comparison of $E^o$ values for $Cu^{2+}/Cu$ and $Zn^{2+}/Zn$ given in Section 7.6 in H&S.

19.4 Relevant sections in H&S from which to draw material are:
- interstitial hydrides: Section 9.7;
- molecular hydride anions: Section 9.7 and Figure 9.10;
- metal borides: Section 12.10 and Table 12.3;
- interstitial carbides: Section 13.7;
- steel (iron carbides and alloys): Section 5.7;
- interstitial nitrides: Section 14.6.

Include examples of syntheses as well as some structural data.

**19.5**  Properties that should be included; information from Section 19.5 of H&S:
- metal atom has one or more occupied *nd* orbitals in the valence shell;
- variable oxidation states possible with largest range being for metals in the middle of the row, e.g. Mn, Fe;
- compounds often coloured – exceptions are $d^0$, e.g. Sc(III), and $d^{10}$, e.g. Zn(II);
- compounds may be diamagnetic or paramagnetic depending on the electronic configuration and the coordination environment;
- complex formation – this is *not* unique to *d*-block metal ions but is a common property; changes in coordination environment can lead to a change in colour, e.g. pink $[Co(H_2O)_6]^{2+}$ to blue $[CoCl_4]^{2-}$, both Co(II).

**19.6**  (a) High coordination numbers are usually not feasible on steric grounds because cations of first row metals are too small to accommodate large numbers of donor atoms. Coordination number of 6 is common, e.g. in aqua ions.

(b) A high oxidation state places a high *formal* charge on the metal centre; by the electroneutrality principle, the distribution of charge in the metal-containing species is such that the actual charge on the metal atom is no greater than $\approx +1$. In $[Ti(NO_3)_4]$, for example, the Ti(IV) centre is 8-coordinate (four didentate $[NO_3]^-$, see Figure 19.4a in H&S) – the ligands remove some of the positive charge from metal centre.

(c) Ligands which are formally $F^-$ or $O^{2-}$ are highly electronegative and can remove positive charge from metal centres in high ox. states, see part (b); also, $F_2$ and $O_2$ are strongly oxidizing and are often associated with the formation of high ox. states.

**19.7**  Refer to Table 19.1 for ground state configurations of metal *atoms*. Refer to Table 6.7 in H&S for ligand abbreviations.

(a) $[Mn(CN)_6]^{4-}$ contains 6 $[CN]^-$ ligands; ox. state of Mn = +2; valence electronic configuration is $d^5$.

(b) $[FeCl_4]^{2-}$ contains 4 $Cl^-$ ligands; ox. state of Fe = +2; valence electronic configuration is $d^6$.

(c) $[CoCl_3(py)_3]$ contains 3 $Cl^-$ and 3 pyridine (neutral *N*-donor) ligands; ox. state of Co = +3; valence electronic configuration is $d^6$.

(d) $[ReO_4]^-$ contains 4 $O^{2-}$ ligands; ox. state of Re = +7; Re is in group 7; valence electronic configuration is $d^0$.

(e) $[Ni(en)_3]^{2+}$ contains 3 en (neutral *N,N'*-donor) ligands; ox. state of Ni = +2; valence electronic configuration is $d^8$.

(f) $[Ti(H_2O)_6]^{3+}$ contains 6 $H_2O$ ligands; ox. state of Ti = +3; valence electronic configuration is $d^1$.

(g) $[VCl_6]^{3-}$ contains 6 $Cl^-$ ligands; ox. state of V = +3; valence electronic configuration is $d^2$.

(h) $[Cr(acac)_3]$ contains 3 $[acac]^-$ (*O,O'*-donor) ligands; ox. state of Cr = +3; valence electronic configuration is $d^3$.

**Table 19.1**  Ground state electronic configurations of M.

| M | Configuration |
| --- | --- |
| Sc | $[Ar]4s^23d^1$ |
| Ti | $[Ar]4s^23d^2$ |
| V | $[Ar]4s^23d^3$ |
| Cr | $[Ar]4s^13d^5$ |
| Mn | $[Ar]4s^23d^5$ |
| Fe | $[Ar]4s^23d^6$ |
| Co | $[Ar]4s^23d^7$ |
| Ni | $[Ar]4s^23d^8$ |
| Cu | $[Ar]4s^13d^{10}$ |
| Zn | $[Ar]4s^23d^{10}$ |

**19.8**  Kepert model (see Section 19.7 in H&S) considers repulsions between ligands, but lone pairs of electrons on the metal centre do *not* influence the shape:
(a) 2-coordinate predicted to be linear;
(b) 3-coordinate predicted to be trigonal planar;
(c) 4-coordinate predicted to be tetrahedral;
(d) 5-coordinate predicted to be trigonal bipyramidal or square-based pyramidal (small energy difference);
(e) 6-coordinate predicted to be octahedral.

**19.9** Follow the method outlined in Chapter 3 of H&S; look at the 'general notes' on p. 30 of this book, and use the flow-chart in Figure 3.9 in H&S. Trigonal bipyramid, **19.3**:

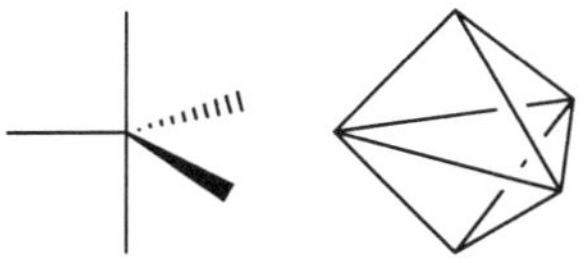

| START ⟹ Is the molecule linear? | No |
| Does it have $T_d$, $O_h$ or $I_h$ symmetry? | No |
| Is there a $C_n$ axis? | Yes: $C_3$ axis |
| Are there 3 $C_2$ axes perpendicular to the principal axis? | Yes |
| Is there a $\sigma_h$ plane? | Yes ⟹ STOP |

Conclusion: the point group is $D_{3h}$.

**(19.3)**

Square-based pyramid, **19.4**:

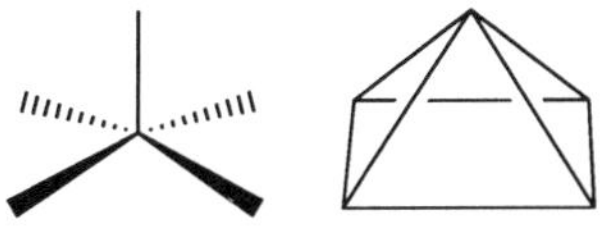

| START ⟹ Is the molecule linear? | No |
| Does it have $T_d$, $O_h$ or $I_h$ symmetry? | No |
| Is there a $C_n$ axis? | Yes: $C_4$ axis |
| Are there 4 $C_2$ axes perpendicular to the principal axis? | No |
| Is there a $\sigma_h$ plane? | No |
| Are there *n* (i.e. 4) $\sigma_v$ planes containing the $C_n$ axis? | Yes ⟹ STOP |

Conclusion: the point group is $C_{4v}$.

**(19.4)**

Square antiprism (two staggered squares form opposite faces), **19.5**:

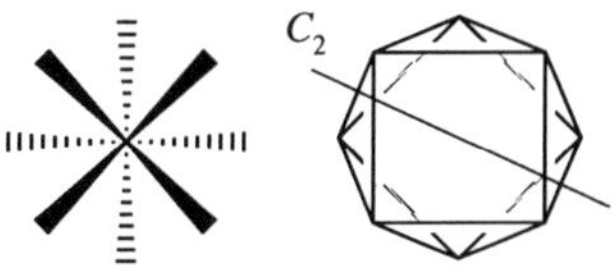

| START ⟹ Is the molecule linear? | No |
| Does it have $T_d$, $O_h$ or $I_h$ symmetry? | No |
| Is there a $C_n$ axis? | Yes: $C_4$ axis (in **19.5**, perpendicular to the plane of the paper) |
| Are there 4 $C_2$ axes perpendicular to the principal axis? | Yes (one $C_2$ axis is shown in **19.5**) |
| Is there a $\sigma_h$ plane? | No |
| Are there *n* (i.e. 4) $\sigma_v$ planes containing the $C_n$ axis? | Yes ⟹ STOP |

Conclusion: the point group is $D_{4d}$.

**(19.5)**

Dodecahedron, **19.6**. Right-hand diagrams in **19.6** are related by rotation about an axis through the * points (using a molecular model or computer modelling package is beneficial).

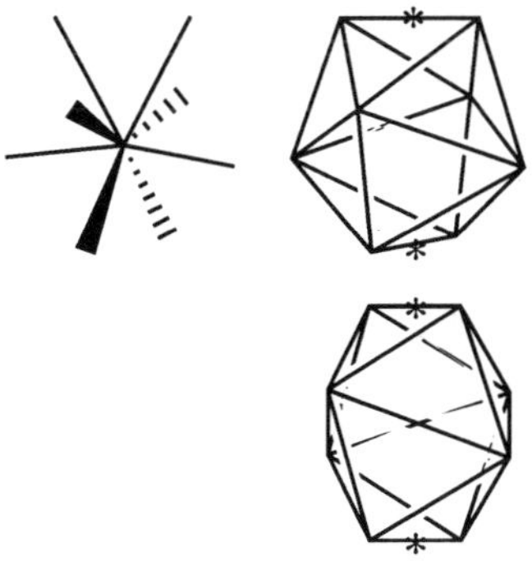

| START ⟹ Is the molecule linear? | No |
| Does it have $T_d$, $O_h$ or $I_h$ symmetry? | No |
| Is there a $C_n$ axis? | Yes: $C_2$ axis (vertical through the * points in **19.6**) |
| Are there 2 $C_2$ axes perpendicular to the principal axis? | Yes (lower right diagram: (i) horizontal, (ii) through plane of paper) |
| Is there a $\sigma_h$ plane? | No |
| Are there *n* (i.e. 2) $\sigma_v$ planes containing the $C_n$ axis? | Yes ⟹ STOP |

Conclusion: the point group is $D_{2d}$.

**(19.6)**

**19.10** (a) Static structure of $Fe(CO)_5$ possesses 2 axial and 3 equatorial CO ligands, and therefore 2 axial and 3 equatorial C environments.

(b) The molecule undergoes a low energy fluxional process – look back at Figure 2.8, p. 24 which should be included in answer 19.10. The process interconverts $CO_{ax}$ and $CO_{eq}$ at a rate that is faster than the $^{13}C$ NMR timescale; only one signal is observed in the spectrum.

**19.11** The structures for this question are redrawn here in **19.7-19.9**. Remember that these are *solid state* data. The bond angles allow you to determine whether the metal is in a planar or non-planar environment: test for planarity is that the sum of bond angles is 360°. Compounds **19.7** and **19.8** have planar metal centres (but not strictly trigonal planar). In **19.9**, Y is non-planar ($\Sigma$ angles = 345°); angles are equal, therefore the coordination environment is *trigonal* pyramidal. The deviation from planarity in **19.9** is probably caused by crystal packing forces and so may be a feature of only the solid state structure. In **19.7**, the steric crowding of the two bulky silyl groups causes $\angle$Si–Fe–Si > $\angle$Si–Fe–Cl. In **19.8**, the bite angle of chelating ligand is a constraint on the N–Cu–N angle.

**(19.8)**

**(19.7)**

**(19.9)**

**19.12** Ligand **19.10** is a tripodal ligand, i.e. three 'arms' with *N*-donors radiate from the central *N*-donor. The complex [Cu(**19.10**)Cl]$^+$ is likely to be trigonal bipyramidal with the tetradentate **19.10** occupying four sites, and the Cl$^-$ ligand occupying one axial site as in **19.11**.

**(19.11)**

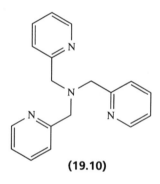

**(19.10)**

**19.13** (a) [Co(NH$_3$)$_5$Br][SO$_4$] contains free [SO$_4$]$^{2-}$ and coordinated Br$^-$, whereas [Co(NH$_3$)$_5$(SO$_4$)]Br contains free Br$^-$ and coordinated [SO$_4$]$^{2-}$. Aqueous solutions of BaCl$_2$ and [Co(NH$_3$)$_5$Br][SO$_4$] give BaSO$_4$ precipitate; aqueous solutions of AgNO$_3$ and [Co(NH$_3$)$_5$(SO$_4$)]Br give AgBr precipitate. Only *free* ions are precipitated:

$$Ba^{2+}(aq) + [SO_4]^{2-}(aq) \rightarrow BaSO_4(s)$$
$$Ag^+(aq) + Br^-(aq) \rightarrow AgBr(s)$$

**(19.12)**

(b) Distinguishing between [CrCl$_2$(H$_2$O)$_4$]Cl·2H$_2$O and [CrCl(H$_2$O)$_5$]Cl$_2$·H$_2$O needs *quantitative* study of precipitation of free Cl$^-$ by Ag$^+$. [CrCl$_2$(H$_2$O)$_4$]Cl·2H$_2$O contains one equivalent of non-coordinated Cl$^-$, but [CrCl(H$_2$O)$_5$]Cl$_2$·H$_2$O contains two.

(c) [Co(NH$_3$)$_5$Br][SO$_4$] and [Co(NH$_3$)$_5$(SO$_4$)]Br are ionization isomers; [CrCl$_2$(H$_2$O)$_4$]Cl·2H$_2$O and [CrCl(H$_2$O)$_5$]Cl$_2$·H$_2$O are hydration isomers (see Section 19.8 in H&S).

(d) [CrCl$_2$(H$_2$O)$_4$]$^+$ is octahedral, therefore *trans* (**19.12**) and *cis* (**19.13**) isomers are possible.

**(19.13)**

19.14    (a) Coordination isomers are salts in which ligands are exchanged between the cation and anion. $[Co(bpy)_3]^{3+}$ contains neutral bpy (**19.14**) in which the *N*-donors can be only mutually *cis*. Coordination isomers of $[Co(bpy)_3]^{3+}[Fe(CN)_6]^{3-}$ are:

$[Co(bpy)_2(CN)_2]^+[Fe(bpy)(CN)_4]^-$
$[Fe(bpy)_2(CN)_2]^+[Co(bpy)(CN)_4]^-$
$[Fe(bpy)_3]^{3+}[Co(CN)_6]^{3-}$

(b) *trans* and *cis*-$[Co(bpy)_2(CN)_2]^+$ in respect of the arrangement of the $CN^-$ ligands; *cis*-$[Co(bpy)_2(CN)_2]^+$ has optical isomers (**19.15** and **19.16**).

*trans* and *cis*-$[Fe(bpy)_2(CN)_2]^+$, and *cis*-$[Fe(bpy)_2(CN)_2]^+$ has optical isomers. $[Fe(bpy)_3]^{3+}$ has optical isomers, (look at Figure 19.12 in H&S for related complex).

**(19.15)**          **(19.16)**

19.15    The chelate rings are 5-membered and for this answer, ignore the conformations of the chelate rings (see answer 19.16).
(a) There are four isomers of $[Pt(H_2NCH_2CHMeNH_2)_2]^{2+}$ depending on the orientations of the Me groups (Figure 19.1a-d).
(b) There are two isomers of $[Pt(H_2NCH_2CMe_2NH_2)(H_2NCH_2CPh_2NH_2)]^{2+}$ depending on the relative positions of the Me and Ph groups (Figure 19.1e-f).

19.16    First, refer to Box 19.2 in H&S for the notation needed for this answer and for additional explanations. $[Co(en)_3]^{3+}$ contains 3 *didentate* ligands and therefore has optical isomers (enantiomers) labelled $\Delta$-$[Co(en)_3]^{3+}$ and $\Lambda$-$[Co(en)_3]^{3+}$. Now consider the chelate ring conformations: using the notation from Box 19.2 in H&S, the $\Delta$-$[Co(en)_3]^{3+}$ enantiomer can exist as $(\delta\delta\delta)$, $(\delta\delta\lambda)$, $(\delta\lambda\lambda)$ or $(\lambda\lambda\lambda)$, and similarly, $\Lambda$-$[Co(en)_3]^{3+}$ can adopt $(\delta\delta\delta)$, $(\delta\delta\lambda)$, $(\delta\lambda\lambda)$ or $(\lambda\lambda\lambda)$ configurations. All the isomers are related as diastereomers except in the cases where every chiral centre has changed configuration, e.g. $\Delta$-$(\delta\delta\lambda)$ and $\Lambda$-$(\lambda\lambda\delta)$.

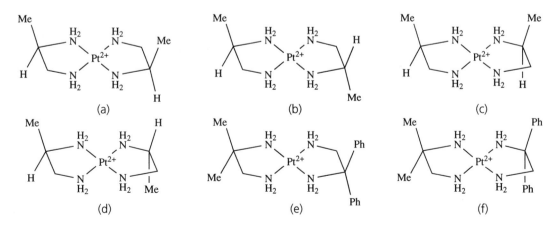

**Figure 19.1** (a)-(d) Isomers of $[Pt(H_2NCH_2CHMeNH_2)_2]^{2+}$; (e)-(f) isomers of $[Pt(H_2NCH_2CMe_2NH_2)(H_2NCH_2CPh_2NH_2)]^{2+}$.

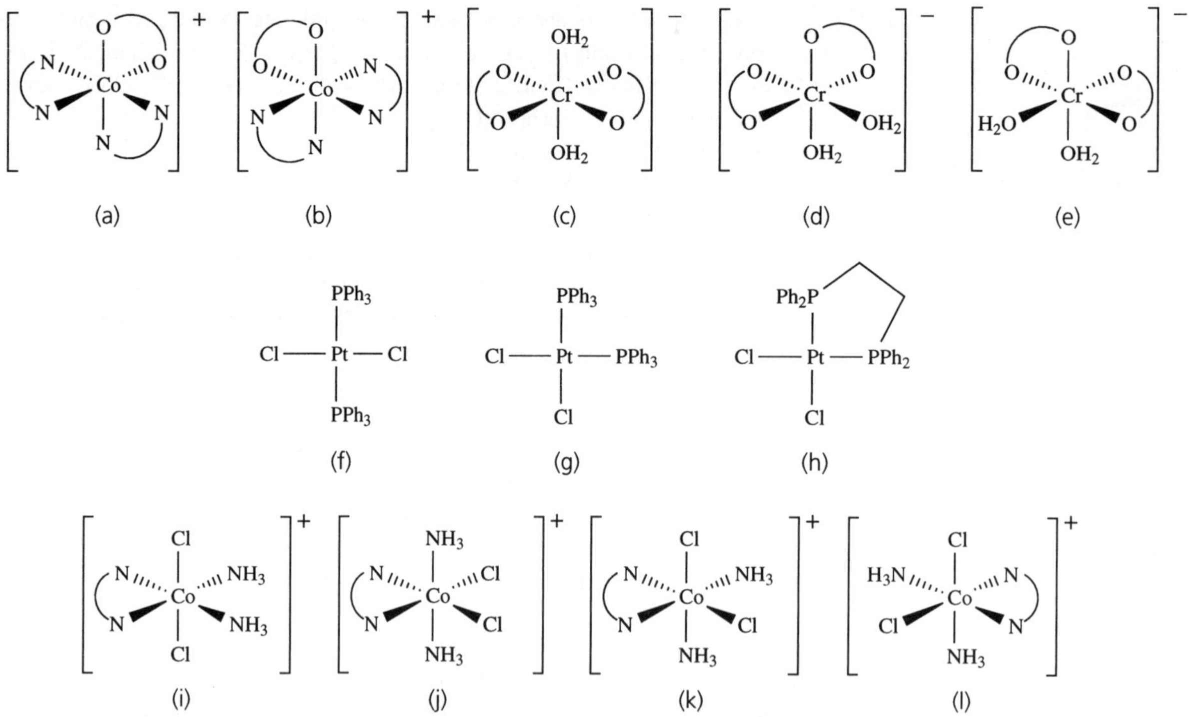

**Figure 19.2** (a)-(b) Isomers of [Co(en)$_2$(ox)]$^+$, (c)-(e) isomers of [Cr(ox)$_2$(H$_2$O)$_2$]$^-$, (f)-(g) isomers of [PtCl$_2$(PPh$_3$)$_2$], (h) the structure of [PtCl$_2$(Ph$_2$PCH$_2$CH$_2$PPh$_2$)], and (i)-(l) isomers of [Co(en)(NH$_3$)$_2$Cl$_2$]$^+$.

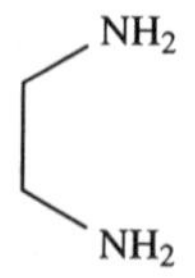

**19.17**   Ligands en and [ox]$^{2-}$ are drawn in **19.17** and **19.18**; both are didentate.
(a) Octahedral [Co(en)$_2$(ox)]$^+$; 3 didentate ligands; optical isomers (Figure 19.2a,b).
(b) Octahedral [Cr(ox)$_2$(H$_2$O)$_2$]$^-$; 2 monodentate and 2 didentate ligands; *cis* and *trans*-isomers; *cis*-isomer has optical isomers (Figure 19.2c-e).
(c) Pt(II), square planar; [PtCl$_2$(PPh$_3$)$_2$] has *trans* and *cis*-isomers (Figure 19.2f,g).
(d) Square planar with one didentate ligand with a short chain length that forces a *cis*-arrangement (Figure 19.2h).
(e) Octahedral [Co(en)(NH$_3$)$_2$Cl$_2$]$^+$ may have *trans*-Cl, *trans*-NH$_3$, or both sets mutually *cis* and for the latter, there are also optical isomers (Figure 19.2i-l).

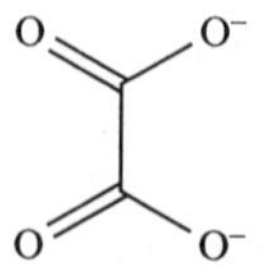

**(19.17)**    **(19.18)**

**19.18**   (a) and (b) Pd(II) and Pt(II) form square planar complexes. Isomers of [PtCl$_2$(PPh$_3$)$_2$] are shown in Figure 19.2f-g. Symmetric *and* asymmetric stretches of PtCl$_2$ unit are IR active for *cis*-isomer; only asymmetric stretch is IR active for *trans*-isomer (see Figure 19.11 in H&S). Also applies to Pd(II) complex. For Pt(II) complex, $^{31}$P NMR spectra show satellite peaks from $^{195}$Pt-$^{31}$P coupling, $J(^{195}\text{Pt}^{31}\text{P})$ *cis* > *trans*.
(c) Use $^{31}$P NMR spectroscopy. *fac*-[RhCl$_3$(PMe$_3$)$_3$] has one P environment, and gives one doublet due to $J(^{103}\text{Rh}^{31}\text{P})$. *mer*-Isomer has two P environments (label them P and P′); gives a doublet of triplets (rel. integral 1) and a doublet of doublets (rel. integral 2) due to $J(^{103}\text{Rh}^{31}\text{P})$ *and* $J(^{31}\text{P}^{31}\text{P}')$.

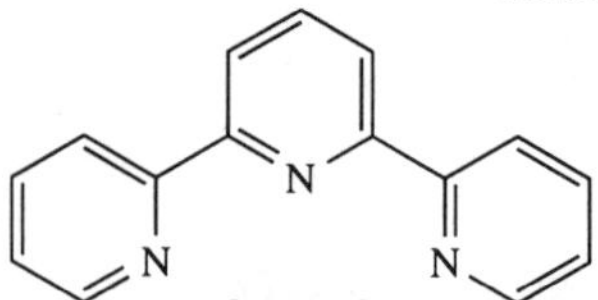

$^{195}$Pt 33.8% $I = \frac{1}{2}$
$^{103}$Rh 100% $I = \frac{1}{2}$
See Box 19.1 in H&S

**19.19**   All are octahedral complexes – see Table 6.7 in H&S for py and bpy abbreviations.
(a) 6 monodentate ligands, therefore *mer* and *fac*-isomers.
(b) 2 monodentate and 2 didentate ligands; *cis* and *trans*-isomers; *cis*-isomer has enantiomers.
(c) tpy (**19.19**) is tridentate and not very flexible; it is constrained to occupying coordination sites in a *mer*-arrangement; no isomers are possible.

**(19.19)**

# 20 *d*-Block chemistry: coordination complexes

**20.1** Refer to Section 20.3 in H&S. Points to include:

- Gas phase metal ion $M^{n+}$ has degenerate $nd$ atomic orbitals, and a valence configuration of $nd^x$.
- Consider formation of octahedral complex $[ML_6]^{n+}$ – crystal field theory treats metal ion and ligands as point charges – repulsions between electrons in $M^{n+}$ $d$ orbitals and L donor electrons.
- Ligands create a 'crystal field' around $M^{n+}$ – in a spherical field, energy of $d$ orbitals is raised with respect to energy in gas phase $M^{n+}$. See left-hand side of Figure 19.2 in H&S.
- Octahedral crystal field leads to splitting of $d$ orbitals into 2 sets (**20.1**): (i) higher energy $d_{z^2}$ and $d_{x^2-y^2}$, and (ii) lower energy $d_{xy}$, $d_{xz}$ and $d_{yz}$. The $d_{z^2}$ and $d_{x^2-y^2}$ orbitals point *directly* at the ligands while $d_{xy}$, $d_{xz}$ and $d_{yz}$ orbitals point *between* the ligands. Therefore, repulsion between ligand electrons and electrons in $d_{z^2}$ and $d_{x^2-y^2}$ is greater than between ligand electrons and electrons in $d_{xy}$, $d_{xz}$ and $d_{yz}$ orbitals. See right-hand side of Figure 19.2 in H&S.
- The raising and lowering of energies is measured with respect to the energy level in the spherical crystal field – the barycentre (see diagram **20.14**).
- Include redrawn Figures 20.2 and 20.3 from H&S in your answer.

The diagram shows an energy-level splitting: upper set $e_g$ with $d_{z^2}$ and $d_{x^2-y^2}$, lower set $t_{2g}$ with $d_{xy}$, $d_{xz}$ and $d_{yz}$, separated by $\Delta_{oct}$, with Energy on the vertical axis.

**(20.1)**

**20.2** Look at Table 19.2 in H&S. $\lambda_{max}$ is the wavelength of the absorption maximum; a value of $\lambda_{max} = 510$ nm corresponds to absorption of green light and transmittance of red and violet, so solutions of $[Ti(H_2O)_6]^{3+}$ actually look purple.

**20.3** (a) en = 1,2-diaminoethane; $N,N'$-donor, **20.2**. Usually didentate; forms 5-membered chelate ring (chelate effect, see Section 6.12 in H&S). Occasionally monodentate.
(b) bpy = 2,2'-bipyridine; $N,N'$-donor, **20.3**. Didentate; 5-membered chelate ring.
(c) Cyanide, $[CN]^-$ (**20.4**); usually $C$-donor, monodentate; sometimes bridges in an M–C≡N–M mode (see examples in Chapter 21 of H&S).
(d) azide, $[N_3]^-$, (**20.5**); usually monodentate $N$-donor; sometimes bridges (e.g. Figure 21.16 in H&S).
(e) CO, (**20.6**); monodentate, $C$-donor (see Section 23.2 in H&S).
(f) phen = 1,10-phenanthroline (**20.7**); $N,N'$-donor; didentate forming 5-membered chelate ring.
(g) $[ox]^{2-}$ = oxalate (**20.8**); $O,O'$-donor; didentate forming 5-membered chelate ring.
(h) $[NCS]^-$ (thiocyanate, **20.9**) can be an $N$- or $S$-donor; usually monodentate but sometimes bridges in an M–N=C=S–M mode.
(i) PMe$_3$ (trimethylphosphine, **20.10**) is monodentate, $P$-donor.

$H_2N$ $\quad$ $NH_2$

**(20.2)**

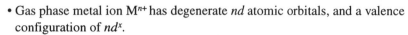

**(20.3)**

$\bar{C}\equiv N$

**(20.4)**

$\bar{N}=\overset{+}{N}=\bar{N}$ $\qquad$ $\bar{C}\equiv\overset{+}{O}$

**(20.5)** $\qquad$ **(20.6)**

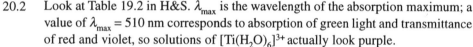

**(20.7)** $\qquad$ **(20.8)** $\qquad$ **(20.9)** $\qquad$ **(20.10)**

**20.4**   The order is as ligands appear in the spectrochemical series (Section 20.3 in H&S):

$$Br^- < F^- < [OH]^- < H_2O < NH_3 < [CN]^-$$

$\xrightarrow{\hspace{5cm}}$

weak field                     strong field

**20.5**   Factors to look for are different oxidation states of metal, different field strengths of ligands, or metals with same oxidation state and in the same triad.
(a) $[Cr(H_2O)_6]^{3+}$ should have larger $\Delta_{oct}$ than $[Cr(H_2O)_6]^{2+}$ (higher ox. state).
(b) $[Cr(NH_3)_6]^{3+}$ should have larger $\Delta_{oct}$ than $[CrF_6]^{3-}$ (both Cr(III), but $NH_3$ is a stronger field ligand than $F^-$).
(c) $[Fe(CN)_6]^{3-}$ is Fe(III), $[Fe(CN)_6]^{4-}$ is Fe(II), therefore $[Fe(CN)_6]^{3-}$ should have larger $\Delta_{oct}$.
(d) $[Ni(en)_3]^{2+}$ should have larger $\Delta_{oct}$ than $[Ni(H_2O)_6]^{2+}$ (en is stronger field ligand).
(e) $[MnF_6]^{2-}$ and $[ReF_6]^{2-}$ both contain M(IV) with M from group 7; $[ReF_6]^{2-}$ should have larger $\Delta_{oct}$ because Re is 3rd row metal, Mn is 1st row.
(f) $[Co(en)_3]^{3+}$ and $[Rh(en)_3]^{3+}$ both contain M(III) with M from group 9; $[Rh(en)_3]^{3+}$ should have larger $\Delta_{oct}$ because Rh is 2nd row metal, Co is 1st row.

**20.6**   (a) Diagram **20.11** shows the $d^8$ configuration in an octahedral field. There is no vacant $e_g$ orbital and so no possibility of promoting an electron from a fully occupied $t_{2g}$ orbital to generate a high-spin configuration. $\therefore$ Only one configuration.
(b) Consider an example, e.g. octahedral $d^4$, that can be low-spin (**20.12**) or high-spin (**20.13**). Preference for high-spin or low-spin configuration depends on which configuration has the lower energy. This, in turn, depends on whether it is energetically preferable to pair the fourth electron (**20.12**) or promote it to the $e_g$ level (**20.13**). Need to consider the energy required to transform two electrons with parallel spins in different degenerate orbitals into spin-paired electrons in the same orbital – the pairing energy, $P$, depends on (i) the loss in the *exchange energy* on pairing the electrons (see Box 1.8 in H&S), and (ii) the coulombic repulsion between the spin-paired electrons.

For high-spin:     $\Delta_{oct} < P$        weak field
For low-spin:      $\Delta_{oct} > P$        strong field

(c) Different numbers of unpaired electrons give rise to different effective magnetic moments ($\mu_{eff}$), e.g. low-spin $d^4$ has 2 unpaired electrons, high-spin $d^4$ has 4. Use the spin-only formula to estimate the magnetic moment for $n$ unpaired electrons:

$$\mu(\text{spin-only}) = \sqrt{n(n+2)}$$

In the case of an octahedral $d^6$ ion, low-spin is diamagnetic, high-spin is paramagnetic.

**(20.11)**

**(20.12)**

**(20.13)**

**20.7**   From diagram **20.14**, CFSE can be found from the equation:

CFSE = $(-0.4\Delta_{oct})$(number of electrons in $t_{2g}$ level)
                    $+ (0.6\Delta_{oct})$(number of electrons in $t_{2g}$ level)

Selected examples:

$d^1$     CFSE = $(-0.4\Delta_{oct})(1) + 0 = -0.4\Delta_{oct}$
$d^4$     CFSE = $(-0.4\Delta_{oct})(4) + 0 = -1.6\Delta_{oct}$          for low-spin
$d^4$     CFSE = $(-0.4\Delta_{oct})(3) + (0.6\Delta_{oct})(1) = -0.6\Delta_{oct}$     for high-spin
$d^5$     CFSE = $(-0.4\Delta_{oct})(5) + 0 = -2.0\Delta_{oct}$          for low-spin
$d^5$     CFSE = $(-0.4\Delta_{oct})(3) + (0.6\Delta_{oct})(2) = 0$        for high-spin

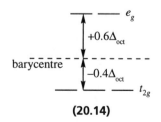

**(20.14)**

**Figure 20.1** Rationalizing numbers of unpaired electrons in problem 20.8. For the ground state electronic configurations of the metal *atoms*, see Table 19.1, p. 156.

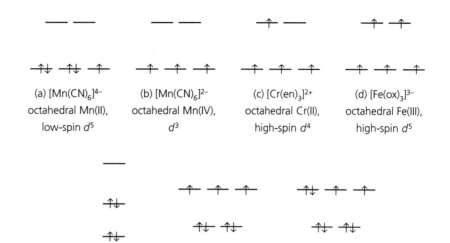

(a) [Mn(CN)$_6$]$^{4-}$
octahedral Mn(II),
low-spin $d^5$

(b) [Mn(CN)$_6$]$^{2-}$
octahedral Mn(IV),
$d^3$

(c) [Cr(en)$_3$]$^{2+}$
octahedral Cr(II),
high-spin $d^4$

(d) [Fe(ox)$_3$]$^{3-}$
octahedral Fe(III),
high-spin $d^5$

(e) [Pd(CN)$_4$]$^{2-}$
$d^8$, square planar Pd(II)

(f) [CoCl$_4$]$^{2-}$
tetrahedral Co(II), $d^7$

(g) [NiBr$_4$]$^{2-}$
tetrahedral Ni(II), $d^8$

20.8    See Figure 20.1.

20.9    (a) Define an axis set; by convention, take the axial ligands to lie on the *z* axis (**20.15** and **20.16**).
***Trigonal bipyramid***: the $d_{z^2}$ orbital points directly at 2 ligands and is destabilized the most (Figure 20.2); the equatorial ligands lie in the *xy* plane, and the $d_{x^2-y^2}$ and $d_{xy}$ orbitals are degenerate and higher in energy than the $d_{xz}$ and $d_{yz}$ orbitals which point between the ligands. ***Square-based pyramid***: 1 ligand lies on the *z* axis, 2 lie ≈ along the *x* axis, 2 lie ≈ on the *y* axis; the $d_{x^2-y^2}$ orbital (points ≈ at the basal ligands) is destabilized the most, the $d_{z^2}$ orbital is destabilized to a lesser extent (Figure 20.2); since the basal ligands lie in the *xy* plane, the $d_{xy}$ orbital lies at higher energy than the $d_{xz}$ and $d_{yz}$ orbitals.

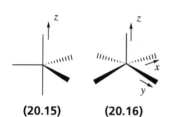

(20.15)    (20.16)

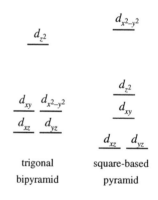

trigonal bipyramid        square-based pyramid

**Figure 20.2** Crystal field splitting diagrams for trigonal bipyramidal and square-based pyramidal fields.

(b) Ni(II) is $d^8$ and both trigonal bipyramidal and square-based pyramidal complexes will be diamagnetic (place 8 electrons in the levels shown in Figure 20.2).

20.10    (a) On going from gaseous M$^{n+}$ to complexed M$^{n+}$, interelectronic repulsion between metal *d* electrons decreases – pairing energies reduced. This is caused by an increase in effective size of metal orbitals – the *nephelauxetic effect* ('cloud expanding'). For a common M$^{n+}$, nephelauxetic effect of ligands follows series:

$$F^- < H_2O < NH_3 < en < [ox]^{2-} < [NCS]^- < Cl^- < [CN]^- < Br^- < I^-$$

For metal ions (with a common ligand) nephelauxetic effect follows series:

$$Mn(II) < Ni(II) \approx Co(II) < Mo(II) < Re(IV) < Fe(III) < Ir(III) < Co(III) < Mn(IV)$$

Parameters for ligands ($h$) and metal ions ($k$) (see Table 20.6 in H&S) are used to estimate the reduction in electron-electron repulsion upon complex formation:

$$\frac{B_0 - B}{B_0} \approx \left(h_{\text{ligands}}\right)\left(k_{\text{metal ion}}\right)$$

where $B$ is the Racah parameter, $B_0$ is interelectronic repulsion in the free ion.
(b) From nephelauxetic series for ligands:    $F^- < H_2O < NH_3 < en < [CN]^- < I^-$

**20.11**    (a) $[CoCl_4]^{2-}$ is Co(II), $d^7$. $[CuCl_4]^{2-}$ is Cu(II), $d^9$.
Tetrahedral $Co^{2+}$ (**20.17**), $t_2$ orbitals are *all* singly occupied; in tetrahedral $Cu^{2+}$ (**20.18**), $t_2$ orbitals are asymmetrically filled. Thus, the complex suffers a Jahn-Teller distortion leading to the observed flattened tetrahedron.

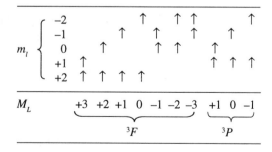

(**20.17**)        (**20.18**)

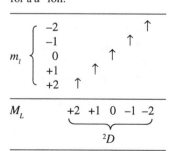

(**20.19**)

(b) Octahedral $[CoF_6]^{3-}$ is Co(III), $d^6$. $F^-$ is weak field ligand; ground state is high-spin (**20.19**). A $d^6$ ion might be expected to give a single absorption in the electronic spectrum (see Orgel diagram in Figure 20.17 in H&S). The excited state of $[CoF_6]^{3-}$ ($t_{2g}^3e_g^3$) suffers a Jahn-Teller effect because the $e_g$ level is asymmetrically filled. Small splitting of $e_g$ level leads to 2 possible transitions: 11 500 and 14 500 cm$^{-1}$.

**20.12**

➤

Term symbols:
see Box 20.5 in H&S

(a) Set up the table of microstates as in Table 20.1. The left-hand column in the top part of Table 20.1 gives $m_l$ values for a $d$ orbital; the row in the lower part of Table 20.1 gives $M_L$ values for a $d^1$ ion. In a tetrahedral field, the $^2D$ term has $E$ and $T_2$ components (see Orgel diagram in Figure 20.17 in H&S). The multiplicity of the term (the 2 in the $^2D$ symbol) is determined from $2S + 1$; for 1 electron, $S = \frac{1}{2}$.
(b) Set up the table of microstates for the $d^2$ ion remembering that the electrons singly occupy orbitals – Table 20.2. The $M_L$ values are obtained by summing the $m_l$ values in each column. Your table may not look identical to Table 20.2 because the columns may come in a different order; the columns here are arranged to give the values of $M_L$ in an order from which the term symbols are easily determined. The $^3P$ term does not split – gives a $T_1$ component in a tetrahedral field, and a $T_{1g}$ component in an octahedral field. In an octahedral field, the components of the $^3F$ term are $A_{2g}$, $T_{2g}$ and $T_{1g}$; in a tetrahedral field, they are $A_2$, $T_2$ and $T_1$.

**Table 20.1** Table of microstates for a $d^1$ ion.

| $m_l$ | | | | | ↑ |
|---|---|---|---|---|---|
| −2 | | | | ↑ | |
| −1 | | | ↑ | | |
| 0 | | ↑ | | | |
| +1 | ↑ | | | | |
| +2 | ↑ | | | | |

| $M_L$ | +2 | +1 | 0 | −1 | −2 |
|---|---|---|---|---|---|
| | | | $^2D$ | | |

**Table 20.2** Table of microstates for a $d^2$ ion.

| $m_l$ | | | ↑ | ↑ ↑ | | ↑ |
|---|---|---|---|---|---|---|
| −2 | | ↑ | ↑ | ↑ | | ↑ |
| −1 | | | | ↑ ↑ | | |
| 0 | ↑ | | | | ↑ ↑ | |
| +1 | ↑ | | | | | ↑ ↑ |
| +2 | ↑ ↑ ↑ ↑ | | | | | |

| $M_L$ | +3 | +2 | +1 | 0 | −1 | −2 | −3 | +1 | 0 | −1 |
|---|---|---|---|---|---|---|---|---|---|---|
| | | | | $^3F$ | | | | | $^3P$ | |

20.13    (a) Equations needed:

**Remember**: *lower wavenumber* corresponds to *longer wavelength*.

$$\text{Wavenumber (in cm}^{-1}) = \frac{1}{\text{Wavelength (in cm)}} \qquad 1 \text{ cm} = 10^7 \text{ nm}$$

$10\,000 \text{ cm}^{-1} = 1000$ nm; $20\,000 \text{ cm}^{-1} = 500$ nm; $30\,000 \text{ cm}^{-1} = 333$ nm.

(b) Visible range $\approx$ 400-700 nm (25 000-14 285 cm$^{-1}$).

Visible range of spectrum and absorption/transmittance of light: see Table 19.2 in H&S

(c) Look at $\lambda$ corresponding to the visible part of spectrum. [Ni(H$_2$O)$_6$]$^{2+}$ absorbs $\approx$660 nm, so appears green; [Ni(NH$_3$)$_6$]$^{2+}$ absorbs $\approx$570 nm, so appears purple.

(d) Consider positions of NH$_3$ and H$_2$O in spectrochemical series: H$_2$O is weaker field ligand than NH$_3$. The relative energies of transitions are estimated from an Orgel diagram (Figure 20.18 in H&S) and for the weaker field ligand, the transition energies are lower, i.e. [Ni(H$_2$O)$_6$]$^{2+}$ < [Ni(NH$_3$)$_6$]$^{2+}$. Since $E$ is proportional to wavenumber, the absorptions for [Ni(H$_2$O)$_6$]$^{2+}$ appear at lower wavenumbers than those for [Ni(NH$_3$)$_6$]$^{2+}$.

20.14    (a) Cr(III) is $d^3$. Sketch an Orgel diagram for an octahedral $d^3$ ion – this corresponds to the left-hand side of Figure 20.18 in H&S. Three absorption transitions are expected:

$$T_{2g} \leftarrow A_{2g} \qquad T_{1g}(F) \leftarrow A_{2g} \qquad T_{1g}(P) \leftarrow A_{2g}$$

centre of symmetry

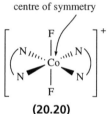

**(20.20)**

(b) *trans*-[Co(en)$_2$F$_2$]$^+$ (**20.20**) has a centre of symmetry, but the *cis*-isomer does not; loss of centre of symmetry permits greater *p-d* mixing and, therefore, a greater probability of transitions, leading to more intense colour for the *cis*-isomer. Comparing chloro and fluoro *trans*-complexes: charge transfer (CT) from Cl$^-$ to Co$^{3+}$ accounts for the more intense colour of chloro complex; CT for F$^-$ is unlikely.

20.15    Use spin-only formula to find number of unpaired electrons, or fit values calculated from the spin-only formula to the experimental values (assume you can ignore magnetic moment associated with orbital angular momentum):

$$\mu(\text{spin-only}) = \sqrt{n(n+2)}$$

**Table 20.3** Calculated values of $\mu$(spin-only) for $n$ unpaired electrons.

| $n$ | $\mu$(spin-only) / $\mu_B$ |
| --- | --- |
| 1 | 1.73 |
| 2 | 2.83 |
| 3 | 3.87 |
| 4 | 4.90 |
| 5 | 5.92 |

From values in Table 20.3:

(a) [VCl$_x$(bpy)] has 1 unpaired electron; $d^1$ corresponds to V(IV), therefore $x = 4$.

(b) K$_x$[V(ox)$_3$] has 2 unpaired electrons; $d^2$ corresponds to V(III); the complex anion is therefore [V(ox)$_3$]$^{3-}$ and so $x = 3$.

(b) [Mn(CN)$_6$]$^{x-}$ has 3 unpaired electrons; $d^3$ corresponds to Mn(IV); since the cyano ligand is [CN]$^-$, the overall charge must be 2–, i.e. $x = 2$.

20.16    For an electron to have orbital angular momentum, it must be possible to transform the orbital containing the electron into an equivalent and degenerate orbital by rotation. For a fuller explanation, refer to Section 20.8 in H&S, the subsection entitled 'Spin and orbital contributions to the magnetic moment'.

20.17    (a) K$_3$[TiF$_6$] contains [TiF$_6$]$^{3-}$, therefore Ti(III), $d^1$.

$$\mu(\text{spin-only}) = \sqrt{n(n+2)} = \sqrt{3} = 1.73 \, \mu_B$$

(b) Take into account spin-orbit coupling. The $[TiF_6]^{3-}$ contains $Ti^{3+}$ ($d^1$) with a $T$ ground term. The simple equation:

$$\mu_{\text{eff}} = \mu(\text{spin - only})\left(1 - \frac{\alpha\lambda}{\Delta_{\text{oct}}}\right)$$

is not applicable to this case; it can only be used for $A$ and $E$ ground states.

20.18   Octahedral $Ni^{2+}$ ($d^8$) should have no orbital contribution, and $\mu_{\text{eff}}$ is expected to be close to spin-only value. Tetrahedral $Ni^{2+}$ has an orbital contribution because ground state configuration is $e^4 t_2^4$, and so spin-orbit coupling occurs. This results in $\mu_{\text{eff}} > \mu(\text{spin-only})$. In a square planar $Ni^{2+}$ complex, all electrons are paired leading to a diamagnetic complex; look back at Figure 20.1e, p. 163 – the promotion energy to the $d_{z^2}$ orbital (highest level) is too large for a high-spin complex to form.

20.19   (a) $r_{\text{ion}}$ is estimated from ionic lattice; $r_{\text{ion}}$ values for high-spin, octahedral ions are:

| $Ti^{2+}$ | $V^{2+}$ | $Cr^{2+}$ | $Mn^{2+}$ | $Fe^{2+}$ | $Co^{2+}$ | $Ni^{2+}$ | $Cu^{2+}$ | $Zn^{2+}$ | |
|---|---|---|---|---|---|---|---|---|---|
| $d^2$ | $d^3$ | $d^4$ | $d^5$ | $d^6$ | $d^7$ | $d^8$ | $d^9$ | $d^{10}$ | |
| $t_{2g}^2 e_g^0$ | $t_{2g}^3 e_g^0$ | $t_{2g}^3 e_g^1$ | $t_{2g}^3 e_g^2$ | $t_{2g}^4 e_g^2$ | $t_{2g}^5 e_g^2$ | $t_{2g}^6 e_g^2$ | $t_{2g}^6 e_g^3$ | $t_{2g}^6 e_g^4$ | |
| 86 | 79 | 80 | 83 | 78 | 75 | 69 | 73 | 74 | pm |

Plot of $r_{\text{ion}}$ against number of $d$ electrons shows an 'inverse' double-humped curve. Radii increase at points in series when electrons enter $d_{z^2}$ or $d_{x^2-y^2}$ orbitals – these point directly at the ligands, and interelectronic repulsion increases. (b) The trend is the inverse of that for lattice energies (Figure 20.27 in H&S): lattice energy is inversely proportional to the internuclear separation and is therefore inversely related to $r_{\text{cation}}$; the double hump is rationalized in terms of variation in LFSE (include a graph or table of LFSE vs $d^n$ electrons, Figure 20.26 in H&S). Hydration enthalpies (Figure 20.28 in H&S) behave similarly to lattice energies.

20.20   Normal spinel has tetrahedral $Ni^{2+}$ ($d^8$) and 2 octahedral $Mn^{3+}$ ($d^4$); inverse spinel has tetrahedral $Mn^{3+}$, octahedral $Mn^{3+}$ and octahedral $Ni^{2+}$. Compare LFSE values:

➤

$\Delta_{\text{tet}} \approx \,^4/_9 \Delta_{\text{oct}}$

LFSE tet. $Ni^{2+}$ + oct. $Mn^{3+}$ = $-(0.8 \times \,^4/_9 \times 8500) - (0.6 \times 21\,000) = -15\,622$ cm$^{-1}$

➤

LFSE oct. $Ni^{2+}$ + tet. $Mn^{3+}$ = $-(1.2 \times 8500) - (0.4 \times \,^4/_9 \times 21\,000) = -13\,933$ cm$^{-1}$

➤

Spinels $AB_2O_4$: see Box 12.6 and Section 20.9 in H&S

Predict normal spinel; factor not taken into account is Jahn-Teller effect for $Mn^{3+}$ ($d^4$); although predict normal spinel by a small margin, the structure is, in practice, an inverse spinel.

20.21   (a) LFSEs can be estimated as in answer 20.7, p. 162 – see also Table 20.3 in H&S. The difference in LFSE on going from oct. $[Co(H_2O)_6]^{2+}$ to tet. $[CoCl_4]^{2-}$ is much less for $Co^{2+}$ ($d^7$) than for $Ni^{2+}$ ($d^8$). Remember that $\Delta_{\text{tet}} \approx \,^4/_9 \Delta_{\text{oct}}$.

➤

Refer to Section 7.3 in H&S

(b) Data are consistent with $H_4[Fe(CN)_6]$ being a weak acid with respect to the 4th dissociation constant; $H^+$ complexing of $[Fe(CN)_6]^{4-}$ makes reduction easier. (c) LFSE plays only a minor part. There is a loss of LFSE on reduction of $Mn^{3+}$

➤

Refer to Section 20.11 in H&S

($d^4$), a gain on reduction of $Fe^{3+}$ ($d^5$), and a loss on reduction of $Cr^{3+}$ ($d^3$); the decisive factor is the large value of the 3rd ionization energy for Mn.

# 21 *d*-Block chemistry: the first row metals

**21.1**

| M | Sc | Ti | V | Cr | Mn | Fe | Co | Ni | Cu | Zn |
|---|---|---|---|---|---|---|---|---|---|---|
| M | $4s^23d^1$ | $4s^23d^2$ | $4s^23d^3$ | $4s^13d^5$ | $4s^23d^5$ | $4s^23d^6$ | $4s^23d^7$ | $4s^23d^8$ | $4s^13d^{10}$ | $4s^23d^{10}$ |
| $M^{2+}$ | $3d^1$ | $3d^2$ | $3d^3$ | $3d^4$ | $3d^5$ | $3d^6$ | $3d^7$ | $3d^8$ | $3d^9$ | $3d^{10}$ |

**21.2**

*See Section 19.2 in H&S* ➤

See Table 19.3 in H&S: range of oxidation states is greatest for metals in the middle of the row with the highest ox. state exhibited by Mn (+7). The +2 state is exhibited by all metals except Sc, and +3 for all metals except Zn. A 'transition' (as opposed to a *d*-block) metal has an incomplete *d*-subshell or gives rise to a cation with an incomplete *d*-subshell. Sc and Zn each shows only one oxidation state in their compounds: $Sc^{3+}$ is $d^0$ (empty *d* shell) and $Zn^{2+}$ is $d^{10}$ (full *d* shell).

**21.3**

$MeOCH_2CH_2OMe$ (**21.1**) is *O*,*O*′-donor, giving a 5-membered chelate ring; $[BH_4]^-$ can be mono-, di- or tridentate (structures **12.3-12.5** in H&S); if the $Ti^{3+}$ centre is 8-coordinate, complex probably contains didentate **21.1** and 3 didentate $[BH_4]^-$.

**(21.1)**

**21.4**

*Solid solution: see Section 5.7 in H&S* ➤

(a) MgO crystallizes with NaCl lattice; therefore to form a continuous range of solid solutions with MgO, $Li_2TiO_3$ must also have the NaCl structure, i.e. it is $[Li^+]_2Ti^{4+}[O^{2-}]_3$. Substitution is possible because $Li^+$, $Ti^{4+}$ and $Mg^{2+}$ are about the same size, and because electrical neutrality can be maintained.

(b) From equation 21.9 and Section 7.2 in H&S:

$$[TiO]^{2+} + 2H^+ + e^- \rightleftharpoons Ti^{3+} + H_2O \qquad E^o = +0.1\ V \quad \text{at pH } 0$$

$$2H^+ + 2e^- \rightleftharpoons H_2 \qquad E^o = 0\ V \qquad \text{at pH } 0$$

From these values, no (or little) reaction at pH 0 occurs *provided everything stays in solution*. But $TiO_2$ is extremely insoluble, and its insolubilty is increased by the presence of $[OH]^-$. Further, the insolubility of $H_2$ in $H_2O$ encourages reaction.

**21.5**

Ammonium vanadate in acidic solution contains $[VO_2]^+$.
Amount of $[VO_2]^+ = 25.00 \times 0.1000 \times 10^{-3} = 2.500 \times 10^{-3}$ moles
After reduction by $SO_2$, vanadium is oxidized back to $[VO_2]^+$ by $[MnO_4]^-$.
Amount of vanadium to be oxidized $= 2.500 \times 10^{-3}$ moles
Amount of $[MnO_4]^- = 25.00 \times 0.0200 \times 10^{-3} = 0.500 \times 10^{-3}$ moles
∴ Ratio $[MnO_4]^-$ : vanadium = 1:5
$[MnO_4]^-$ undergoes a 5-electron reduction, and therefore vanadium undergoes a 1-electron oxidation.

$$[MnO_4]^- + 8H^+ + 5e^- \rightleftharpoons Mn^{2+} + 4H_2O$$

$$[VO_2]^+ + 2H^+ + e^- \rightleftharpoons [VO]^{2+} + H_2O$$

Hence, reduction of $[VO_2]^+$ by $SO_2$ gives $[VO]^{2+}$.
Reaction of $[VO_2]^+$ with Zn reduces it to $V^{n+}$ – the need to add this immediately to $[VO_2]^+$ arises from its instability in air. Oxidation of $V^{n+}$ to $V^{m+}$ takes place, and at

the same time, some of the excess $[VO_2]^+$ is reduced to either $[VO]^{2+}$ or $V^{3+}$ which is then oxidized by $[MnO_4]^-$. The uncertainty about these vanadium oxidation states does not matter since the *net change* in oxidation states *equals* the change when $[VO_2]^+$ is reduced to $V^{n+}$.

Amount of $[VO_2]^+$ initially $= 2.500 \times 10^{-3}$ moles $=$ amount of $V^{n+}$ to be oxidized.

Amount of $[MnO_4]^- = 75.4 \times 0.0200 \times 10^{-3} = 1.49 \times 10^{-3}$ moles

∴ Ratio $[MnO_4]^-$ : vanadium $= 3:5$

$[MnO_4]^-$ undergoes a 5-electron reduction, and therefore vanadium undergoes a 3-electron oxidation, and $V^{n+}$ is $V^{2+}$ (i.e. $[V(H_2O)_6]^{2+}$).

> Alternative method is to notice that in 2nd titration, volume of $KMnO_4$ is 3× that in the first titration, and so, in the 2nd titration, V undergoes a 3-electron reduction

$$[MnO_4]^- + 8H^+ + 5e^- \rightleftharpoons Mn^{2+} + 4H_2O$$

$$[VO_2]^+ + 4H^+ + 3e^- \rightleftharpoons V^{2+} + 2H_2O$$

**21.6**  Heating $VBr_3$ causes disproportionation:  $2VBr_3 \rightarrow VBr_4 + VBr_2$

> See V(III) and V(IV) subsections in Section 21.6 in H&S

$VBr_4$ formed decomposes on heating:  $2VBr_4 \rightarrow 2VBr_3 + Br_2$

If $Br_2$ is removed, the final product will be $VBr_2$.

**21.7**  $[NH_4]V(SO_4)_2 \cdot 12H_2O$ is an alum. Use $\mu_{eff} = 2.8 \, \mu_B$ to find the number of unpaired electrons and, hence, the oxidation state of V:

$$\mu(\text{spin-only}) = \sqrt{n(n+2)}$$

Value of $n = 2$ gives $\mu(\text{spin-only}) = 2.83 \, \mu_B$.

Ground state electronic configuration of $V = 4s^2 3d^3$, therefore 2 unpaired electrons corresponds to $V^{3+}$ ($d^2$). The alum contains octahedral $[V(H_2O)_6]^{3+}$. Sketch an Orgel diagram for an octahedral $d^2$ ion – this corresponds to the right-hand side of Figure 20.18 in H&S. Three absorption transitions are expected:

$$T_{2g} \leftarrow T_{1g}(F) \qquad T_{1g}(P) \leftarrow T_{1g}(F) \qquad A_{2g} \leftarrow T_{1g}(F)$$

**(21.2)**

**21.8**  The ligand **L** is a hexadentate $N,N',N'',O,O',O''$-donor. It is likely to form a 6-coordinate, octahedral complex with $Cr^{3+}$, and the complex is therefore neutral $[Cr(\mathbf{L})]$. The three N atoms are restricted by being in the macrocycle, and only the *fac*-isomer **21.2** is possible.

**L**

**21.9**  Use Appendix 11 for data. Half-equations to consider:

$$Cr^{2+} + 2e^- \rightleftharpoons Cr \qquad\qquad E^\circ = -0.91 \text{ V}$$
$$Cr^{3+} + 3e^- \rightleftharpoons Cr \qquad\qquad E^\circ = -0.74 \text{ V}$$
$$Cr^{3+} + e^- \rightleftharpoons Cr^{2+} \qquad\qquad E^\circ = -0.41 \text{ V}$$
$$2H^+ + 2e^- \rightleftharpoons H_2 \qquad\qquad E^\circ = 0 \text{ V}$$
$$O_2 + 4H^+ + 4e^- \rightleftharpoons 2H_2O \qquad\qquad E^\circ = +1.23 \text{ V}$$
$$[Cr_2O_7]^{2-} + 14H^+ + 6e^- \rightleftharpoons 2Cr^{3+} + 7H_2O \qquad\qquad E^\circ = +1.33 \text{ V}$$

Aqueous $HClO_4$ is kinetically very inert to reduction; look back at the discussion accompanying equations 16.67 and 16.68 in H&S. Oxidation of Cr *by* $[ClO_4]^-$ can therefore be ignored; the purpose of $HClO_4$ is to make the solution acidic. Cr is oxidized by $H^+$, and it will be oxidized to $Cr^{3+}$ rather than $Cr^{2+}$. Even though $E^o$ for oxidation to $Cr^{2+}$ is larger than for oxidation of Cr to $Cr^{3+}$, $\Delta G^o$ per mole of Cr is more negative for oxidation to $Cr^{3+}$. Look back to worked example 7.3 in H&S and include a similar calculation in your answer.

In air, $Cr^{3+}$ will *not* be further oxidized.

21.10    (a) The reaction occurring is:

$$2[MnO_4]^- + 16H^+ + 5[C_2O_4]^{2-} \rightarrow 2Mn^{2+} + 8H_2O + 10CO_2$$

Follow the reaction by colorimetry (loss of absorbance from $[MnO_4]^-$) or gas evolution (production of $CO_2$).

(b) Reaction rate increases after $Mn^{2+}$ begins to be present in solution; this is *autocatalysis* – $Mn^{2+}$ catalyses the reaction between $[MnO_4]^-$ and $[C_2O_4]^{2-}$.

21.11    Consider the modes of bonding *within the complexes* listed. Other modes may be possible in other complexes.

$O=PPh_3$ (**21.3**) is *O*-bound, monodentate.

$[N_3]^-$ (azide, **21.4**), monodentate.

$[Se_4]^{2-}$ (**21.5**) didentate; donor atoms are *cis*, giving 5-membered chelate ring.

$[pc]^{2-}$ (**21.6**) has a non-flexible framework (related to a porphyrin); $Mn^{2+}$ is coordinated approximately within the plane of the central four *N*-donors.

$OC(NHMe)_2$ (**21.7**) is monodentate through *O*-donor.

$N(CH_2CH_2NMe_2)_3$ (**21.8**) is tripodal and tetradentate, i.e. 3 'arms' radiating from central *N*-donor; in a trigonal bipyramidal complex, **21.8** favours occupying axial and 3 equatorial sites as in **19.11**, p. 158.

THF (**21.9**) is monodentate, *O*-bound.

Hpz (**21.10**) is monodentate, *N*-bound.

bpy (**21.11**), didentate, N atoms must be mutually *cis*.

$[NCS]^-$ may be *N*- or *S*- bonded, but $[NCS-N]^-$ shows it coordinates through N.

$HOCH_2CH_2OH$, didentate and the O atoms must be mutually *cis*.

tpy (**21.12**), tridentate; *N*-donors are *mer.*

$[EDTA]^{4-}$ (**21.13**) hexadentate but relatively flexible allowing octahedral coordination.

12-crown-4 is a tetradentate macrocycle but is conformationally restricted.

(21.6)

(21.12)

(21.13)

(21.3)

(21.4)

(21.5)

(21.7)

(21.8)

(21.9)

(21.10)

(21.11)

21.12    (a) Fe(II) and Fe(III) can be distinguished by using Mössbauer spectroscopy (refer to Section 2.12 in H&S for discussion of this technique).

(b) For hard and soft metal ions, refer to Table 6.9 in H&S. Look also at the stability constant data in Table 6.8 in H&S. It can be shown that $Fe^{3+}$ is a hard cation by observing that $Fe^{3+}$(aq) changes colour at high [Cl$^-$] *and* changes colour if Cl$^-$ is displaced on addition of F$^-$. Both Cl$^-$ and F$^-$ are hard anions; hardness F$^-$ > Cl$^-$.

(c) Mn(V) disproportionates unless it is in strongly alkaline solution. Treat a known amount of the precipitate with acid to give $MnO_2$ (as a precipitate) and $[MnO_4]^-$ (in solution). Separate $MnO_2$ and $[MnO_4]^-$ by filtration, and determine the amount of each by titrating against oxalic acid in strongly acidic ($H_2SO_4$) solution. Results will allow determination of oxidation state changes and hence confirm Mn(V).

21.13    (a) and (b) See beginning of Section 21.9 in H&S for reactions of halogens with Fe: $Cl_2$ oxidizes Fe to Fe(III), but $I_2$ oxidizes it to Fe(II):

$$2Fe + 3Cl_2 \rightarrow 2FeCl_3$$

$$Fe + I_2 \rightarrow FeI_2$$

(c) Concentrated $H_2SO_4$ acts as oxidizing agent, Fe(II) $\rightarrow$ Fe(III):

$$2FeSO_4 + 2H_2SO_4 \rightarrow Fe_2(SO_4)_3 + SO_2 + 2H_2O$$

(d) [SCN]$^-$ acts as a ligand; $Fe^{3+}$ is a hard metal ion and therefore expect that [SCN]$^-$ will coordinate through hard *N*-donor rather than soft *S*-donor:

$$[Fe(H_2O)_6]^{3+} + [SCN]^- \rightarrow [Fe(H_2O)_5(SCN\text{-}N)]^{2+} + H_2O$$

(e) $K_2C_2O_4$ provides oxalate ligand which is didentate (**21.14**) with 2 hard *O*-donors; on standing, $Fe^{3+}$ will oxidize oxalate ions:

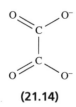

**(21.14)**

$$[Fe(H_2O)_6]^{3+} + 3[C_2O_4]^{2-} \rightarrow [Fe(C_2O_4)_3]^{3-} + 6H_2O$$

$$2Fe^{3+} + [C_2O_4]^{2-} \rightarrow 2Fe^{2+} + 2CO_2$$

(f) FeO is basic and reacts with acids to give salt and water:

$$FeO + H_2SO_4 \rightarrow FeSO_4 + H_2O$$

(g) Addition of alkali to Fe(II) salts precipitates white hydroxide which is oxidized in air to give mixed Fe(II)Fe(III) and Fe(III) oxides. The reaction with alkali is:

$$FeSO_4 + 2NaOH \rightarrow Fe(OH)_2(s) + Na_2SO_4$$

21.14    (a) Figure 21.1 shows the trend in lattice energies for first row metal(II) chlorides. A similar plot applies to fluorides. The 'double hump' is attributed to crystal (or ligand) field stabilization energy contributions. For 0, 5 and 10 *d* electrons in high-spin configurations, CFSE = 0, and interpolation from these points on the graph allows an estimate of (lattice energy – CFSE). The CFSE for $FeF_2$ is therefore estimated from the difference between the lattice energy value determined from a

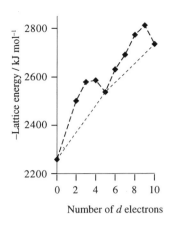

**Figure 21.1** Trend in lattice energies for first row MCl₂.

Born-Haber cycle, and a value interpolated from the lattice energies of $MnF_2$ (high-spin $d^5$) and $ZnF_2$ ($d^{10}$).

(b) Construct an appropriate thermochemical cycle:

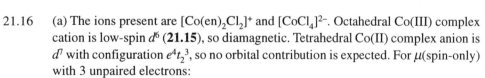

Values of $\Delta G^\circ_2$ and $\Delta G^\circ_4$ can be found from the $E^\circ$ values in the question:

$$\Delta G^\circ_2 = -zFE^\circ = -(1)(96\,485)(+1.92) \times 10^{-3} = -185 \text{ kJ mol}^{-1}$$
$$\Delta G^\circ_4 = -zFE^\circ = -(1)(96\,485)(+0.11) \times 10^{-3} = -11 \text{ kJ mol}^{-1}$$

Use the overall formation constant for $[Co(NH_3)_6]^{2+}$ to find $\Delta G^\circ_3$:

$$\Delta G^\circ_3 = -RT \ln K = -(8.314)(298)\ln(10^5) \times 10^{-3} = -28.5 \text{ kJ mol}^{-1}$$

From the cycle above:  $\Delta G^\circ_1 = \Delta G^\circ_2 + \Delta G^\circ_3 - \Delta G^\circ_4 = -202.5 \text{ kJ mol}^{-1}$

For the overall formation constant of $[Co(NH_3)_6]^{3+}$:

$$K = e^{-\frac{\Delta G^\circ}{RT}} = e^{-\frac{(-202.5)}{(8.314 \times 10^{-3})(298)}}$$

$$K \approx 10^{35}$$

21.15    Preferences between spinel structures: see end of Section 20.9 in H&S. The distribution of metal ions between tetrahedral and octahedral sites in a spinel can be rationalized in terms of LFSEs. In a *normal* spinel $A^{II}B^{III}_2O_4$, tetrahedral sites are occupied by the $A^{2+}$ and octahedral sites by $B^{3+}$. An inverse spinel has metal ions arranged $(B^{III})_{tet}(A^{II}B^{III})_{oct}O_4$. Now consider $Co_3O_4$: rewrite as $Co^{II}Co^{III}_2O_4$. $Co^{3+}$ is $d^6$ and low-spin, therefore LFSE is greater if the $Co^{3+}$ ions occupy octahedral rather than tetrahedral sites. This means that a *normal* spinel is favoured.

21.16    (a) The ions present are $[Co(en)_2Cl_2]^+$ and $[CoCl_4]^{2-}$. Octahedral Co(III) complex cation is low-spin $d^6$ (**21.15**), so diamagnetic. Tetrahedral Co(II) complex anion is $d^7$ with configuration $e^4t_2{}^3$, so no orbital contribution is expected. For $\mu$(spin-only) with 3 unpaired electrons:

$$\mu\text{(spin-only)} = \sqrt{n(n+2)} = 3.87\ \mu_B$$

This is only a little larger than the observed value of 3.71 $\mu_B$.

(b) Value of $\mu_{eff} = 5.01\ \mu_B$ for $[CoI_4]^{2-}$ is significantly greater than $\mu$(spin-only) due to spin-orbit coupling. Also greater than $\mu_{eff}$ for $[CoCl_4]^{2-}$, e.g. value in part (a). Although the following equation is only applicable to $A$ and $E$ ground terms:

$$\mu_{eff} = \mu\text{(spin-only)}\left(1 - \frac{\alpha\lambda}{\Delta_{oct}}\right) \qquad \text{(where } \lambda \text{ is negative for a } d^7 \text{ ion)}$$

it shows that $\mu_{eff}$ is inversely related to ligand field strength, and $\Delta_{oct}$ for $Cl^- > I^-$.

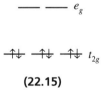

(22.15)

21.17   (a) The reaction to give the green precipitate is:

$$Ni^{2+}(aq) + 2[CN]^- \rightarrow Ni(CN)_2(s) \qquad \text{(forms as hydrated compound)}$$

Dissolution to give the yellow and then red solutions is due to complex formation; note that the hexacyano ion does *not* form:

$$Ni(CN)_2 + 2[CN]^- \rightarrow [Ni(CN)_4]^{2-} \qquad \text{(yellow)}$$

$$[Ni(CN)_4]^{2-} + [CN]^- \rightarrow [Ni(CN)_5]^{3-} \qquad \text{(red)}$$

(b) Yellow salt from part (a) is $K_2[Ni(CN)_4]$. Sodium in liquid $NH_3$ is a reducing reagent. Ni(II) may be reduced to Ni(I) or Ni(0) forming $K_4[Ni_2(CN)_6]$ or $K_4[Ni(CN)_4]$.

21.18   An aqueous solution of $NiCl_2$ contains $[Ni(H_2O)_6]^{2+}$. This reacts with the ligand to give a complex which still contains $H_2O$ and is paramagnetic; value of 3.30 $\mu_B$ (larger than spin-only value) corresponds to 2 unpaired electrons ($d^8$). The complex is likely to be octahedral *trans*-$[Ni(\mathbf{21.16})_2(H_2O)_2]$. This loses $2H_2O$ to give square planar $[Ni(\mathbf{21.16})_2]$ which is diamagnetic (analogous to $Pd^{2+}$ in Figure 20.1e, p. 163). Isomers of $[Ni(\mathbf{21.16})_2]$ can arise because of different orientations of the Ph groups of the ligand. Note that isomerization does *not* involve square planar-tetrahedral interconversion; a tetrahedral complex would not be diamagnetic.

**(21.16)**

21.19   (a) Addition of alkali to Cu(II) salts precipitates blue copper(II) hydroxide:

$$CuSO_4 + 2NaOH \rightarrow Cu(OH)_2(s) + Na_2SO_4$$

(b) Reduction of Cu(II) in presence of $Cl^-$ leads to copper(I) chloride:

$$CuO + Cu + 2HCl \rightarrow 2CuCl + H_2O$$

(c) Concentrated $HNO_3$ is an oxidant:

$$Cu + 4HNO_3(conc) \rightarrow Cu(NO_3)_2 + 2H_2O + 2NO_2$$

(d) Complex formation with $NH_3$ as the ligand, and dissolution of Cu(II) complex:

$$Cu(OH)_2 + 4NH_3 \rightarrow [Cu(NH_3)_4]^{2+} + 2[OH]^-$$

(e) Addition of alkali gives a precipitate of $Zn(OH)_2$, but this dissolves in the presence of excess $[OH]^-$ as $[Zn(OH)_4]^{2-}$ forms:

$$ZnSO_4 + 2NaOH \rightarrow Zn(OH)_2(s) + Na_2SO_4$$

$$Zn(OH)_2(s) + 2NaOH \rightarrow Na_2[Zn(OH)_4]$$

(f) Action of acid on a metal sulfide:

$$ZnS + 2HCl \rightarrow H_2S + ZnCl_2$$

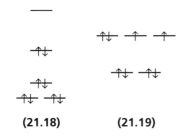

**(21.17)**

21.20 (a) Structure **21.17** shows $H_2$dmg (dimethylglyoxime). Refer to Figures 21.25 and 21.28b in H&S which show the solid state structures of $[Ni(Hdmg)_2]$ and $[Cu(Hdmg)_2]$. In $[Ni(Hdmg)_2]$, there is strong hydrogen bonding between the two $[Hdmg]^-$ ligands and this helps to stabilize both the square planar coordination environment and planarity of the ligand. Molecules stack in the solid state with relatively close Ni····Ni contacts. In $[Cu(Hdmg)_2]$, there is also association between the ligands through hydrogen bonding, but in the solid state, dimers exist with each Cu(II) being 5-coordinate.

(b) Ni and Pd are in group 8. Pd(II) usually has a strong preference for square planar coordination (large CFSE for $d^8$). Therefore one expects $[Pd(Hdmg)_2]$ to be structurally similar to $[Ni(Hdmg)_2]$, exhibiting stacks of planar molecules in the solid state. This is actually observed with close Pd····Pd contacts of 325 pm.

21.21 $SO_2$ will be oxidized to sulfate and the relevant half-equation is:

$$[SO_4]^{2-} + 4H^+ + 2e^- \rightleftharpoons SO_2 + 2H_2O$$

➤

See Section 7.2 in H&S for the effects of pH on $E^o$ values, and Section 7.3 in H&S for the effects of complex formation on $E^o$ values

but $E^o$ for this reduction will be pH-dependent. Conc. HCl also provides $Cl^-$ which acts as a ligand to $Cu^{2+}$ forming, for example, $[CuCl_4]^{2-}$. The reduction potential for $Cu^{2+}/Cu$ of +0.34 V refers to aqueous $Cu^{2+}$. Complex formation alters the $E^o$ value. To test this reasoning, try the effect of replacing HCl by (i) saturated LiCl or any other very soluble chloride which will alter the concentration of $Cl^-$ present, and (ii) $HClO_4$ or another very strong acid, the anion of which is a very poor complexing agent and difficult to reduce (see answer 21.9, p. 168).

21.22 Steric factors will always favour tetrahedral coordination. In the absence of any steric control, electronic factors prevail.

(a) Pd(II) is $d^8$ and a square planar arrangement maximizes LFSE: Pd(II) and also Pt(II) complexes are usually square planar.

(b) and (c) Cu(I) and Zn(II) are $d^{10}$ and so LFSE = 0 for all crystal fields. Therefore, there is no electronic driving force for a particular structure and tetrahedral coordination is favoured over square planar on steric grounds.

Ni(II) is $d^8$. Consider the crystal field splitting of the $d$ orbitals: **21.18** and **21.19** show that a square planar complex is diamagnetic, but a tetrahedral complex is paramagnetic.

**(21.18)**    **(21.19)**

21.23 In giving your answer, check that the oxidation states match those in the question.

(a) $[MnO_4]^-$ (also called permanganate ion)    (b) $[MnO_4]^{2-}$
(c) $[Cr_2O_7]^{2-}$    (d) $[VO]^{2+}$
(e) $[VO_4]^{3-}$ (*ortho*); $[VO_3]^-$ (*meta*)    (f) $[Fe(CN)_6]^{3-}$

21.24 Detailed information is given in Chapter 21 in H&S. Points to include:
- Variation in oxidation states across row – how this affects numbers of oxides formed by each metal.
- Basic (e.g. CuO), amphoteric (e.g. $V_2O_5$) or acidic (e.g. $CrO_3$) behaviour; reactions of oxides with water, alkalis and acids.
- Oxidizing properties of $CrO_3$ and $MnO_2$.
- Highest ox. state metal oxide is $Mn_2O_7$ (it is highly explosive).
- Mixed oxidation state oxides, e.g. $Fe_3O_4$, $Co_3O_4$ (spinels).
- Hydrated oxides, e.g. $Fe_2O_3 \cdot H_2O$.
- Example of non-stoichiometric metal(II) oxides (see Section 27.2 in H&S).

**21.25**   Detailed information is given in Chapter 21 in H&S. Points to include:
- Sc chemistry limited; hard $Sc^{3+}$ forms discrete $[ScF_6]^{3-}$ and $[ScF_7]^{4-}$.
- Halo anions for oxidation states > +4 are very rare indeed (e.g. $[VF_6]^-$).
- Octahedral anions common, e.g. $[TiCl_6]^{2-}$, $[TiCl_6]^{3-}$, $[VF_6]^{3-}$, $[CoF_6]^{3-}$, $[NiF_6]^{2-}$; fluoro ligands often associated with high ox. states.
- Tetrahedral anions for the later metal ions, e.g. $[FeCl_4]^{2-}$, $[FeBr_4]^{2-}$, $[CoCl_4]^{2-}$, $[NiCl_4]^{2-}$.
- Halo bridges lead to multinuclear anions, e.g. $[V_2Cl_9]^{3-}$ which has octahedral coordination.
- Halo bridges may lead to chain structures or 3D-lattices in examples such as $CsNiCl_3$, $Na_2MnF_5$, $[Me_2NH_2][MnCl_3]$, Cr(II) halo anions; octahedral metal sites often observed.

**21.26**   Start with the analysis:      **X** contains 11.4% Fe and 53.7% $ox^{2-}$

$ox^{2-} = [C_2O_4]^{2-}$   ➤      Ratio Fe : $ox^{2-}$ = $\dfrac{11.4}{55.85} : \dfrac{53.7}{88.02} = 0.20 : 0.61 = 1 : 3$

This suggests a complex $[Fe(ox)_3]^{n-}$ where $n = 4$ or 3.
Since the reaction conditions are $Fe(ox)_2 + H_2O_2$ (oxidizing agent) + $H_2ox$ (source of $ox^{2-}$) + $K_2ox$ (source of $ox^{2-}$ and $K^+$ cation), it is reasonable to suggest that Fe(II) is oxidized to Fe(III) giving $K_3[Fe(ox)_3]$. The remaining 34.9% (look at the analysis results) is $K^+$ plus (most likely) water of crystallization:

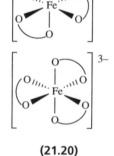

$$K_3[Fe(ox)_3]\cdot 3H_2O \quad 11.4\%\ Fe,\ 53.7\%\ ox^{2-},\ 23.9\%\ K,\ 11.0\%\ H_2O = 100\%$$

Therefore, **X** is $K_3[Fe(ox)_3]\cdot 3H_2O$. Its reactions with alkali (precipitation of metal hydroxide with alkali) and photochemical decomposition are:

$$[Fe(ox)_3]^{3-} + 3[OH]^- \rightarrow Fe(OH)_3\ (\text{i.e. } Fe_2O_3\cdot 3H_2O) + 3ox^{2-}$$

$$2K_3[Fe(ox)_3] \rightarrow 2Fe[ox] + 3K_2[ox] + 2CO_2$$

**(21.20)**

$[Fe(ox)_3]^{3-}$ is chiral (3 didentate ligands, octahedral, **21.20**), *but* the reaction with $[OH]^-$ suggests that $[Fe(ox)_3]^{3-}$ may be too labile to be resolved into enantiomers.

**21.27**   DMSO (**21.21**) acts as a ligand. With $Co^{2+}$ (the $ClO_4^-$ anion will act an as a counterion), DMSO forms $[Co(DMSO)_6]^{2+}$, and compound **A** is the 1:2 electrolyte $[Co(DMSO)_6][ClO_4]_2$. With $CoCl_2$, DMSO will give the same complex cation but the $Cl^-$ ions also act as ligands; **B** is the 1:1 electrolyte $[Co(DMSO)_6][CoCl_4]$:

**(21.21)**

$$2CoCl_2 + 6DMSO \rightarrow [Co(DMSO)_6][CoCl_4]$$

**21.28**   $H_2S$ with aqueous $Cu^{2+}$ precipitates CuS which has a very low solubility ($K_{sp} = 6.0 \times 10^{-37}$); it precipitates even in acidic medium:

$$Cu^{2+} + H_2S \rightarrow CuS(s) + 2H^+$$

➤

See Sections 7.2 and 7.3 in
H&S for factors that influence
$E^o$ values

With hot conc. $H_2SO_4$, Cu is oxidized and S reduced. High $[H^+]$ and low solubility of CuS affect the $E^o$ values to an extent that makes *both* of the following reductions possible:

$$[SO_4]^{2-} + 4H^+ + 2e^- \rightarrow SO_2 + 2H_2O$$
$$[SO_4]^{2-} + 8H^+ + 8e^- \rightarrow S^{2-} + 4H_2O \qquad \text{(which gives CuS)}$$

# 22 *d*-Block chemistry: the second and third row metals

22.1 (a) The aim of this question is to give you an easy way to learn the positions of the metals in the periodic table: *learn by building up each triad from the first row.*

| Group: | 3 | 4 | 5 | 6 | 7 | 8 | 9 | 10 | 11 | 12 |
|---|---|---|---|---|---|---|---|---|---|---|
| | Sc | Ti | V | Cr | Mn | Fe | Co | Ni | Cu | Zn |
| | Y | Zr | Nb | Mo | Tc | Ru | Rh | Pd | Ag | Cd |
| | La | Hf | Ta | W | Re | Os | Ir | Pt | Au | Hg |

(b) Lanthanoid series comes between La and Hf.

22.2 Data for this answer are found in Tables 5.2, 21.1 and 22.1 in H&S.
(a) The trends in metallic radii for 12-coordination are plotted in Figure 22.1; although not all the metals adopt close-packed structures at 298 K (see below), radii for different coordination numbers can be adjusted as shown in Section 5.5 in H&S. Figure 22.1 shows that, in general, values of metallic radii:
- show little variation across a given row of the *d*-block;
- are greater for second and third row metals than for first row metals:
- are similar for the second and third row metals in a given triad.

The last fact is due to the *lanthanoid contraction* (see Section 24.3 in H&S).

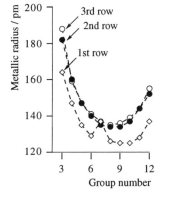

**Figure 22.1** The trends in metallic radii of the *d*-block metals on crossing the periods.

(b) First note that comparisons between values of $\Delta_a H^\circ$ are valid if the metals are structurally similar. Close-packed lattices (hcp or ccp) are the norm in the *d*-block, except for group 5 and 6 (bcc) and group 12 metals. There is a rough correlation between the number of unpaired electrons and the value of $\Delta_a H^\circ$, with the maximum values being for middle *d*-block metals. The trend in values of $\Delta_a H^\circ$ for the third row metals was shown in Figure 5.3 (p. 44); this general trend is followed for each row. Down each triad, the general trend is $\Delta_a H^\circ(\text{third row}) > \Delta_a H^\circ(\text{second row}) > \Delta_a H^\circ(\text{first row})$; the differences are greatest for the middle group metals.

22.3 (a) Use data from the Appendices in H&S.
The question gives a value for $\Delta_f H^\circ(\text{CrCl}_2)$ and so this gives a clue as to the starting point. Assume that $\text{CrCl}_2$ and $\text{WCl}_2$ (both group 6 metal(II) chlorides) have the same lattice type. First, calculate $\Delta_{\text{lattice}} H^\circ(\text{CrCl}_2)$ from a Born-Haber cycle:

$$\Delta_{\text{lattice}} H^\circ = \Delta_f H^\circ(\text{CrCl}_2, \text{s}) - \Delta_a H^\circ(\text{Cr}, \text{s}) - IE_1 - IE_2 - D(\text{Cl}_2, \text{g})$$

For a reminder of the method of calculation, see answer 5.15, p. 48

$$-2\Delta_{\text{EA}} H(\text{Cl}, \text{g})$$

$$= -397 - 397 - 652.9 - 1591 - 242 - 2(-349)$$

$$= -2582 \text{ kJ mol}^{-1}$$

A value of $\Delta_{\text{lattice}} H^\circ(\text{WCl}_2)$ can be estimated by using the fact that $\Delta U$ ($\Delta U \approx \Delta_{\text{lattice}} H^\circ$) is inversely proportional to internuclear distance:

$$\frac{\Delta_{\text{lattice}} H^\circ(\text{WCl}_2)}{\Delta_{\text{lattice}} H^\circ(\text{CrCl}_2)} = \frac{r(\text{Cr}^{2+}) + r(\text{Cl}^-)}{r(\text{W}^{2+}) + r(\text{Cl}^-)}$$

Since $r(Cl^-) \gg r(Cr^{2+})$ or $r(W^{2+})$, the difference between the radii of $Cr^{2+}$ and $W^{2+}$ will not affect the ratio of internuclear distances very much. Therefore:

$$\Delta_{lattice}H^\circ(WCl_2) \approx \left( \frac{r(Cr^{2+}) + r(Cl^-)}{r(W^{2+}) + r(Cl^-)} \right) \Delta_{lattice}H^\circ(CrCl_2) \approx \Delta_{lattice}H^\circ(CrCl_2)$$

$$\approx 1 \text{ (or slightly larger than 1)}$$

The radius of $W^{2+}$ will be slightly larger than that of $Cr^{2+}$, so $\Delta_{lattice}H^\circ(WCl_2)$ will be a little less negative than $\Delta_{lattice}H^\circ(CrCl_2)$. A reasonable estimate is $\Delta_{lattice}H^\circ(WCl_2)$ $\approx -2450$ to $-2500$ kJ mol$^{-1}$. Now use these values in a Born-Haber cycle to estimate $\Delta_f H^\circ(WCl_2, s)$. For $\Delta_{lattice}H^\circ(WCl_2) = -2450$ kJ mol$^{-1}$:

$$\Delta_f H^\circ(WCl_2, s) = \Delta_{lattice}H^\circ(WCl_2, s) + \Delta_a H^\circ(W, s) + IE_1 + IE_2 + D(Cl_2, g)$$

$$+ 2\Delta_{EA}H(Cl, g)$$

$$= -2450 + 850 + 758.8 + 1700 + 242 + 2(-349)$$

$$= +403 \text{ kJ mol}^{-1}$$

Similarly, a value of $\Delta_{lattice}H^\circ(WCl_2) = -2500$ kJ mol$^{-1}$ gives $\Delta_f H^\circ(WCl_2, s) = +353$ kJ mol$^{-1}$.

> See Section 22.7 in H&S for the strucure of WCl$_2$

(b) The high endothermic value of $\Delta_f H^\circ(WCl_2, s)$ shows that an *ionic* compound is *unlikely* to form. The factors that contribute to the endothermic formation reaction (as opposed to the exothermic formation of $CrCl_2$) are the higher ionization energies of W than Cr, and the higher enthalpy of atomization of W versus Cr.

22.4 (a) Hf(IV) and Zr(IV) oxides are isostructural and have structures in which Zr and Hf are 7-coordinate. The lanthanoid contraction results in the Hf and Zr centres being approximately the same size, and the unit cell sizes for $MO_2$ are the same. The big difference in densities (i.e. $\rho(HfO_2)$ is almost double $\rho(ZrO_2)$) is because the atomic weight of Hf $\gg$ Zr (178.5 versus 91.2); there are 14 lanthanoid elements between La and Hf in the periodic table.

> $\rho = \dfrac{M}{V}$

(b) Nb(IV) is $d^1$, and in NbF$_4$ (**22.1**) each Nb(IV) centre has an unpaired electron and the compound is paramagnetic. In NbX$_4$ (**22.2**, X = Cl or Br), the Nb centres (and therefore the odd electrons) are paired up with Nb–Nb interactions; the compound is therefore diamagnetic.

**(22.1)**

> For an alternative way of representing the structure of NbCl$_4$ or NbBr$_4$, see structure **22.10** in H&S

**(22.2)**

22.5 (a) CsBr is ionic and provides Br$^-$; NbBr$_5$ acts as a bromide acceptor:

$$CsBr + NbBr_5 \rightarrow Cs[NbBr_6]$$

(b) Molten KF contains K⁺ and F⁻; TaF₅ acts as a fluoride acceptor. Under these conditions (and given that high coordination numbers are possible for the heavier, early *d*-block metals, and F is *not* sterically demanding) the most likely reactions are:

$$2KF + TaF_5 \rightarrow K_2[TaF_7] \qquad \text{or} \qquad 3KF + TaF_5 \rightarrow K_3[TaF_8]$$

rather than:

$$KF + TaF_5 \rightarrow K[TaF_6]$$

(c) Several products are possible; the addition of bpy is likely:

$$NbF_5 + bpy \rightarrow [Nb(bpy)F_5]$$

(d) $MF_5$ (M = Nb, Ta) is tetrameric (**22.3**) and $NbBr_5$ is a dimer (**22.4**).

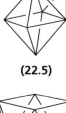

**(22.5)**

**(22.6)**

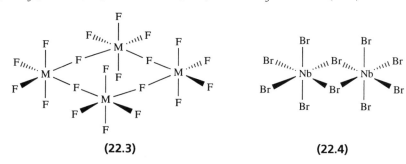

**(22.3)**　　　　　　　**(22.4)**

Products: $[NbBr_6]^-$ is octahedral; $[TaF_7]^{2-}$ is a monocapped octahedron (ligand polyhedron is shown in **22.5**); $[TaF_8]^{3-}$ is a square antiprism (ligand polyhedron **22.6**); $[Nb(bpy)F_5]$ is 7-coordinate; bpy is a didentate ligand with a fairly restricted bite angle, so a pentagonal bipyramid with bpy in the equatorial plane is plausible.

22.6　Chloro-bridges support the dinuclear framework in $[Cr_2Cl_9]^{3-}$ (**22.7**) and there is no Cr–Cr bonding. Each Cr(III) is octahedral $d^3$, and has a magnetic moment characteristic of 3 unpaired electrons. The tendency for metal-metal bonding increases down the triad. In $[W_2Cl_9]^{3-}$ (which also has 3 chloro bridges), the 6 electrons from the two W(III) centres pair up to give a W≡W bond and salts of $[W_2Cl_9]^{3-}$ are diamagnetic.

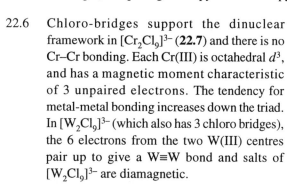

**(22.7)**

22.7　(a) The formula $[Mo_6Cl_8]Cl_2Cl_{4/2}$ means that the compound can be formally 'broken down' into the units:
- $[Mo_6Cl_8]^{4+}$ core (the 4+ charge balances the $(2 + {}^4/_2)$ Cl⁻);
- 2 terminal Cl;
- 4 Cl bridges (2 Cl 'belonging' to each $Mo_6$ unit), each bridging between 2 $Mo_6$ units giving *sheet* structure **22.8**.

The $[Mo_6Cl_8]^{4+}$ core has 8 triangular faces, and the 8 Cl atoms are in $\mu_3$-modes.

$$\therefore [Mo_6Cl_8]Cl_2Cl_{4/2} = [Mo_6Cl_8]Cl_{2+2} = Mo_6Cl_{12} = MoCl_2$$

(b) How many valence electrons (ve) are available?　W = $s^2d^4$ = 6 ve, Cl = 1 ve

Total ve = 36 (from 6W) + 8 (from 8Cl) – 4 (for 4+ charge) = 40

16 ve are used for 8 M–Cl (remember that this assumes nothing about the bonding modes of the Cl atoms), leaving 24 ve for 12 W–W bonds, i.e. bond order = 1.

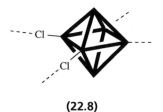

**(22.8)**

22.8    Refer to Section 22.8 in H&S for more details. Points to include:
- Oxohalides: tetrahedral $TcOCl_3$ and $ReOX_3$ (X = F, Cl, Br).
- Large numbers of Tc(V) and Re(V) complexes with one oxo ligand; often square-based pyramidal with oxo group in apical site, e.g. $[TcOCl_4]^-$, $[TcO(ox)_2]^-$.
- Examples of octahedral complexes: $[ReOCl_5]^{2-}$, *trans*-$[TcO_2(py)_4]^+$; the *trans*-dioxo group appears in a range of complexes; protonation equilibria such as:

$$\textit{trans-}[ReO_2(en)_2]^+ \underset{-H^+}{\overset{H^+}{\rightleftharpoons}} \textit{trans-}[ReO(OH)(en)_2]^{2+} \underset{-H^+}{\overset{H^+}{\rightleftharpoons}} \textit{trans-}[Re(OH)_2(en)_2]^{3+}$$

- $ReOCl_3(PPh_3)_2$ is an important starting material in Re(V) chemistry; made from $[ReO_4]^-$ and $PPh_3$ in alcohol in presence of HCl.
- Tc(V) oxo complexes are important in development of medical imaging agents.

22.9    Refer to Sections 21.8 and 22.8 in H&S for greater detail. Points to include:
- Mn occurs naturally ($\beta$-$MnO_2$ is major ore); Tc is a man-made element and is a $\beta$-emitter; thus, chemistry of Mn is more widely studied than that of Tc.
- Range of Mn oxidation states is 0 to +7, with +2 and +7 being of major importance; range for Tc also 0 to +7, but higher oxidation states (e.g. +5) more important for Tc than Mn; there is a wide range of Tc(V) complexes.
- $[TcO_4]^-$ more stable with respect to reduction than $[MnO_4]^-$; $[TcO_4]^-$ is an important starting material in Tc chemistry.
- Chemistry of Mn-containing cations exceeds that of Tc-containing cations.
- Tc–Tc bond formation more important than Mn–Mn bond formation.

22.10    Diagram **22.9** shows the structure of $[Re_2Cl_8]^{2-}$, with eclipsed ligands and unsupported (i.e. no $\mu$-ligands) metal–metal bond as the important features. Take the Re atoms to lie on the $z$ axis; each Re atom uses four atomic orbitals ($s$, $p_x$, $p_y$, $d_{x^2-y^2}$) to form Re–Cl bonds. Mixing of $p_z$ and $d_{z^2}$ orbitals gives two hybrids pointing along $z$ axis; each Re atom uses four orbitals for Re–Re bonding: $d_{xz}$, $d_{yz}$, $d_{xy}$ and one $p_zd_{z^2}$ hybrid (the other $p_zd_{z^2}$ hybrid is non-bonding). The bonding can be considered in terms of the interactions shown in Figure 22.2. Each Re(III) provides 4 electrons for Re–Re bond formation, giving a $\sigma^2\pi^4\delta^2$ configuration, i.e. quadruple bond. The presence of the $\delta$-component forces the two $ReCl_4$-units to be eclipsed.

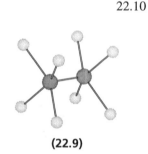

**(22.9)**

22.11    The shortest Re–Re distances (224 pm) are in $[Re_2Cl_8]^{2-}$ (quadruple bond, see above) and $[Re_2Cl_4(\mu\text{-}Ph_2PCH_2CH_2PPh_2)_2]$. This contains two Re(II) and has 10 electrons for Re–Re bonding and a $\sigma^2\pi^4\delta^2\delta^{*2}$ configuration, therefore Re≡Re

$d_{xy}$ (or $d_{x^2-y^2}$)    $\delta$ interaction

$d_{xz}$ (also $d_{yz}$)    2 $\pi$ interactions

$p_zd_{z^2}$    $\sigma$ interaction

—Re——Re→$_z$

Energy

— $\sigma^*$

— — $\pi^*$

— $\delta^*$

⥮ $\delta$

⥮  ⥮ $\pi$

⥮ $\sigma$

**Figure 22.2** Formation of an Re≡Re quadruple bond in $[Re_2Cl_8]^{2-}$.

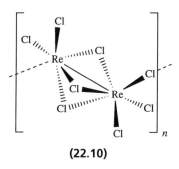

**(22.10)**

bond. The fact that the bond length is the same in $[Re_2Cl_8]^{2-}$ ($\{Re_2\}^{6+}$ core with quadruple bond) and $[Re_2Cl_4(\mu\text{-}Ph_2PCH_2CH_2PPh_2)_2]$ ($\{Re_2\}^{4+}$ core with triple bond) may be an effect of the bridging ligands 'clamping' the metals together and opposing the lowering in bond order. Next in the bond length sequence is $Re_3Cl_9$ (249 pm); triangular $Re_3$ unit with Re=Re double bonds – $\{Re_3\}^{9+}$ core with 12 valence electrons for Re–Re bonding. The longest bonds (273 pm in $ReCl_4$ and 270 pm in $[Re_2Cl_9]^-$) are single bonds; $ReCl_4$ is polymeric (**22.10**) and $[Re_2Cl_9]^-$ has an $\{Re_2\}^{8+}$ core and a triple bond would be suggested. The bond length appears long (given that there are 3 $\mu$-Cl) and suggests a single bond, with two electrons per Re not used for Re–Re bonding.

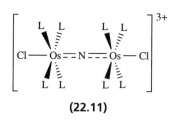

**(22.11)**

22.12  Inspection of the formula of the Os-containing species suggests (i) '$N_9H_{24}$' replaces $3Cl^-$, and (ii) two Os units come together when **A** is formed. The reaction with HI indicates that there are two types of chloride, probably 3 non-coordinated $Cl^-$ and 2 coordinated $Cl^-$. Now consider the reaction of **A** with KOH: since 1 equivalent of **A** contains 9 equivalents of N, the results show that all the N is liberated as $NH_3$. The group '$N_9H_{24}$' could contain $8NH_3 + N^{3-}$. The nitride ligand is likely to be the linking group between the two Os centres and **22.11** is a proposed structure of the cation: **A** is [**22.11**]$Cl_3$. The vibrational spectra show that the cation has a centre of symmetry, and so the bridge is linear. This is consistent with bridging $N^{3-}$; bonding is analogous to that described in Figure 22.15 in H&S for a bridging $O^{2-}$ ligand; **A** contains Os(III), $\pi$-bond order = 0.5.

22.13  Refer to Section 22.9 in H&S for greater detail. Points to include:
 • For oxidation states of $\geq +6$, only $RuF_6$ and $OsF_6$ (octahedral monomer).
 • For M(V): $RuF_5$ and $OsF_5$ (both tetrameric), and $OsCl_5$ (dimeric).
 • For M(IV): $RuF_4$, $OsF_4$, $OsCl_4$ and $OsBr_4$ (all polymeric).
 • For M(III), ruthenium halides outnumber osmium halides: $RuF_3$, $RuCl_3$, $RuBr_3$, $RuI_3$, $OsCl_3$, $OsI_3$; this is the lowest ox. state for which binary halides are well established.
 • Give syntheses and some properties of each compound mentioned.

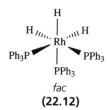

*fac*
**(22.12)**

22.14  (a) Section 22.10 in H&S contains details needed. Points to include:
 • Neutral Rh(IV) and Ir(IV) halides include only fluorides; $RhF_4$ made from $RhBr_3$ or $RhCl_3$ with $BrF_3$; $IrF_4$ made from $IrF_6$ or $IrF_5$ and Ir.
 • M(IV) halo anions $[MX_6]^{2-}$ are known for M = Rh, X = F, Cl and M = Ir, X = F, Cl, Br.
 • $[RhCl_6]^{3-}$ is fluorinated and oxidized to $[RhF_6]^{2-}$ by $BrF_3$.
 • $[RhCl_6]^{2-}$ is prepared from $[RhCl_6]^{3-}$ and $Cl_2$.
 • $[IrF_6]^{2-}$ made by treating $[IrF_6]^-$ with water.
 • $Na_2[IrCl_6]$ (a useful synthon) is made from NaCl, Ir and $Cl_2$; $Br^-$ for $Cl^-$ converts $[IrCl_6]^{2-}$ to $[IrBr_6]^{2-}$.

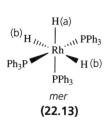

*mer*
**(22.13)**

(b) Octahedral $[IrH_3(PPh_3)_3]$ has *fac*- and *mer*-isomers (**22.12** and **22.13**). Use $^{31}P\{^1H\}$ NMR spectroscopy as one method of distinguishing between them; *fac*-isomer shows a singlet (one P environment) while the *mer*-isomer has two P environments and exhibits a triplet and doublet ($J_{PP}$). (Remember that the notation '$^{31}P\{^1H\}$' means 'proton decoupled $^{31}P$'). $^1H$ NMR spectroscopy could also be used: hydride signals are diagnostic – *fac*-isomer shows one resonance with $J_{PH}$ to one *trans* and two *cis* P, while *mer*-isomer exhibits two resonances with both $J_{HH}$ and $J_{PH}$. To work out these latter couplings, consider H(a) in **22.13**: it couples to two H(b), to one *trans* P and to two *cis* P. (To what does H(b) couple?)

**22.15**

Look at answer 2.13, p. 19

**X** is a non-electrolyte, and so a neutral molecule. The IR spectroscopic data indicate that an Rh–H bond is present in **X**, the H originating from $H_3PO_2$. Use of $D_3PO_2$ produces an Rh–D bond, the vibrational wavenumber of which is lower than that of the Rh–H bond. The role of the $MePh_2As$ is to act as a ligand. Therefore **X** is $RhBr_2H(AsMePh_2)_3$. Reaction with $Br_2$:

$$RhBr_2H(AsMePh_2)_3 + Br_2 \rightarrow RhBr_3(AsMePh_2)_3 + HBr$$

Treatment with mineral acid regenerates $RhBr_3$ and free $MePh_2As$.

**22.16**    (a) $\beta$-$PdCl_2$ is a hexamer (**22.14**) with Pd atoms arranged in an octahedron but at *non-bonded* separations. The $[Nb_6Cl_{12}]^{2+}$ unit is present in $Nb_6Cl_{14}$ (Figure 22.6 in H&S) and contains an octahedron of Nb atoms in which there is Nb–Nb bonding (**22.15**); along each Nb–Nb edge there is a $\mu$-Cl. The arrangement of Cl atoms is the same in both **22.14** and **22.15**.

(**22.14**)

(**22.15**)

(b) For details, see Section 22.11 in H&S. Points to include:
- Compounds which appear to contain Pd(III) tend to be mixed valence species, e.g. '$PdF_3$' is $Pd[PdF_6]$ with Pd(II) and Pd(IV).
- Examples of dinuclear complexes formally containing $\{Pd_2\}^{6+}$ core with Pd–Pd bond could be classed as Pd(III), but Pd–Pd bonding makes ox. state assignment ambiguous.

(c) $Ni^{2+}$, $Pd^{2+}$ and $Pt^{2+}$ are $d^8$ and CFSE effects mean that a square planar complex is favoured on electronic grounds, especially for $Pd^{2+}$ (2nd row metal) and $Pt^{2+}$ (3rd row) for which field strengths are large. $Ni^{2+}$ shows examples of both tetrahedral and square planar complexes depending on field strengths and steric effects of ligands (e.g. $[Ni(CN)_4]^{2-}$ is square planar, $[NiCl_4]^{2-}$ is tetrahedral). Steric effects always favour tetrahedral over square planar coordination sphere, but competing electronic effects (e.g. strong field $[CN]^-$) can tip the balance in favour of square planar.

For more about CFSE, look back at Chapter 20

**22.17**    (a) The results of a single crystal X-ray diffraction study are definitive, but it is not always possible to grow suitable crystals. *cis*- and *trans*-$[PtCl_2(NH_3)_2]$ can be distinguished by their dipole moments (*cis* is polar, *trans* is non-polar) and by IR spectra. Because the *trans*-isomer is non-polar, only the asymmetric $PtCl_2$ stretch is IR active, but for the *cis*-isomer, both symmetric and asymmetric stretches are IR active. $[Pt(NH_3)_4][PtCl_4]$ contains the ions $[Pt(NH_3)_4]^{2+}$ and $[PtCl_4]^{2-}$ and so is a 1:1 electrolyte. $[PtCl_2(NH_3)_2]$ is a non-electrolyte.

(b) $[(H_3N)_2Pt(\mu\text{-}Cl)_2Pt(NH_3)_2]Cl_2$ would contain the ions $[(H_3N)_2Pt(\mu\text{-}Cl)_2Pt(NH_3)_2]^{2+}$ and $2Cl^-$ and would be a 1:2 electrolyte. It also does *not* contain terminal Pt–Cl bonds and, therefore, no associated IR absorptions are observed.

**22.18**    (a) Halide exchange reaction, with *excess* $I^-$ :

$$K_2[PtCl_4] + 4KI \rightarrow K_2[PtI_4] + 4KCl$$

$[PtI_4]^{2-}$ is square planar ($d^8$).

(b) $NH_3$ substitutes for $Cl^-$, but only for *two* Cl which must be in *cis* positions because of the *trans effect* (see Chapter 25); the product is square planar:

$$[PtCl_4]^{2-} + 2NH_3(aq) \rightarrow cis\text{-}[PtCl_2(NH_3)_2] + 2Cl^-$$

(c) phen (**22.16**) is didentate and binds *cis*; only two $Cl^-$ are displaced (see above) and the product is square planar:

$$[PtCl_4]^{2-} + phen \rightarrow cis\text{-}[PtCl_2(phen)] + 2Cl^-$$

(d) tpy (**22.17**) is tridentate With the chelate effect in its favour, it displaces three $Cl^-$ from $[PtCl_4]^{2-}$; product cation is square planar:

$$[PtCl_4]^{2-} + tpy \rightarrow [PtCl(tpy)]^+Cl^- + 2Cl^-$$

[PtCl(tpy)]Cl can also be made by reacting $PtCl_2$ with tpy.

(e) $[CN]^-$ (strong field ligand) displaces all four $Cl^-$; product $[Pt(CN)_4]^{2-}$ is square planar and, in the solid state, forms stacks with Pt–Pt interactions:

$$K_2[PtCl_4] + 4[CN]^- \rightarrow K_2[Pt(CN)_4] + 4Cl^-$$

22.19    Complexes of type $[PtCl_2(R_2P(CH_2)_nPR_2)]$ (and $[PdCl_2(R_2P(CH_2)_nPR_2)]$) contain square planar M(II) centres. The coordination mode of the didentate bisphosphine ligand depends on the length of the $(CH_2)_n$ backbone. For a small chain, a chelating mode, as shown in **22.18** for $n = 2$, will force a *cis* arrangement of chloro ligands. For *trans*-$[PtCl_2(R_2P(CH_2)_nPR_2)]$ to form, the $(CH_2)_n$ backbone must be long, e.g. $n = 12$ would give a long enough chain to 'reach' over between the two *trans* coordination sites. For intermediate chains, e.g. $n = 7$ or 9, it is likely that steric factors will hinder the formation of the *cis*-monomer, and a dimer will form instead. Of the possible *cis* or *trans* arrangements, the *trans*-form is favoured as shown in **22.19**.

22.20    (a) $[Pt(NH_3)_4][PtCl_4]$ (Magnus's green salt) forms chains of alternating $[Pt(NH_3)_4]^{2+}$ and $[PtCl_4]^{2-}$ ions (**22.20**) with a degree of Pt–Pt interaction that affects the electronic spectrum of $[PtCl_4]^{2-}$. In $[Pt(EtNH_2)_4][PtCl_4]$, the absorption spectrum shows the presence of $[Pt(EtNH_2)_4]^{2+}$ and $[PtCl_4]^{2-}$ ions, indicating that the discrete ions are present and the salt is structurally dissimilar from $[Pt(NH_3)_4][PtCl_4]$. The $EtNH_2$ ligand is significantly more sterically demanding than $NH_3$, and the more bulky amine ligand forces the Pt atoms so far apart that interaction between them is impossible.

(b) Both AgCl and AgI are sparingly soluble in water. The increased solubility of AgI in saturated $AgNO_3$ must be due to complex formation. $Ag^+$ is a soft metal centre, as is $I^-$, while $Cl^-$ is hard (see Table 6.9 in H&S). In the presence of an excess of $Ag^+$ in solution, AgI forms $[Ag_2I]^+$, thereby enhancing the solubility of AgI. An analogous complex for AgCl is not favoured and AgCl remains sparingly soluble in saturated $AgNO_3$ solution.

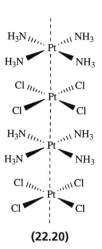

(**22.20**)

(c) If the equilibrium is: $\qquad$ $Hg^{2+} + Hg \rightleftharpoons 2Hg^+$

the equilibrium constant is: $\qquad$ $K = \dfrac{[Hg^+]^2}{[Hg^{2+}]}$

and the ratio of [Hg(I)]/[Hg(II)] depends on [Hg(II)]. Mercury(I) forms the dinuclear species $[Hg_2]^{2+}$. If the equilibrium is:

$$Hg^{2+} + Hg \rightleftharpoons [Hg_2]^{2+}$$

the equilibrium constant is: $\qquad$ $K = \dfrac{[Hg_2^{2+}]}{[Hg^{2+}]}$

and so the ratio [Hg(I)]/[Hg(II)] is always constant.

22.21    Refer to Sections 21.12 and 22.12 in H&S for detailed discussion. Points to include:
- Cu(I) and Cu(II) are the important oxidation states for Cu; in solution, Cu(I) is generally unstable with respect to disproportionation to Cu(II) and Cu.
- A few complexes of Cu(III) and Cu(IV) are known, e.g. $[CuF_6]^{3-}$, $[CuF_6]^{2-}$.
- Ag(I) dominates in the chemistry of Ag.
- Au(I) and Au(III) important for Au, Au(III) being the more stable; this is rationalized in terms of relativistic effects (see Box 12.2 in H&S).
- Compare $E^o$ values for $Ag^+/Ag$, $Cu^+/Cu$, $Cu^{2+}/Cu^+$, $Au^+/Au$ couples.
- High oxidation states: Au(V) and Ag(III).

22.22    Refer to Section 22.13 in H&S for detailed discussion. Points to include:
- Group 12 metals are Zn, Cd, Hg; each metal has an $ns^2nd^{10}$ configuration.
- Hg atypical of *d*-block because liquid at 298 K; amalgams (e.g. Na/Hg) important.
- For Zn and Cd, only the +2 oxidation states ($d^{10}$) are stable, but for Hg, Hg(I) is stable in the form of $[Hg_2]^{2+}$.
- Zn and Cd are unlike metals in groups 4 to 11 because they do not exhibit variable oxidation states.
- In triads of groups 4-11, 2nd and 3rd row metals similar as a consequence of lanthanoid contraction, but Cd and Hg are somewhat unlike each other, the effects of lanthanoid contraction being relatively unimportant.
- Stability of halide complexes: $F^-$ complex most stable for hard $Zn^{2+}$, but $I^-$ complexes most stable for soft $Cd^{2+}$ and $Hg^{2+}$

22.23    Refer to Chapters 19 and 22 in H&S for relevant examples/discussions. Outline answer:
- Metal-metal bonding: high values of $\Delta_a H^o$ (see answer 22.2, p. 175) indicate a tendency to form strong M–M bonds; M–M bond strength increases down a triad and 2nd and 3rd row metals form more compounds with M–M bonds than do 1st row metals; multiple bonds are also important (see answers 22.10 and 22.11, p. 178).
- High coordination numbers: 2nd and 3rd row metal ions larger than 1st row congeners; examples of high coordination number include complexes with macrocyclic ligands which occupy equatorial plane in, e.g. pentagonal bipyramid.
- Metal halo clusters: focus on those with M–M bonding; show formation of 2D-layered structures, 3D-networks and discrete clusters; illustrate electron counting schemes to work out M–M bond orders (see answer 22.7, p. 177).
- Polyoxometallates: within 2nd and 3rd rows, you should discuss Nb, Ta, Mo and W-containing species; of these, Mo and W species have the wider chemistries; show solution equilibria, structural families including Lindqvist, α-Keggin and α-Dawson anions.

# 23 Organometallic compounds of *d*-block elements

23.1    (a) The μ notation refers to 'bridging' ligands; η notation gives the 'hapticity' of the ligand, i.e. the number of atoms in the ligand which interact with the metal (see Box 18.1 in H&S).

μ-CO bridges *two* M atoms (**23.1**); $\mu_4$-PR bridges *four* M (**23.2**); $\eta^5$-$C_5Me_5$ has all five C atoms interacting with M (**23.3**, Me groups omitted); $\eta^4$-$C_6H_6$ is coordinated so only *four* of the C atoms interact with M (**23.4**); $\mu_3$-H bridges *three* M (**23.5**).

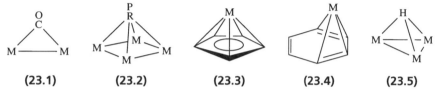

    **(23.1)**      **(23.2)**      **(23.3)**      **(23.4)**      **(23.5)**

(b) The Cp$^-$ ligand can coordinate through any number of C atoms and therefore can adopt an $\eta^1$, $\eta^2$, $\eta^3$, $\eta^4$ or $\eta^5$-mode. CO can be terminal or bridging, and the bridging mode may be μ- or $\mu_3$. Other modes are also possible for CO, e.g. **23.6**.
(c) PPh$_3$ always coordinates in a terminal mode and so is not versatile.

**(23.6)**

23.2    A synergic effect is one in which there is cooperation between two (or more) effects. With respect to metal carbonyl bonding, the synergic effect (or Dewar-Chatt-Duncanson model) refers to the two complementary components of the M–CO interaction (scheme **23.7**):

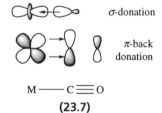

σ-donation

π-back donation

$M \longrightarrow C \equiv O$

**(23.7)**

- donation of a lone pair of electrons from CO to vacant M orbital (e.g. $d_{z^2}$); this provides the metal with an excess of electron density; electroneutrality principle indicates that this charge distribution is unsatisfactory;
- donation of electrons from filled M orbital (e.g. $d_{xz}$) to vacant CO $\pi^*$ orbital; this reduces the negative charge on the metal, and at the same time strengthens the M–C interaction.

23.3    Apply the bonding model in answer 23.2, remembering that back donation weakens the C–O bond because electrons occupy the CO $\pi^*$ orbital.
(a) $[V(CO)_6]^-$ and $Cr(CO)_6$ are isoelectronic, but the formal negative charge in $[V(CO)_6]^-$ leads to greater back donation, thereby weakening the C–O bond. The vibrational wavenumber is related to the force constant, *k*, of the bond:

$$\bar{v} = \frac{1}{2\pi c}\sqrt{\frac{k}{\mu}}$$

and the weaker the bond, the lower the wavenumber. The increased back bonding strengthens the M–C bond, and the vibrational wavenumber shifts to a higher value on going from $Cr(CO)_6$ to $[V(CO)_6]^-$.
(b) PPh$_3$ is shown in **23.8**. The Tolman cone angle gives a measure of the steric demands of the ligand when coordinated to the metal. Taking PPh$_3$ as a starting point, the introduction of Me groups in the 4-positions does not affect the steric demands of the coordinated ligand, but if Me groups are in the 2-positions, the ligand becomes more bulky, as is shown by the increased value of the cone angle.

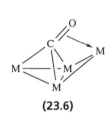

**(23.8)**

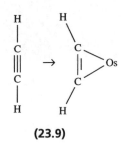

**(23.9)**

(c) MeCN is a more labile ligand than CO and the replacement of one or two CO in $Ru_3(CO)_{12}$ by MeCN gives preferential sites of substitution for $PPh_3$. Direct reaction of $Ru_3(CO)_{12}$ with MeCN does not occur; $Me_3NO$ acts as an oxidizing agent:

$$Me_3NO + CO \rightarrow Me_3N + CO_2$$

creating a vacant coordination site which is occupied by MeCN, the latter being readily replaced by $PPh_3$.

(d) Free $HC{\equiv}CH$ is linear with *sp* hybridized C atoms; $\eta^2$-coordination results in back donation from Os to the CC $\pi^*$ orbital reducing the C-C bond order (**23.9**) and making the C atoms more $sp^2$-like. $\therefore$ The Os–C–H bond angle increases.

**23.4**  (a) See structure **23.10**; the prefix $\mu$ tells you the ligand is bridging; other ligands are terminal.

(b) Negative $\delta$ is consistent with a metal hydride and the $\delta$ –10.2 signal can be assigned to the Ru–H–Ru hydride. $^1H$ nucleus couples to 4 equivalent $^{31}P$ nuclei (100%, $I = {}^1/_2$) to give a binomial quintet.

**(23.10)**

**23.5**  (a) The lengthening from 135 to 151 pm is a significant change and indicates substantial back donation into the CC $\pi^*$ orbital. The CN substituents are electron withdrawing and encourage back donation , i.e. back donation for $C_2(CN)_4 > C_2H_4$.

> See Figure 23.5 in H&S

(b) THF is a labile ligand and is readily replaced by $PPh_3$:

$$Mo(CO)_5(THF) + PPh_3 \xrightarrow{h\nu} Mo(CO)_5(PPh_3) + THF$$

In the $^{31}P$ NMR spectrum, the signal at $\delta$ –6 is assigned to free $PPh_3$; usually, on coordination, the signal shifts to higher frequency (more +ve $\delta$) and the signal at $\delta$ +37 is assigned to $Mo(CO)_5(PPh_3)$.

(c) Absorptions around 2000 cm$^{-1}$ are assigned to CO stretching modes. $PPh_3$ is a poorer $\pi$-acceptor than CO and, compared to $Fe(CO)_5$, in $Fe(CO)_3(PPh_3)_2$ more charge is available on Fe for back donation to the remaining 3 COs. This lowers the C–O bond strength and lowers $\bar{\nu}$ (CO). The fact there are 3 bands in $Fe(CO)_3(PPh_3)_2$ but only 2 in $Fe(CO)_5$ reflects a change in molecular symmetry.

**23.6**  The bonding scheme is shown in Figure 23.1. The allyl ligand has 3 $\pi$ orbitals, $\psi_1$ (Figure 23.1) is the lowest in energy and is occupied; electrons are donated to the metal. In $C_3H_5$, MO $\psi_2$ (non-bonding) is singly occupied; $\psi_3$ is vacant and back donation into this MO weakens the C–C bonds.

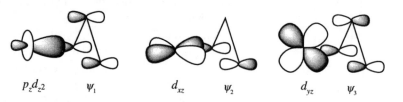

**Figure 23.1**  Orbital interactions in a metal $\pi$-allyl complex.

**23.7** 

**Each metal is taken as being in oxidation state zero; all ligands are treated as being neutral**

(a) $(\eta^5\text{-Cp})Rh(\eta^2\text{-}C_2H_4)(PMe_3)$ — Electron count $= 5 + 9 + 2 + 2 = 18$

(b) $(\eta^3\text{-}C_3H_5)_2Rh(\mu\text{-Cl})_2Rh(\eta^3\text{-}C_3H_5)_2$ — (bridging Cl is a 3-electron donor)
Electron count per Rh $= (2 \times 3) + 9 + 3 = 18$

(c) $Cr(CO)_4(PPh_3)_2$ — Electron count $= 6 + (4 \times 2) + (2 \times 2) = 18$

(d) $Fe(CO)_3(\eta^4\text{-}CH_2=CHCH=CH_2)$ — Electron count $= 8 + (3 \times 2) + 4 = 18$

(e) $Fe_2(CO)_9$ (structure **23.11**) — Electron count per Fe $= 8 + (3 \times 2) + 3 + 1 = 18$
(1 electron per metal from an M–M bond)

(f) $[HFe(CO)_4]^-$ — Electron count $= 1 + 8 + (4 \times 2) + 1 = 18$
(1 electron from the –ve charge)

(g) $[(\eta^5\text{-Cp})CoMe(PMe_3)_2]^+$ — Electron count $= 5 + 9 + 1 + (2 \times 2) - 1 = 18$
(subtract 1 electron for +ve charge)

(h) $RhCl(H)_2(\eta^2\text{-}C_2H_4)(PPh_3)_2$ — Electron count $= 9 + 1 + (2 \times 1) + 2 + (2 \times 2) = 18$
(terminal Cl is a 1-electron donor)

**(23.11)**

**23.8** Raman data show an *unbridged* Fe–Fe bond is present. Each Fe centre will obey the 18 electron rule and so the anion will be symmetrical and each Fe will have the same number of CO ligands: $[(CO)_xFe\text{–}Fe(CO)_x]^{2-}$. To find $x$, apply the 18 electron rule, allocating one electron from the –ve charge per Fe:

$$\text{Electron count per Fe} = 18 = 8 + 1 + 2x + 1 \qquad x = 4$$

$[A]^{2-}$ is $[Fe_2(CO)_8]^{2-}$. Likely structure is **23.12** with staggered ligands.

**(23.12)**

**23.9** NO can be a 1- or 3-electron donor, but to achieve an 18 electron count per Fe, each NO must act as a 3-electron donor. **23.13** and **23.14** are possible structures in which each Fe obeys the 18 electron rule. They could be distinguished by using IR spectroscopy; terminal and $\mu$-NO can be distinguished in the same way as one tells terminal and $\mu$-CO apart, i.e. $\mu$-NO at lower wavenumber.

**(23.13)**          **(23.14)**

**23.10** (a) The vibrational wavenumber of 1652 cm$^{-1}$ reflects the strength of the C=C bond in the free ligand. Coordination to the metal in $[PtCl_3(\eta^2\text{-MeCH=CH}_2)]^-$ involves donation of $\pi$-electrons and back donation of metal electrons into the C=C $\pi^*$ orbital (**23.15**). This weakens the CC interaction, shifting the vibrational wavenumber to 1504 cm$^{-1}$.

$\pi$-electron donation

$\pi$-back donation

**(23.15)**

(b) In $(\eta^5\text{-Cp})(\eta^1\text{-Cp})Fe(CO)_2$ (**23.16**), the 5 protons of the $\eta^5$-Cp ring are equivalent and give one signal in the $^1$H NMR spectrum. The $\eta^1$-Cp ring in the *static* structure contains 3 H environments, but in solution at 303 K, the observation of only one proton signal for this Cp ring indicates that the $\{(\eta^5\text{-Cp})Fe(CO)_2\}$-unit is 'hopping' from one CH unit to the next in the $\eta^1$-Cp ring at a rate that is faster than the NMR timescale.

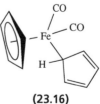

**(23.16)**

**(23.17)**

**(23.18)**

23.11    Using Wade's rules (see also p. 105 for application in boranes), break the cluster formula down into $Os(CO)_3$ units, extra (or fewer) CO ligands, and the charge (if applicable). Each $Os(CO)_3$ unit provides 2 electrons for cluster bonding, and each extra CO, 2 electrons.

$Os_7(CO)_{21}$        $7\,Os(CO)_3 = (7 \times 2) = 14$ electrons $= 7$ pairs

7 pairs of electrons are consistent with a 'parent' deltahedron with 6 vertices – an octahedron. There are 7 Os atoms to be accommodated, and so this is a capped *closo* structure. (No extra electrons are associated with the addition of the capping unit.) A capped octahedron (**23.17**) is suggested for $Os_7(CO)_{21}$.

$[Os_8(CO)_{22}]^{2-}$        $8\,Os(CO)_3 = (8 \times 2) = 16$ electrons
        Subtract 2 CO $= -4$ electrons
        2– charge $= 2$ electrons
        Total $= 16 - 4 + 2 = 14$ electrons $= 7$ pairs

7 pairs of electrons are consistent with a 'parent' octahedron. There are 8 Os atoms to be accommodated, and so this is a bicapped *closo* structure. Structure **23.18** is suggested for $[Os_8(CO)_{22}]^{2-}$; this is one of three possible isomers depending on the positions of the two capping units.

23.12    Total valence electron (ve) counts for selected low oxidation state metal cluster shapes are listed in Table 23.1.

**Table 23.1** Total valence electron (ve) counts for $M_x$ polyhedra.

| Cluster shape | ve |
|---|---|
| Triangle | 48 |
| Tetrahedron | 60 |
| Butterfly | 62 |
| Square | 64 |
| Trigonal bipyramid | 72 |
| Square-based pyramid | 74 |
| Octahedron | 86 |
| Trigonal prism | 90 |

(a) $[Ru_6(CO)_{18}]^{2-}$        Total ve $= (6 \times 8) + (18 \times 2) + 2 = 86$
(b) $H_4Ru_4(CO)_{12}$        Total ve $= (4 \times 1) + (4 \times 8) + (12 \times 2) = 60$
(c) $Os_5(CO)_{16}$        Total ve $= (5 \times 8) + (16 \times 2) = 72$
(d) $Os_4(CO)_{16}$        Total ve $= (4 \times 8) + (16 \times 2) = 64$
(e) $Co_3(CO)_9(\mu_3\text{-CCl})$        Total ve $= (3 \times 9) + (9 \times 2) + 3 = 48$
(f) $H_2Os_3(CO)_9(\mu_3\text{-PPh})$        Total ve $= (2 \times 1) + (3 \times 8) + (9 \times 2) + 4 = 48$
(g) $HRu_6(CO)_{17}B$        Total ve $= 1 + (6 \times 8) + (17 \times 2) + 3 = 86$
                        (interstitial B)
(h) $Co_3(\eta^5\text{-Cp})_3(CO)_3$        Total ve $= (3 \times 9) + (3 \times 5) + (3 \times 2) = 48$
(i) $Co_3(CO)_9Ni(\eta^5\text{-Cp})$        Total ve $= (3 \times 9) + (9 \times 2) + 10 + 5 = 60$

23.13    (a) Number of valence electrons *available* in $Os_5(CO)_{18} = (5 \times 8) + (18 \times 2) = 76$
Now consider the $Os_5$ framework shown in **23.19**.

**(23.19)**

Number of valence electrons *required* for a framework of 3 edge-sharing triangles $= (3 \times 48) - (2 \times 34) = 76$

                for 3          for 2 shared
                triangles        edges

Therefore the number of valence electrons required for this raft cluster is consistent with the number available.

(b) $[Ir_8(CO)_{22}]^{2-}$ (**23.20**) consists of two tetrahedra connected by an Ir–Ir bond which is a *localized* 2c-2e interaction. Write the formula as $[\{Ir_4(CO)_{11}\}_2]^{2-}$ and work out an electron count for each sub-cluster (the two sub-clusters are identical).

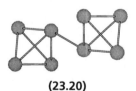

**(23.20)**

Total ve per sub-cluster $= (4 \times 9) + (11 \times 2) + 1 + 1 = 60$

        from the          from the Ir–Ir bond
        charge per        between the sub-
        sub-cluster        clusters

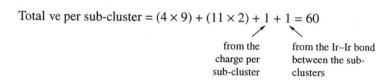

23.14    (a) Ligand substitution reaction, ethene displacing CO:

$$Fe(CO)_5 + C_2H_4 \xrightarrow{h\nu} Fe(CO)_4(\eta^2\text{-}C_2H_4)$$    Product structure **23.21**.

**(23.21)**

(b) Na/Hg is a reducing agent; Re–Re bond in $Re_2(CO)_{10}$ is cleaved:

$$Re_2(CO)_{10} \xrightarrow{Na/Hg} 2Na[Re(CO)_5]$$

$[Re(CO)_5]^-$ is isoelectronic and isostructural with $Fe(CO)_5$, i.e. trigonal bipyramidal.

(c) ONCl provides $[NO]^+$; it reacts with $[Mn(CO)_5]^-$ to give a nitrosyl complex:

$$Na[Mn(CO)_5] + ONCl \rightarrow NaCl + Mn(CO)_4(NO) + CO$$

If NO is a 3-electron donor, CO must be lost so that Mn obeys the 18 electron rule; product is trigonal bipyramidal, one of two isomers **23.22**.

**(23.22)**

(d) The carbonyl anion is protonated by phosphoric acid:

$$Na[Mn(CO)_5] + H^+ \rightarrow HMn(CO)_5 + Na^+$$    Product structure **23.23**.

(e) Ligand substitution reaction. More than one CO could be displaced, but on steric grounds, more than two may be unfavourable:

$$Ni(CO)_4 + PPh_3 \rightarrow Ni(CO)_3(PPh_3) \quad or \quad Ni(CO)_2(PPh_3)_2$$

**(23.23)**

$Ni(CO)_4$ is tetrahedral; the phosphine substituted products will also be tetrahedral.

23.15    Starting from the labelled compound shown in Figure 23.15 in H&S, migration of the 'inserted' CO could take place to a *cis* position with concomitant loss of a terminal CO. One of the *cis* sites contains $^{13}CO$ while three sites contain unlabelled CO:

The product distribution is therefore 75% **B** and 25% **A**: the Me group can never be *trans* to the labelled CO ligand (compare Figure 23.15 in H&S).

23.16    For specific examples to include in your answer, see Section 23.7 in H&S.
(a) Oxidative addition involves, for example, addition of XY with cleavage of the X–Y single bond, addition of a multiply bonded species with reduction in the bond order and formation of a metallacycle, addition of a C–H bond in an orthometallation step. The metal increases its oxidation state by 2 units and its coordination number by 2. Initially, a *cis* product results from oxidative addition; rearrangement may occur.
(b) Reductive elimination is the reverse of oxidative addition.
(c) The position of α-H atoms with respect to M are illustrated in **23.24**. Abstraction of one α-H atom from $L_nMCHR_2$ gives a carbene complex ($M=CR_2$) and abstraction of two α-H atoms from $L_nMCH_2R$ gives a carbyne complex ($M\equiv CR$).

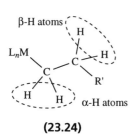

**(23.24)**

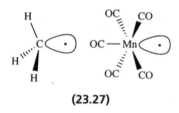

**(23.25)**

(d) The position of β-H atoms with respect to M are shown in **23.24**. β-Hydrogen elimination involves transfer of a β-H atom from (for example) an alkyl group to M, and the conversion of the σ-alkyl group to a π-bonded alkene.

(e) Alkyl, R, migration involves the migration of an R group to a *cis* CO ligand and formation of an acyl group; the reaction is facilitated by an incoming ligand (e.g. CO) which fills the coordination site vacated by R. The process is concerted.

(f) Orthometallation is a type of oxidative addition reaction involving the *ortho* C–H bond of a phenyl (or similar) ring. A cyclized product such as **23.25** results.

23.17  (a) Iron tricarbonyl complexes of 1,3-dienes are stable under various reaction conditions, and the Fe(CO)$_5$ precursor is cheap. The Fe(CO)$_3$ group acts as a protecting group for the diene group and this permits reactions to be carried out on other parts of the organic molecule (e.g. **23.26**); the diene group can be

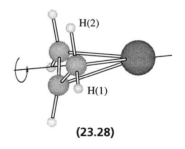

**(23.26)**

deprotected in a later step in the reaction sequence. The presence of the Fe(CO)$_3$ group facilities reactions of the 1,3-diene with nucleophiles with stereochemical control; the nucleophile is only able to attack the coordinated diene at the side away from the metal centre.

More examples in Section 23.10 in H&S

(b) Fullerenes (of which C$_{60}$ and C$_{70}$ have been the most studied) are carbon cages with C–C and C=C bonds. The C=C bonds function like alkenes, forming η$^2$-complexes, e.g.

$$Pt(\eta^2\text{-}C_2H_4)(PPh_3)_2 + C_{60} \rightarrow C_2H_4 + Pt(\eta^2\text{-}C_{60})(PPh_3)_2$$

(c) Two fragments are isolobal if they exhibit frontier MOs with the same symmetries, approximately the same energies, and containing the same number of electrons. CH$_3$ and Mn(CO)$_5$ are isolobal: each has one frontier MO of σ-symmetry, containing 1 electron (**23.27**). Combination of two CH$_3$ fragments gives C$_2$H$_6$ with a C–C single bond (overlap of σ-orbitals and pairing of the 2 electrons). Similarly, two Mn(CO)$_5$ fragments combine to give Mn$_2$(CO)$_{10}$ with an Mn–Mn single bond.

**(23.27)**

23.18  Diagram **23.28** illustrates the bonding mode of an η$^3$-allyl ligand to a metal centre. The plane containing the C and H atoms is approximately perpendicular to an axis connecting M and the centroid of the C$_3$-unit. Rotation of the allyl ligand about this axis (as in **23.28**) can never exchange atoms H(1) and H(2). Since $^1$H NMR spectra show these protons to be equivalent on the NMR time-scale, the rotation shown in **23.28** cannot be the mechanism for fluxionality.

H(2)

H(1)

**(23.28)**

23.19  The two C$_5$ rings in Cp$_2$Fe could be eclipsed or staggered; in this answer, they are taken as being eclipsed (as in the gas phase). The five π-MOs of C$_5$H$_5$ are drawn in Box 18.2 in H&S. For a bonding scheme in ferrocene, consider the orbital

**Figure 23.2** Orbital interactions between atomic orbitals of an Fe atom and the ligand group orbitals (LGOs) of a set of two eclipsed Cp ligands. Each of LGOs $\psi_3$, $\psi_4$, $\psi_7$ and $\psi_8$ is one of a degenerate pair of orbitals; $\psi_5$, $\psi_6$ and $\psi_9$ interact with the Fe $p_y$, $d_{yz}$ and $d_{x^2-y^2}$ atomic orbitals, while $\psi_{10}$ (like $\psi_8$) has no match.

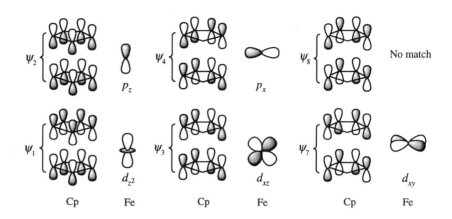

interactions between the Fe atom and a pair of Cp ligands – draw out the 10 ligand group orbitals that arise from taking combinations of the MOs of two Cp's. The left-hand column in Figure 23.2 shows the in-phase and out-of-phase combinations of the lowest energy $\pi$-MOs of the Cp ligands, and matches each (by symmetry) to one of the atomic orbitals of Fe. This operation is repeated using pairs of Cp $\pi$-MOs (see Box 18.2 in H&S). In Figure 23.2, the middle column shows in-phase and out-of-phase combinations of the second $\pi$-MO of Cp and matches each (by symmetry) to an Fe atomic orbital; see also figure caption. Of the last two LGOs shown in Figure 23.2, only one ($\psi_7$) can be matched by symmetry to an Fe atomic orbital; LGO $\psi_8$ becomes non-bonding in ferrocene; see also figure caption.

23.20    (a) $FeCl_3$ oxidizes Fe(0) to Fe(I) forming $[(\eta^5\text{-}Cp)_2Fe]^+[FeCl_4]^-$ and $FeCl_2$.
(b) Friedel-Crafts acylation of the Cp ring to give **23.29**, or of both rings to give $(\eta^5\text{-}C_5H_4C(O)Ph)_2Fe$.
(c) In the presence of Al and $AlCl_3$, toluene displaces a Cp ligand. Since toluene provides 6 $\pi$-electrons, the product is cationic with Fe retaining an 18-electron centre: product is $[(\eta^5\text{-}Cp)Fe(\eta^6\text{-}C_6H_5Me)]^+[AlCl_4]^-$ containing cation **23.30**.
(d) NaCl is eliminated, and a Co–Fe bond forms, giving $(\eta^5\text{-}Cp)(CO)_2Fe\text{–}Co(CO)_4$. This actually has two $\mu$-CO ligands, but from the data given, this cannot be deduced.

**(23.29)**

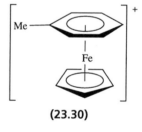

**(23.30)**

23.21    The singly substituted product is structurally related to **23.29** with Ph replaced by Me; in the doubly substituted product, both ligands are identical. Use $^1H$ NMR spectroscopy to distinguish between the products. The $\eta^5$-Cp ring gives a singlet ($\approx \delta$ 5); $\eta^5$-$C_5H_4C(O)$Me gives a singlet for the Me group and two multiplets for the ring protons (coupling between H(a) or H(a)' and H(b) and H(b)', see **23.31**). The compounds could also be distinguished by using $^{13}C$ NMR spectroscopy, or mass spectrometry; elemental analysis would distinguish compositions.

**(23.31)**

23.22    $C_6Me_6$ is a $\pi$-ligand, donating 6 (or 4 or 2) electrons to Ru. In $[(C_6Me_6)RuCl_2]_2$, use the 18-electron rule to work out the bonding modes of the Cl atoms and whether or not there is an Ru–Ru bond. Per Ru: Ru (8 ve), $\eta^6$-$C_6Me_6$ (6 ve), one terminal Cl (1 ve), 2 $\mu$-Cl (3 ve) gives 18 ve, suggesting structure **23.32** for **A**.

**(23.32)**

Product **B** contains $[(C_6Me_6)_2Ru]^{2+}$. Use 18-electron rule to work out the hapticity of the ligands: Ru (8 ve) needs 10 more electrons so, given there is a 2+ overall charge, each $C_6Me_6$ must provide 6 ve and be $\eta^6$-bonded (structure **23.33**). Conversion of **B** to **C** is reduction step (Na in liquid $NH_3$), suggesting **C** is $(C_6Me_6)_2Ru$. Apply 18-electron rule: Ru (8 ve) needs 10 more electrons so one ligand is $\eta^6$- and one is $\eta^4$-bonded (**23.34**).

**(23.33)**          **(23.34)**

23.23    (a) 2,5-Norbornadiene could act as a 2- or 4-electron donor although it is conformationally restricted and usually binds in an $\eta^4$-mode. Use the 18-electron rule to confirm that this is satisfactory in $Fe(CO)_3L$: Fe (8 ve), 3CO (6 ve) leaving 4 ve to be contributed by L. Structure **23.35** is suggested. Similarly, for $Fe(CO)_3L$ where L = heptatriene, an $\eta^4$-mode is expected on the basis of the 18-electron rule. Structure **23.36** is proposed.

(b) Moving from Fe to Mo reduces by two the number of electrons contributed by the metal. Therefore, to maintain an 18-electron count at Mo, heptatriene must provide an extra 2 electrons and is $\eta^6$.

(c) The reagent $[Ph_3C]^+$ abstracts $H^-$ from the organic ligand, converting heptatriene to $[C_7H_7]^+$. Neutral $\eta^7$-$C_7H_7$ is a 7 $\pi$-electron ligand and the product is $[(\eta^7$-$C_7H_7)Mo(CO)_3]^+$ – check the electron count at Mo: Mo (6 ve), $\eta^7$-$C_7H_7$ (7 ve), 3CO (6 ve), +ve charge (–1 ve) = 18 electrons.

**(23.35)**

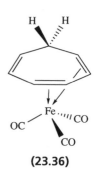

**(23.36)**

23.24    Draw out the frontier MOs of $C_{3v}$ $Fe(CO)_3$ and cyclobutadiene ($C_4H_4$) ligand (Figure 23.3). The $Fe(CO)_3$ unit provides 2 electrons and the $C_4H_4$ ligand has 4 $\pi$-electrons. Match fragment orbitals by looking at their symmetries: the Fe $p_z d_{z^2}$ hybrid overlaps with $\psi_1$ of the organic ligand, and the Fe $d_{xz}$ and $d_{yz}$ orbitals overlap with $\psi_2$ and $\psi_3$ of the ligand. Orbital $\psi_4$ becomes non-bonding in $(\eta^4$-$C_4H_4)Fe(CO)_3$. In the complex, there are 3 bonding MOs involving $Fe$–$C_4H_4$ character and the 6 electrons fully occupy these MOs – electrons are paired and the complex is diamagnetic.

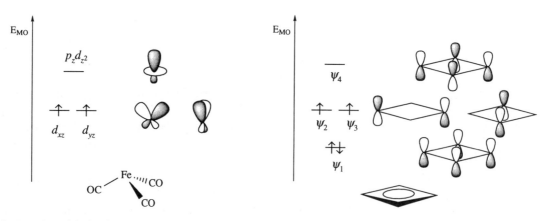

**Figure 23.3** Frontier orbitals of a $C_{3v}$ $Fe(CO)_3$ unit and of a cyclobutadiene ligand.

# 24 The *f*-block elements: lanthanoids and actinoids

**24.1** (a) Lanthanoids (Ln) are the 14 elements which follow lanthanum in the periodic table; they lie between La and Hf. Their valence electrons occupy the 4*f* level. Lanthanum, $[Xe]6s^2 5d^1 4f^0$, is strictly not a lanthanoid but is often classed with these metals. Physical and chemical properties of the lanthanoids tend to be similar; chemistry is generally that of the +3 oxidation state.

(b) See Section 24.5 in H&S for details – points to include:
- $Ln^{3+}$ ion sizes are similar, as are their properties, so separation is difficult.
- Separation is by solvent extraction using $(^nBuO)_3PO$ or cation-exchange; solution containing $Ln^{3+}$ ions is poured on to a cation-exchange resin column, where $Ln^{3+}$ exchanges with $H^+$ or $Na^+$ ions; resin-bound $Ln^{3+}$ removed using, for example, $[EDTA]^{4-}$ because formation constants of $[Ln(EDTA)]^-$ complexes (unlike those of resin-bound $Ln^{3+}$ which are nearly constant) increase regularly from $La^{3+}$ to $Lu^{3+}$, and the complexes can be eluted preferentially.

**24.2** $Ce^{3+}$ is $4f^1$. For the *f* orbital, quantum number $l = 3$ and the single electron may occupy one of 7 atomic orbitals. For the metal ion, $L = 3$ (see answer 20.12, p. 164 and apply the same method of working). For a single electron, spin quantum number $S = 1/2$ and the spin multiplicity is $(2S+1) = 2$. Quantum number $J$ takes values $(L+S)$ ... $|L-S|$; for $L = 3$ and $S = 1/2$, $J = 5/2$ or $7/2$; the ground state has the lower value of $J$. The full term symbol for the ground state is $^2F_{5/2}$.

> See also Box 20.5 in H&S

For calculating the magnetic moment, the spin-only formula is *not* applicable to lanthanoid ions. The appropriate formula is:

$$\mu_{eff} = g_J \sqrt{J(J+1)} \qquad g_J = 1 + \frac{S(S+1) - L(L+1) + J(J+1)}{2J(J+1)}$$

$$= 1 + \frac{\left(\frac{1}{2} \times \frac{3}{2}\right) - (3 \times 4) + \left(\frac{5}{2} \times \frac{7}{2}\right)}{2\left(\frac{5}{2} \times \frac{7}{2}\right)} = \frac{6}{7}$$

$$\mu_{eff} = g_J \sqrt{J(J+1)} = \frac{6}{7}\sqrt{\frac{5}{2} \times \frac{7}{2}} = 2.54 \ \mu_B$$

**24.3** Consider the reaction: $\qquad 3LnX_2 \rightarrow 2LnX_3 + Ln$

> See answers 5.17b and 10.10 for related lattice energy discussions

For a given metal, the relative stabilities of $LnX_2$ and $LnX_3$ depend on the difference between the lattice energies of $3LnX_2$ and $2LnX_3$. For a given lanthanoid, the variable that is of importance is the anion radius (lattice energy is inversely proportional to internuclear separation) and the difference in lattice energies is smallest for the largest anion, i.e. $I^-$.

**24.4** A 'saline' iodide $LnI_2$ would be of the form $Ln^{2+}(I^-)_2$ whereas a 'metallic' iodide is formulated $Ln^{3+}(e^-)(I^-)_2$. The electrical conductivity of the metallic iodide is higher than expected for a saline formulation. Saline iodides form for Ln = Sm, Eu, Yb, and metallic diiodides for Ln = La, Ce, Pr, Gd.

24.5     (a) The reaction to consider is:

$$[Ln(H_2O)_y]^{3+} + [EDTA]^{4-} \rightarrow [Ln(EDTA)(H_2O)_x]^- + (y-x)H_2O$$

The negligible variation in values of $\Delta H^\circ$ for this reaction as $Ln^{3+}$ varies arises because of the similarity in size of the metal ions across the first row of the *f*-block; given a constant set of ligands as here, and a fairly constant cation size, then the metal ion-ligand interactions will be similar. The fact that $\Delta H^\circ \approx 0$ indicates that solvent-ligand and metal ion-ligand interactions are similar for both reactants and products.

➤ (b) In *aqueous* solution, the equilibrium to be considered is:

See also Section 7.3 in H&S

$$Ce^{4+}(aq) + e^- \rightleftharpoons Ce^{3+}(aq)$$

but in acidic solution (at constant pH), anions are present which may act as ligands to the metal centres. Complex formation by anions follows the series $Cl^- > [SO_4]^{2-} > [NO_3]^- > [ClO_4]^-$, and the differences in coordinating ability affect the ease of reduction of $Ce^{4+}$.

(c) Perovskite is $CaTiO_3$ – a mixed metal oxide (see Figures 5.7 and 5.8, p. 47) containing $Ca^{2+}$, $Ti^{4+}$ and $O^{2-}$. $BaCeO_3$ must also be a mixed metal oxide and contain $Ba^{2+}$, $Ce^{4+}$ and $O^{2-}$ – it is *not* a salt of type $Ba^{2+}[CeO_3]^{2-}$.

24.6     For details, see Section 24.4 in H&S. Points to include:
- Spin-orbit coupling of greater importance than crystal field splitting; terms differing only in $J$ value are sufficiently different in energy to be separated in the electronic spectrum; high values of $L$ for some $f^n$ ions; large number of electronic transitions possible, so a large number of lines in the spectrum.
- 4*f* electrons are well shielded and little influenced by ligand environment (in contrast to *d* electrons); this leads to sharp lines associated with *f*-*f* transitions, with wavenumbers similar to those of the gas phase $Ln^{3+}$ ions.
- Weak absorptions reflect fact that probabilities of *f*-*f* transitions are low (little *d*-*f* mixing – needed for selection rule $\Delta l = \pm 1$).
- Absorptions due to 4*f*-5*d* transitions are broad and influenced by ligand environment.

24.7     For further details, see Sections 24.3 and 24.7 in H&S; points to include:
- $Ln^{3+}$ ions are large and high coordination numbers (>6) are possible.
- Coordination number is controlled by steric effects, not $f^n$ configuration.
- $Ln^{3+}$ ions are hard and favour hard donors such as *O*.
- Coordination numbers typically range from 6 to 12; aqua ions typically 9-coordinate; highest numbers tend to involve didentate ligands, e.g. $[NO_3]^-$.
- Low coordination numbers can be stabilized using e.g. $[R_2N]^-$ ligands, but R must be sterically demanding.

24.8     First, note that $[NCS]^-$ could coordinate through N or S, but since $Ln^{3+}$ ions are hard, *N*- rather than *S*-coordination is expected. $[Ln(NCS)_6]^{3-}$ is likely to be octahedral.

➤ $[Ln(NCS)_7(H_2O)]^{4-}$ is 8-coordinate and could be dodecahedral, square antiprismatic,

See Figures 19.7 and 19.8 in H&S

cubic or a distorted version of one of these; a hexagonal bipyramid is less likely for monodentate ligands – tends to be imposed by a macrocyclic ligand (but see answer 24.13a). 7-coordinate structures are hard to predict and $[Ln(NCS)_7]^{4-}$ could be pentagonal bipyramidal, capped octahedral or a distorted variant of one of these.

24.9    (a) For further details, see Section 24.8 in H&S. Points to include:
- Lithium alkyls generally used to introduce $\sigma$-bonded alkyls:

$$3LiR + LnCl_3 \rightarrow 3LiCl + LnR_3$$

In coordinating solvents (e.g. THF, DME or TMED), $LnR_3$ can react with excess LiR to give $[LiL_n]_x[LnR_{3+x}]$ where L is the solvent.
- Cp derivatives prepared by general reaction:

$$3Na[Cp] + LnCl_3 \rightarrow 3NaCl + Cp_3Ln$$

or $Cp_2LnCl$ or $CpLnCl_2$ may be formed if stoichiometry of reactants is altered.
- $\eta^5$-Cp mode usual, but complexes may be monomeric, e.g. $Cp_3Tm$, or polymeric, e.g. $Cp_3Pr$, but in presence of a donor solvent, monomeric product may be isolated, e.g. $Cp_3Pr(NCMe)_2$.

(b) $K_2[C_8H_8]$ provides the $[C_8H_8]^{2-}$ ligand which can coordinate in an $\eta^8$-mode, forming sandwich complexes. Reactions proposed are:

$$SmCl_3 + 2K_2[C_8H_8] \rightarrow 3KCl + K[(\eta^8\text{-}C_8H_8)_2Sm]$$

$$SmI_2 + 2K_2[C_8H_8] \rightarrow 2KI + K_2[(\eta^8\text{-}C_8H_8)_2Sm]$$

THF    DME    TMED

24.10    (a) Figure 24.6 in H&S shows potential diagrams for U, Np, Pu and Am. The $E^o$ values show that $Am^{3+}$ is much more difficult to oxidize than U, Np or Pu in oxidation states +3 to +5. The first step in separation could therefore be to oxidize U, Np, Pu to $[MO_2]^{2+}$ using $Ce^{4+}$ (see **24.1**) and precipitate Am(III) as $AmF_3$ along with $CeF_3$. To separate $AmF_3$ from $CeF_3$, $Am^{3+}$ could be oxidized by $[S_2O_8]^{2-}$ and the product extracted. Alternatively, $Am^{3+}$ could be separated from $Ce^{3+}$ using a cation-exchange column, eluting with $H_4EDTA$ (see answer 24.1b).

$$Ce^{4+} + e^- \rightleftharpoons Ce^{3+}$$
$$E^o = +1.72 \text{ V}$$
**(24.1)**

(b) $NpO_2(ClO_4)_2$ contains $[NpO_2]^{2+}$. Zn amalgam is a good reducing agent and should reduce $[NpO_2]^{2+}$ to $Np^{3+}$ (see Figure 24.6 in H&S). At pH 0 (i.e. 1 M $HClO_4$), $O_2$ (see **24.2**) should oxidize $Np^{3+}$ to $[NpO_2]^+$ (with some being oxidized to $[NpO_2]^{2+}$) although oxidation might be slow.

$$O_2 + 4H^+ + 4e^- \rightleftharpoons 2H_2O$$
$$E^o = +1.23 \text{ V}$$
**(24.2)**

24.11    Solution **X** contains 21.4 g of U(VI) per $dm^3$

$$\therefore \text{ Amount of U} = \frac{21.4}{238.03} = 9.00 \times 10^{-2} \text{ mol dm}^{-3}$$

After Zn amalgam reduction followed by $O_2$ oxidation to $U^{n+}$, the 25.00 $cm^3$ aliquot is oxidized to $[UO_2]^{2+}$ by 37.5 $cm^3$ of 0.1200 mol $dm^{-3}$ $Ce^{4+}$ (see **24.1**).

Amount of $U^{n+}$ in 25.00 $cm^3$ = $25.00 \times 10^{-3} \times 9.00 \times 10^{-2} = 2.25 \times 10^{-3}$ moles

Amount of $Ce^{4+}$ needed to oxidize $U^{n+}$ to U(VI) = $37.5 \times 0.1200 \times 10^{-3}$
$$= 4.50 \times 10^{-3} \text{ moles}$$

Ratio of moles $U^{n+} : Ce^{4+} = 1 : 2$

The $Ce^{4+}$ reduction to $Ce^{3+}$ is a 1-electron process, and therefore the oxidation of $U^{n+}$ to U(VI) must be a 2-electron process. Hence, $U^{n+}$ is $U^{4+}$.

In the next set of reactions, 100 $cm^3$ of **X** is converted to $U^{4+}$ :

Amount of $U^{4+} = 0.1 \times 9.00 \times 10^{-2} = 9.00 \times 10^{-3}$ moles

Treatment with aq. KF precipitates $UF_4$, deduced from the amount of substance:

$$\text{Amount of } UF_4 = \frac{2.826}{238.03 + (19.00 \times 4)} = 9.00 \times 10^{-3} \text{ moles}$$

UF$_4$ is now heated at 1070 K with O$_2$ and yields 1.386 g of product, an aqueous solution of which contains F$^-$, precipitated as 2.355 g PbClF.

$$\text{Amount of F}^- = \text{amount of PbClF} = \frac{2.355}{207.19 + 35.45 + 19.00}$$

$$= 9.00 \times 10^{-3} \text{ moles}$$

∴ Not all the fluorine present in the products from $9.00 \times 10^{-3}$ moles of UF$_4$ is present as F$^-$ in solution. Possible products from the reaction of UF$_4$ with O$_2$ are UF$_6$ (a volatile solid which would vaporize at 1070 K) and UO$_2$F$_2$ (ionizes to [UO$_2$]$^{2+}$ and 2F$^-$ in solution). $9.00 \times 10^{-3}$ moles F$^-$ could originate from $4.50 \times 10^{-3}$ moles UO$_2$F$_2$. This would suggest that $4.50 \times 10^{-3}$ moles UF$_6$ were also formed, but were not collected in the solid product (see above). Check the amount of substance:

Mass of $4.50 \times 10^{-3}$ moles UO$_2$F$_2$ = $4.50 \times 10^{-3} \times \{238.03 + (2 \times 19.00) +$

$$(2 \times 15.99)\}$$

$$= 1.386 \text{ g}$$

This matches that observed, and so the reaction of UF$_4$ with O$_2$ is:

$$2UF_4 + O_2 \rightarrow UF_6 + UO_2F_2$$

24.12   (a) F$_2$ oxidizes U(IV) to U(VI) on heating:

$$UF_4 + F_2 \rightarrow UF_6$$

(b) SOCl$_2$ acts as a chlorinating agent, then H$_2$ as a reducing agent:

$$Pa_2O_5 + 5SOCl_2 \rightarrow 2PaCl_5 + 5SO_2$$

$$2PaCl_5 + H_2 \rightarrow 2PaCl_4 + 2HCl$$

(c) H$_2$ reduces UO$_3$ on heating:

$$UO_3 + H_2 \rightarrow UO_2 + H_2O$$

(d) UCl$_5$ disproportionates on heating:

$$2UCl_5 \rightarrow UCl_4 + UCl_6$$

(e) NaOC$_6$H$_2$-2,4,6-Me$_3$ is a source of an aryloxide ligand (24.3) that is sterically demanding, and reaction with UCl$_3$ gives a 3-coordinate complex:

$$UCl_3 + 3NaOC_6H_2\text{-}2,4,6\text{-}Me_3 \rightarrow 3NaCl + U(OC_6H_2\text{-}2,4,6\text{-}Me_3)_3$$

**(24.3)**

24.13   (a) Cs$_2$[NpO$_2$(acac)$_3$] contains Cs$^+$ and [NpO$_2$(acac)$_3$]$^{2-}$ ions; [acac]$^-$ is didentate, and [NpO$_2$(acac)$_3$]$^{2-}$ is 8-coordinate: hexagonal bipyramidal with axial oxo-ligands.
(b) Np(BH$_4$)$_4$ contains 4 [BH$_4$]$^-$ ligands which can coordinate through one, two or three H atoms (i.e. $\eta^1$, $\eta^2$ or $\eta^3$-modes). Given that Np$^{4+}$ can tolerate a high coordination number, suggest that Np(BH$_4$)$_4$ contains four $\eta^3$-[BH$_4$]$^-$ ligands (24.4).
(c) Discrete guanidinium cations and [ThF$_3$(CO$_3$)$_3$]$^{5-}$ anions are present. Planar [CO$_3$]$^{2-}$ acts as a didentate ligand, giving 9-coordinate [ThF$_3$(CO$_3$-$O$,$O'$)$_3$]$^{5-}$.
(d) [Li(DME)]$_3$[LuMe$_6$] contains [Li(DME)]$^+$ and octahedral [LuMe$_6$]$^{3-}$ ions. Likely to be cation-anion interactions as shown in Figure 24.7a in H&S.

**(24.4)**

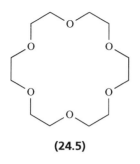

**(24.5)**

(e) $Sm\{CH(SiMe_3)_2\}_3$ is a 3-coordinate organometallic complex with sterically demanding R groups. Trigonal planar might be expected, but by analogy with similar complexes, trigonal pyramidal is possible in the solid state.

(f) By analysis, the compound is $[UO_2][CF_3SO_3]_2 \cdot 2(18\text{-crown-}6) \cdot 5H_2O$. It is reasonable to assume $[CF_3SO_3]^-$ ions are present. This leaves $[UO_2]^{2+}$ which will *not* be present as a discrete entity; it will have *trans*-oxo ligands and could be coordinated by the 18-crown-6 (**24.5**) giving a hexagonal bipyramidal complex with $5H_2O$ of crystallization, or by the $H_2O$ ligands with the crown ether as solvate of crystallization. The complex cation is actually $[UO_2(H_2O)_5]^{2+}$.

24.14   Look back to answer 2.7, p. 17, to check details of the types of decay.
(a) Step 1: gain a neutron: $\mathbf{A} = {}^{239}U$; step 2: lose a β-particle: $\mathbf{B} = {}^{239}Np$; step 3: lose a β-particle: $\mathbf{C} = {}^{239}Pu$.
(b) Step 3: lose a β-particle: $\mathbf{F} = {}^{242}Cm$; step 2: gain a neutron: $\mathbf{E} = {}^{241}Am$; step 1: lose a β-particle: $\mathbf{D} = {}^{241}Pu$.

24.15   Balance each equation using the values of $Z$ and mass numbers.

(a)   ${}^{253}_{99}Es + {}^{4}_{2}He \rightarrow {}^{256}_{101}Md + {}^{1}_{0}n$

(b)   ${}^{244}_{94}Pu + {}^{16}_{8}O \rightarrow {}^{255}_{102}No + 5{}^{1}_{0}n$

(c)   ${}^{249}_{98}Cf + {}^{11}_{5}B \rightarrow {}^{256}_{103}Lr + 4{}^{1}_{0}n$

(d)   ${}^{248}_{96}Cm + {}^{18}_{8}O \rightarrow {}^{261}_{104}Rf + 5{}^{1}_{0}n$

(e)   ${}^{249}_{98}Cf + {}^{18}_{8}O \rightarrow {}^{263}_{106}Sg + 4{}^{1}_{0}n$

24.16   (a) The formulae suggest Th(II), Th(III) and Th(IV) compounds, but $ThI_2$, $ThI_3$ and $ThI_4$ are all Th(IV) iodides. $ThI_4$ is a saline halide; $ThI_2$ and $ThI_3$ are metallic halides formulated as $Th^{4+}(I^-)_2(e^-)_2$ and $Th^{4+}(I^-)_3(e^-)$. See also answer 24.4.
(b) In the solid state, discrete $[UO_2]^{2+}$ (**24.6**) are not present but are complexed by further ligands around the equatorial plane, e.g. $[UO_2(H_2O)_5]^{2+}$.
(c)
$$UCl_4 + 4NaOR \rightarrow 4NaCl + U(OR)_4$$

**(24.6)**

Product is usually of the form $U(OR)_4$ only if R is very bulky, e.g. $R = 2,6\text{-}^tBu_2C_6H_3$; with less bulky R groups, donor solvents, L, may coordinate giving e.g. $U(OR)_4L_x$.

24.17   (a) In each example, NaCl or LiCl is eliminated as the driving force for reaction:

(i)   $(\eta^5\text{-Cp})_3ThCl + Na[Ru(CO)_2(\eta^5\text{-Cp})] \rightarrow NaCl + (\eta^5\text{-Cp})_3Th–Ru(CO)_2(\eta^5\text{-Cp})$

(ii)   $(\eta^5\text{-Cp})_3ThCl + LiCHMeEt \rightarrow LiCl + (\eta^5\text{-Cp})_3ThCHMeEt$

(iii)   $(\eta^5\text{-Cp})_3ThCl + LiCH_2Ph \rightarrow LiCl + (\eta^5\text{-Cp})_3ThCH_2Ph$

(b) $(\eta^5\text{-Cp})_2ThCl_2$ can undergo a redistribution reaction:

$$2(\eta^5\text{-Cp})_2ThCl_2 \xrightarrow{THF} (\eta^5\text{-Cp})_3ThCl + (\eta^5\text{-Cp})ThCl_3(THF)_2$$

Replacing H atoms by Me groups gives the more sterically hindered compound $(\eta^5\text{-C}_5Me_5)_2ThCl_2$ which does not undergo an analogous redistribution reaction.

(c) Elimination of KI occurs; $[C_8H_8]^{2-}$ coordinates in an $\eta^8$-mode. Likely that the coordination sphere will be too crowded if 3 THF ligands are retained. Propose that the product is $(\eta^5\text{-}C_5Me_5)(\eta^5\text{-}C_8H_8)U(THF)_x$ with $x = 1$ or 2. In practice, $x = 1$.

24.18    (a) This is a U(IV) complex and so start with $UCl_4$. Treatment with the Grignard reagent $C_3H_5MgCl$ in an ether solvent:

$$UCl_4 + 4C_3H_5MgCl \xrightarrow{Et_2O} 4MgCl_2 + U(\eta^3\text{-}C_3H_5)_4$$

(b) Treatment with $H^+$ removes propene, leaving a vacant coordination site to be filled by $Cl^-$:

$$U(\eta^3\text{-}C_3H_5)_4 + HCl \rightarrow U(\eta^3\text{-}C_3H_5)_3Cl + CH_3CH=CH_2$$

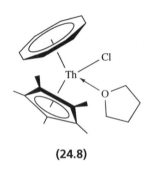

**(24.8)**

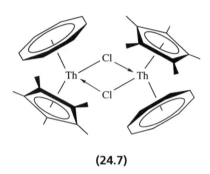

**(24.7)**

(c) In dimer **24.7**, the presence of the *bridging* chloro ligands increases the Th coordination number with respect to a monomer $(\eta^5\text{-}C_5Me_5)(\eta^8\text{-}C_8H_8)ThCl$. In the presence of a coordinating solvent such as THF, the monomer is stabilized because the solvent can fill a vacant coordination site as in **24.8**. This is a general strategy for stabilizing monomeric organothorium and uranium complexes.

24.19    (a) In a non-stoichiometric oxide, vacant $O^{2-}$ sites need to be countered by an appropriate decrease in oxidation state of some metal ions so that the solid retains electrical neutrality. An excess of $O^{2-}$ ions can similarly be offset by the presence

➤ **Non-stoichiometric oxides: see Section 27.2 in H&S**

of some higher oxidation state metal ions. For actinoids, a range of oxidation states with similar stabilities is available, making the formation of non-stoichiometric oxides common. For lanthanoids, the +3 oxidation state dominates, with the result that variation in oxidation state is not favourable.

(b) $[NpO_6]^{5-}$ contains Np(VII). This needs to be stabilized against reduction which involves loss of oxo ligands *in the presence of* $H^+$, e.g.

$$[NpO_6]^{5-} + 8H^+ + e^- \rightleftharpoons [NpO_2]^{2+} + 4H_2O$$

In strongly alkaline solution, no $H^+$ is available, and $[NpO_6]^{5-}$ is stabilized with respect to reduction.

➤ **Disproportionation: see Section 7.4 in H&S**

(c) Figure 24.6 in H&S shows that aqueous solutions of $Pu^{4+}$ should contain significant concentrations of disproportionation products. The fact that this does not occur in the presence of an excess of molar $H_2SO_4$ indicates that $[SO_4]^{2-}$ must stabilize $Pu^{4+}$ more than it does $[PuO_2]^{2+}$, $[PuO_2]^+$ or $Pu^{3+}$ – a consequence of charge effects.

24.20    Refer to Sections 24.8 and 24.11 in H&S for specific details. Major differences between *d*- and *f*-block organometallics to highlight are:
- CO complexes are extremely common within the *d*-block, but rare within the *f*-block.
- Large size of the *f*-block metals allows higher coordination numbers, e.g. $(\eta^5\text{-}Cp)_4Th$ but $(\eta^5\text{-}Cp)_2Fe$, and involvment of larger organic ligands in metallocenes, e.g. $(\eta^8\text{-}C_8H_8)_2U$.

# 25 *d*-Block metal complexes: reaction mechanisms

25.1 (a) A complete reaction pathway from reactants to products usually consists of a series of *elementary steps*, e.g. in a free radical reaction, the pathway involves initiation, propagation and termination steps.

(b) The elementary step with the highest activation energy is the rate determining (or slow) step. The molecularity of this step determines the observed kinetics of the overall reaction.

(c) The activation energy, $E_a$, is the energy barrier that must be overcome for a reaction, or step in a reaction pathway, to proceed (see Figure 25.1).

(d) An intermediate lies in a local energy minimum along a reaction pathway (Figure 25.1); it can be detected (e.g. by spectroscopic methods) and, possibly, isolated.

(e) A transition state lies at an energy maximum along a reaction pathway (Figure 25.1) and cannot be isolated.

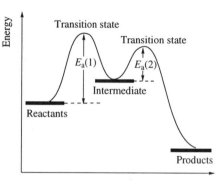

**Figure 25.1** Reaction profile for a two-step pathway; $E_a$ = activation energy.

(f) A rate equation shows the dependence of the rate of reaction on the concentrations of species in the reaction; usually expressed in terms of rate of loss of a starting material or rate of formation of a product.

(g) For a reaction: $\quad A \rightarrow$ products

a zero order dependence on A means that the rate of reaction is not affected by variation in [A]:

$$-\frac{d[A]}{dt} = k \qquad \text{where } k = \text{rate constant}$$

The integrated form of the rate law is:

$$[A] = [A]_0 - kt \qquad \text{where } [A]_0 = [A] \text{ at time } t = 0$$

If the rate is first order with respect to A:

$$-\frac{d[A]}{dt} = k[A] \qquad \text{or} \qquad \ln[A] = \ln[A]_0 - kt$$

If the rate is second order with respect to A:

$$-\frac{d[A]}{dt} = k[A]^2 \qquad \text{or} \qquad \frac{1}{[A]} = \frac{1}{[A]_0} + kt$$

(h) A nucleophile is a species which donates electrons (e.g. $Cl^-$), being attracted to a centre such as a metal which is $\delta^+$.

**25.2**   The profiles should be of the same form as shown in Figure 25.1 on the previous page. For the dissociative process, the reactants are $ML_xX$ and Y, and the intermediate is $ML_x$ (with Y still as a reactant and X having departed); the products are $ML_xY$ and X. For the associative process, the reactants are $ML_xX$ and Y, and the intermediate is $ML_xXY$; the products are $ML_xY$ and X.

**25.3**   For further details, see Section 25.3 in H&S. Points to include and discuss are:
- Negative values of $\Delta V^{\ddagger}$ and $\Delta S^{\ddagger}$.
- Rate constants for substitution of $H_2O$ for $Cl^-$ in $[PtCl_{4-x}(NH_3)_x]^{(2-x)}$ are about the same.
- Solvent effects (see equation 25.13 in H&S).
- Effects of steric demands of entering and leaving groups.
- Retention of stereochemistry.

**25.4**   Reaction is substitution by py in $[Rh(cod)(PPh_3)_2]^+$, and from the general rate law for a square planar complex:

$$-\frac{d[Rh(cod)(PPh_3)_2{}^+]}{dt} = k_1[Rh(cod)(PPh_3)_2{}^+] + k_2[Rh(cod)(PPh_3)_2{}^+][py]$$

where: $k_{obs} = k_1 + k_2[py]$

Data suggest that the pathways (where S = solvent) are:

$$[Rh(cod)(PPh_3)_2]^+ + py \xrightarrow{k_2} [Rh(cod)(PPh_3)(py)]^+ + PPh_3$$

competes with:

$$\left\{ \begin{array}{l} [Rh(cod)(PPh_3)_2]^+ + S \xrightarrow{k_1} [Rh(cod)(PPh_3)S]^+ + PPh_3 \\ [Rh(cod)(PPh_3)S]^+ + py \xrightarrow{fast} [Rh(cod)(PPh_3)(py)]^+ + S \end{array} \right.$$

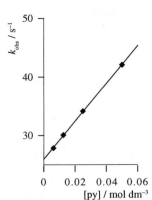

**Figure 25.2** Plot of the kinetic data for question 25.4.

Plot values of $k_{obs}$ against [py] (Figure 25.2) and confirm a linear relationship between $k_{obs}$ and [py]. This corresponds to the equation for $k_{obs}$ given above; gradient of the line in Figure 25.2 = $k_2$ = 322 dm$^3$ mol$^{-1}$ s$^{-1}$; intercept = $k_1$ = 25 s$^{-1}$.

**25.5**   (a) The chosen routes to *cis* and *trans*-$[PtCl_2(NH_3)(NO_2)]^-$ depend on the relative *trans*-effects of the $Cl^-$, $NH_3$ and $[NO_2]^-$ ligands. The relative abilities of the ligands to direct *trans*-substitution are $[NO_2]^- > Cl^- > NH_3$. To prepare the *cis*-isomer, $NH_3$ must be substituted first so that $Cl^-$ directs substitution *trans* to itself and *cis* to $NH_3$:

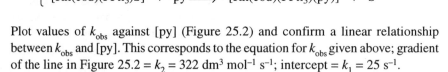

To prepare the *trans*-isomer, $[NO_2]^-$ must be substituted first so that this ligand directs substitution *trans* to itself:

(b) The first step is formation of $[PtCl_3(PEt_3)]^-$. Of $Cl^-$ and $PEt_3$, $PEt_3$ has the greater *trans*-effect, and directs substitution of the next $PEt_3$ ligand *trans* to the first $PEt_3$:

$$\begin{bmatrix} & Cl & \\ & | & \\ Cl-&Pt&-Cl \\ & | & \\ & Cl & \end{bmatrix}^{2-} \xrightarrow[-Cl^-]{PEt_3} \begin{bmatrix} & PEt_3 & \\ & | & \\ Cl-&Pt&-Cl \\ & | & \\ & Cl & \end{bmatrix}^- \xrightarrow[-Cl^-]{PEt_3} \begin{array}{c} PEt_3 \\ | \\ Cl-Pt-Cl \\ | \\ PEt_3 \end{array}$$

**25.6**    (a) Proposed associative mechanism for *trans*-$[PtL_2Cl_2]$ going to *trans*-$[PtL_2ClY]^+$:

$$\begin{array}{c} L \\ | \\ Cl-Pt-Cl \\ | \\ L \end{array} \xrightarrow{Y} \begin{array}{c} L \\ | \\ Cl-Pt^{\cdots\cdots Cl} \\ | \searrow Y \\ L \end{array} \xrightarrow{-Cl} \begin{bmatrix} L \\ | \\ Cl-Pt-Y \\ | \\ L \end{bmatrix}^+$$

intermediate

(b) If the 5-coordinate intermediate is sufficiently long-lived, it may be stereochemically non-rigid in solution (Berry pseudo-rotation); this would prevent the stereoselectivity shown above and both *cis* and *trans*-isomers would form.

**25.7**    $$[Co(NH_3)_5(H_2O)]^{3+} + X^- \rightarrow [Co(NH_3)_5X]^{2+} + H_2O$$

Rate law is:    $$\frac{d[Co(NH_3)_5X^{2+}]}{dt} = k_{obs}[Co(NH_3)_5(H_2O)^{3+}][X^-]$$

The fact that $\Delta V^{\ddagger}$ is positive suggests a dissociative ($D$ or $I_d$) mechanism, but the the rate law suggests an associative mechanism. Application of Eigen-Wilkins mechanism rationalizes this apparent contradiction. An 'encounter complex' is formed between $[Co(NH_3)_5(H_2O)]^{3+}$ and $X^-$ in a pre-equilibrium step (equilibrium constant $K_E$); $H_2O$ then leaves in the rate-determining step. Apply equations 25.23 to 25.30 in H&S to establish that second order kinetics (i.e. the rate equation above) hold at low concentrations of $X^-$ when $K_E[X^-] << 1$:

$$\frac{d[Co(NH_3)_5X^{2+}]}{dt} = \frac{kK_E[Co(NH_3)_5(H_2O)^{3+}]_{total}[X^-]}{1 + K_E[X^-]}$$

$$= kK_E[Co(NH_3)_5(H_2O)^{3+}]_{total}[X^-]$$

where $kK_E = k_{obs}$

**25.8**    (a) First step involves monosubstitution in octahedral complex in which all $H_2O$ ligands are equivalent; there is only one possible product (**25.1**). In the next step, the product is determined by the fact that the *trans*-effect of $Cl^- > H_2O$. Substitution of $H_2O$ by $Cl^-$ in **25.1** is directed by the coordinated $Cl^-$ and the product is *trans*-$[RhCl_2(H_2O)_4]^+$. For the next step, all the $H_2O$ ligands are equivalent and the product has to be the *mer*-isomer:

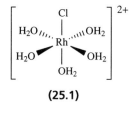

**(25.1)**

$$\begin{bmatrix} & Cl & \\ H_2O_{\prime\prime\prime}, & | & ,,\backslash\backslash OH_2 \\ & Rh & \\ H_2O^{\blacktriangleright} & | & {}^{\blacktriangleleft}OH_2 \\ & Cl & \end{bmatrix}^+ \xrightarrow[-H_2O]{Cl^-} \begin{array}{c} Cl \\ H_2O_{\prime\prime\prime}, \overset{*}{\underset{Rh}{|}} ,,\backslash\backslash OH_2 \\ H_2O^{\blacktriangleright} | {}^{\blacktriangleleft}Cl \\ Cl \end{array}$$

In the last step, the site of substitution is controlled by the *trans*-effect of the $Cl^-$ ligand *trans* to a $H_2O$ ligand (marked by * above) giving *trans*-$[RhCl_4(H_2O)_2]^-$.

(b) $[RhCl_5(H_2O)]^{2-}$ can be prepared from:

$$trans\text{-}[RhCl_4(H_2O)_2]^- + Cl^- \rightarrow [RhCl_5(H_2O)]^{2-} + H_2O$$

or:

$$[RhCl_6]^{3-} + H_2O \rightarrow [RhCl_5(H_2O)]^{2-} + Cl^-$$

Preparation of $cis\text{-}[RhCl_4(H_2O)_2]^-$ cannot start from $mer\text{-}[RhCl_3(H_2O)_3]$ because of the stronger *trans*-effect of $Cl^-$ with respect to $H_2O$. A suitable reaction is:

$$[RhCl_5(H_2O)]^{2-} + H_2O \rightarrow cis\text{-}[RhCl_4(H_2O)_2]^- + Cl^-$$

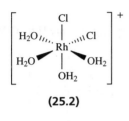

**(25.2)**

Similarly, $fac\text{-}[RhCl_3(H_2O)_3]$ cannot be prepared by treating $cis\text{-}[RhCl_2(H_2O)_4]^+$ **(25.2)** with $Cl^-$ because the *trans*-effect of coordinated $Cl^-$ will direct substitution to give the *mer*-isomer. A suitable synthesis is:

$$cis\text{-}[RhCl_4(H_2O)_2]^- + H_2O \rightarrow fac\text{-}[RhCl_3(H_2O)_3] + Cl^-$$

25.9    Anation is the substitution of an uncharged ligand (here $H_2O$) by an anionic ligand such as $Cl^-$. Co, Rh and Ir are all group 9 metals, and $M^{3+}$ is octahedral $d^6$. CFSE increases down a triad consistent with the rate of substitution following the trend $[Co(H_2O)_6]^{3+} > [Rh(H_2O)_6]^{3+} > [Ir(H_2O)_6]^{3+}$.

25.10

$$[Co(NH_3)_5X]^{2+} + [OH]^- \overset{K}{\rightleftharpoons} [Co(NH_3)_4(NH_2)X]^+ + H_2O$$

$$[Co(NH_3)_4(NH_2)X]^+ \overset{k_2}{\longrightarrow} [Co(NH_3)_4(NH_2)]^{2+} + X^-$$

$$[Co(NH_3)_4(NH_2)]^{2+} + H_2O \overset{fast}{\longrightarrow} [Co(NH_3)_5(OH)]^{2+}$$

$$K = \frac{[Co(NH_3)_4(NH_2)X^+][H_2O]}{[Co(NH_3)_5X^{2+}][OH^-]} \qquad ([H_2O] \approx 1 \text{ in dilute solution})$$

$$\therefore \quad [Co(NH_3)_4(NH_2)X^+] = K[Co(NH_3)_5X^{2+}][OH^-]$$

As $[Co(NH_3)_4(NH_2)X]^+$ is formed, it is used in the rate-determining step, but is also reformed in the pre-equilibrium. The rate can be written in terms of $k_{obs}$ or $k_2$:

$$\frac{d[Co(NH_3)_5(OH)^{2+}]}{dt} = k_{obs}\left([Co(NH_3)_5X^{2+}] + [Co(NH_3)_4(NH_2)X^+]\right)$$

$$= k_2[Co(NH_3)_4(NH_2)X^+]$$

Substituting for $[Co(NH_3)_4(NH_2)X^+]$ from above and rearranging gives:

$$k_{obs} = \frac{k_2 K[Co(NH_3)_5X^{2+}][OH^-]}{[Co(NH_3)_5X^{2+}] + K[Co(NH_3)_5X^{2+}][OH^-]}$$

$$= \frac{k_2 K[Co(NH_3)_5X^{2+}][OH^-]}{[Co(NH_3)_5X^{2+}]\left(1 + K[OH^-]\right)}$$

$$\therefore \quad \frac{d[Co(NH_3)_5(OH)^{2+}]}{dt} = \frac{k_2 K[OH^-][Co(NH_3)_5X^{2+}]_{total}}{1 + K[OH^-]}$$

25.11    $NR_1R_2R_3$ is trigonal pyramidal and, therefore, optically active; the left-hand diagram below shows enantiomers **A** and **B**. Inversion at N is generally facile and this process interconverts **A** and **B** (right-hand diagram) preventing resolution of enantiomers.

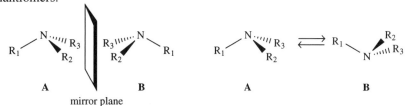

A                              B                              A                              B
mirror plane

25.12    $L^-$ derived from HL (see question 25.12 in H&S) are acac-type ligands. Look at scheme 25.45 in H&S which shows racemization occurring through a 5-coordinate intermediate. The chelating ligands in the question have the possibility of coordination through 3 different pairs of $O,O'$-donors. A mechanism involving dissociation of one end of chelate and reformation of Co-O bond will exchange $C(O)CH_3$ and $C(O)CD_3$ groups, allowing isomerization as well as racemization.

25.13    From the reaction scheme:

$$[Fe(H_2O)_6]^{3+} + [SCN]^- \underset{k_{-1}}{\overset{k_1}{\rightleftharpoons}} [Fe(H_2O)_5(SCN)]^{2+} + H_2O$$

$$\Big\Updownarrow K_1 \qquad\qquad\qquad \Big\Updownarrow K_2$$

$$[Fe(H_2O)_5(OH)]^{2+} + H^+ + [SCN]^- \underset{k_{-2}}{\overset{k_2}{\rightleftharpoons}} [Fe(H_2O)_4(OH)(SCN)]^+ + H^+ + H_2O$$

Let [Fe] represent $[Fe(H_2O)_6^{3+}]$, and [Fe(SCN)] represent $[Fe(H_2O)_5(SCN)^{2+}]$.

$$-\frac{d[SCN^-]}{dt} = k_1[Fe][SCN^-] - k_{-1}[Fe(SCN)][H_2O]$$

$$+ k_2[Fe(H_2O)_5(OH)^{2+}][SCN^-] - k_{-2}[Fe(H_2O)_4(OH)(SCN)^+][H_2O]$$

$$K_1 = \frac{[Fe(H_2O)_5(OH)^{2+}][H^+]}{[Fe]} \qquad \therefore [Fe(H_2O)_5(OH)^{2+}] = \frac{K_1[Fe]}{[H^+]}$$

$$K_2 = \frac{[Fe(H_2O)_4(OH)(SCN)^+][H^+]}{[Fe(SCN)]} \qquad \therefore [Fe(H_2O)_4(OH)(SCN)^+] = \frac{K_2[Fe(SCN)]}{[H^+]}$$

Make substitutions into the rate equation:

$$-\frac{d[SCN^-]}{dt} = k_1[Fe][SCN^-] - k_{-1}[Fe(SCN)][H_2O]$$

$$+ k_2\left(\frac{K_1[Fe]}{[H^+]}\right)[SCN^-] - k_{-2}\left(\frac{K_2[Fe(SCN)]}{[H^+]}\right)[H_2O]$$

> Take $[H_2O] \approx 1$, which is approximately true for dilute solutions

$$= \left(k_1 + \frac{k_2K_1}{[H^+]}\right)[Fe][SCN^-] - \left(k_{-1} + \frac{k_{-2}K_2}{[H^+]}\right)[Fe(SCN)]$$

25.14

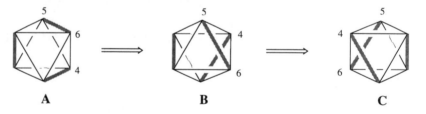

The unlabelled $H_2O$ is essentially $H_2{}^{16}O$, i.e. natural abundance O. The scheme is consistent with $H_2{}^{16}O$ and $H_2{}^{18}O$ being present in the product in a 1:1 ratio.

25.15    Look at the product enantiomers in Figure 25.8 in H&S; each blue line represents an edge of the octahedron involved in a chelate ring in M(L–L)$_3$. Start with the product in Figure 25.8b (shown in **A** below with a numbering scheme to match that in H&S). By rotating this structure about a horizontal axis in the plane of the paper (i.e. apex 5 moves towards you), the enantiomer can be drawn in a new orientation (diagram **B**). Now rotate this structure about a vertical axis in the plane of the paper to generate the structure in orientation **C**, i.e. a structure identical to that of the product enantiomer in Figure 25.8a in H&S. Since rotations merely take the molecule from one orientation to another, the scheme below confirms that the enantiomers formed in Figures 25.8a and b in H&S are the same.

**A**          **B**          **C**

25.16    $L^{2-}$ (**25.5**) is an *N,N'*-donor, and the Fe(II) complex $[FeL_3]^{4-}$ is a tris-chelate. The two sets of experimental data are the *same* within experimental error, meaning that the rate constants for dissociation and racemization are the same. To find $\Delta H^{\ddagger}$ and $\Delta S^{\ddagger}$, construct an Eyring plot. Figure 25.3 shows this for values of $k_r$; an almost identical plot is obtained using $k_d$ data.

(a) Gradient $= -15\,300 = -\dfrac{\Delta H^{\ddagger}}{R}$

$\therefore \quad \Delta H^{\ddagger} = 15\,300 \times 8.314 \times 10^{-3}$
$= 128\ \text{kJ mol}^{-1}$

**(25.5)**

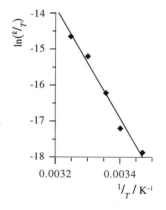

**Figure 25.3** Plot of the racemization data for answer 25.16.

Extrapolation of the line in Figure 25.3 allows you to find the intercept when $^1/_T = 0$.

$$\text{Intercept} = 35.2 = \ln\left(\frac{k'}{h}\right) + \frac{\Delta S^{\ddagger}}{R} \qquad \text{where } k' = \text{Boltzmann constant}$$
$$h = \text{Planck constant}$$

$$\therefore \quad \Delta S^{\ddagger} = R\left\{35.2 - \ln\left(\frac{k'}{h}\right)\right\} = 8.314\left\{35.2 - \ln\frac{1.381\times10^{-23}}{6.626\times10^{-34}}\right\}$$

$$= 95 \text{ J K}^{-1} \text{ mol}^{-1}$$

(b) $k_r = k_d$, so data are consistent with dissociative mechanism for racemization.

**25.17** In liquid ammonia, self-ionization gives:  $2NH_3 \rightleftharpoons [NH_4]^+ + [NH_2]^-$

> Liquid $NH_3$: see Sections 8.4 and 8.6 in H&S

$KNH_2$ in liquid $NH_3$ provides $[NH_2]^-$ and is analogous to the presence of $[OH]^-$ in aqueous solution. The fact that $[NH_2]^-$ catalyses the displacement of $Cl^-$ by $NH_3$ in $[Cr(NH_3)_5Cl]^{2+}$ is therefore consistent with a conjugate base (*Dcb*) mechanism in which $[NH_2]^-$ is the base, removing $H^+$ from coordinated $NH_3$.

**25.18** Reaction 25.48 in H&S provides an example of an inner-sphere mechansm:

$$[Co(NH_3)_5Cl]^{2+} + [Cr(H_2O)_6]^{2+} + 5[H_3O]^+ \rightarrow [Co(H_2O)_6]^{2+} + [Cr(NH_3)_5Cl]^{2+} + 5[NH_4]^+$$

(a) Bridge formation to form A has highest $E_a$.

(b) Electron transfer in B has highest $E_a$ (most common).

(c) Bridge cleavage in complex C has highest $E_a$.

> P = reactants
> P = products
> A, B and C = transition states

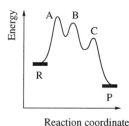
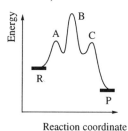
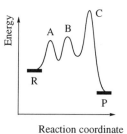

**25.19** Refer to Section 25.5 in H&S for details. Points to include:

- Outer-sphere mechansm does *not* involve covalent bond formation between reactants; occurs when metal centres are kinetically inert; mechanism is important for long-range electron transfer in biological systems.
- In an inner-sphere mechansm, electron transfer occurs between reactants across a covalently-bound bridging ligand.
- Agreement with Marcus-Hush theory is test for outer-sphere mechanism – this relates data for self-exchange reactions and the corresponding cross reaction.
- Self-exchange reaction is electron transfer between like metal centres in different oxidation states, e.g. $[Ru(bpy)_3]^{3+}$ with $[Ru(bpy)_3]^{2+}$. For such a reaction, $\Delta G^{\circ} = 0$. Free energy of activation, $\Delta G^{\ddagger}$, for self-exchange is related to the rate constant by:

> For more detail, see equations 25.56 and 25.58 in H&S

$$k = \kappa Z e^{\left(-\Delta G^{\ddagger}/RT\right)} \qquad \text{where } \kappa \approx 1; Z \approx 10^{11} \text{ dm}^3 \text{ mol}^{-1} \text{ s}^{-1}$$

**25.20** **I** and **III** are self-exchanges, and **II** is the corresponding cross-reaction. In **I**, $k$ is relatively large and indicates fast electron transfer – 2nd row metal has relatively large $\Delta_{oct}$ and both $Ru^{3+}$ ($d^5$) and $Ru^{2+}$ ($d^6$) are low-spin, differing only in an extra non-bonding electron in a $t_{2g}$ orbital; for ground state $Ru^{3+}$ and $Ru^{2+}$ complexes, the Ru–N bond distances are similar. In **III**, $[Co(NH_3)_6]^{3+}$ is low-spin $d^6$, but $[Co(NH_3)_6]^{2+}$ is high-spin $d^7$ and has longer Co–N bonds. Electron transfer between

See Figure 25.9 in H&S

$[Co(NH_3)_6]^{3+}$ and $[Co(NH_3)_6]^{2+}$ occurs between vibrationally excited states having similar Co–N bond lengths, i.e. an 'encounter complex'. The greater the changes in bond lengths needed to establish the encounter complex, the slower the rate of electron transfer. In **III**, value of $k$ indicates relatively slow reaction. In cross reaction **II**, $Co^{3+}$ to $Co^{2+}$ involves low to high-spin change; unlike self-exchanges **I** and **III** where $\Delta G° = 0$, the cross reaction is assisted by the free energy change of reaction.

25.21   (a)   **A:**   $[Fe^{III}(CN)_6]^{3-} + [Fe^{II}(CN)_6]^{4-} \rightarrow [Fe^{II}(CN)_6]^{4-} + [Fe^{III}(CN)_6]^{3-}$
        **B:**   $[Mo^{V}(CN)_8]^{3-} + [Mo^{IV}(CN)_8]^{4-} \rightarrow [Mo^{IV}(CN)_8]^{4-} + [Mo^{V}(CN)_8]^{3-}$
        **C:**   $[Mo^{V}(CN)_8]^{3-} + [Fe^{II}(CN)_6]^{4-} \rightarrow [Mo^{IV}(CN)_8]^{4-} + [Fe^{III}(CN)_6]^{3-}$

(b) Let rate constants for **A**, **B** and **C** be $k_{11}$, $k_{22}$ and $k_{12}$ and let equilibrium constant for **C** be $K_{12}$. Equation needed:

$$k_{12} = \sqrt{k_{11}k_{22}K_{12}f_{12}} \qquad \text{where} \qquad \log f_{12} = \frac{\left(\log K_{12}\right)^2}{4\log\left(\dfrac{k_{11}k_{22}}{Z^2}\right)}$$

Substituting in values from question gives:

$$\log f_{12} = \frac{\left(\log 1.0 \times 10^2\right)^2}{4\log\left(\dfrac{7.4\times10^2 \times 3.0\times10^4}{10^{22}}\right)} = -0.07 \qquad f_{12} = 0.85$$

$$k_{12} = \sqrt{7.4\times10^2 \times 3.0\times10^4 \times 1.0\times10^2 \times 0.85} = 4.3\times10^4 \text{ dm}^3 \text{ mol}^{-1} \text{ s}^{-1}$$

This agrees well with experimental value implying an outer-sphere mechanism.

25.22   (a) By summation of equations in the mechanism, the overall reaction is

$$2[Fe(CN)_6]^{3-} + H_2A \rightarrow 2[Fe(CN)_6]^{4-} + A + 2H^+$$

Compare with +ve values of $E°$ for a spontaneous overall reaction.

(b) For the rate-determining step, $E°_{cell} = -0.26$ V. From $E°_{cell}$, find $\Delta G°$ and $K_{12}$:

$$\Delta G° = -zFE° = -(1)(96\ 485)(-0.26) \times 10^{-3} = +25 \text{ kJ mol}^{-1}$$

$$K_{12} = e^{-\frac{\Delta G°}{RT}} = e^{-\frac{(+25)}{(8.314 \times 10^{-3})(298)}} = 4.1\times10^{-5}$$

From data given:   $k_{11} = 5.8 \times 10^4 \text{ dm}^3 \text{ mol}^{-1} \text{ s}^{-1}$   $k_{22} = 3.0 \times 10^5 \text{ dm}^3 \text{ mol}^{-1} \text{ s}^{-1}$

See answer 25.21 for equations

$$\log f_{12} = \frac{\left(\log 4.1\times10^{-5}\right)^2}{4\log\left(\dfrac{5.8\times10^4 \times 3.0\times10^5}{10^{22}}\right)} = -0.41 \qquad f_{12} = 0.39$$

$$k_{12} = \sqrt{5.8\times10^4 \times 3.0\times10^5 \times 4.1\times10^{-5} \times 0.39} = 530 \text{ dm}^3 \text{ mol}^{-1} \text{ s}^{-1}$$

This value for $k_{12}$ is the same order of magnitude as the experimental value, and so the values are in good agreement – supports an outer-sphere mechanism.

# 26 Homogeneous and heterogeneous catalysis

26.1 (a)

See Section 23.7 in H&S for discussion of reaction types

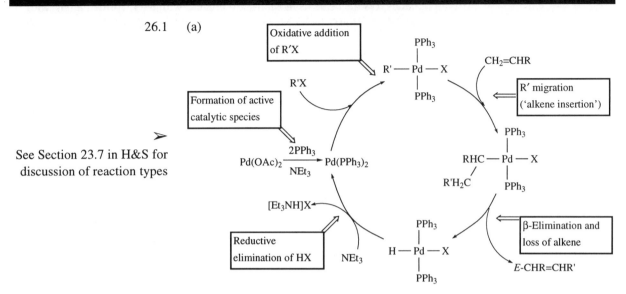

(b) When R′ is vinyl, benzyl or aryl, there is no β-hydrogen present in the R′ group. See Section 23.7 in H&S for further discussion.

26.2 (a) A catalyst precursor is not the catalytically active species but is one added to the system; it undergoes e.g. loss of CO or another ligand to form the active catalyst. $HCo(CO)_4$ contains an 18-electron Co and is not catalytically active; loss of CO generates $HCo(CO)_3$ with a 16-electron centre – coordinatively unsaturated.

(b)

See Section 23.7 in H&S for discussion of reaction types

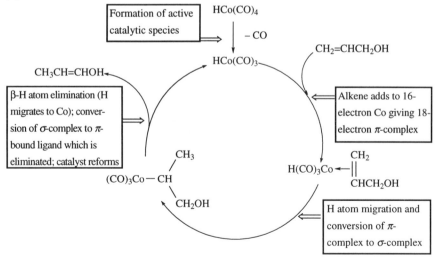

26.3 For complete details, see Section 26.3 in H&S. Points to include:
 • Monsanto process converts MeOH to $MeCO_2H$; catalyst is *cis*-$[Rh(CO)_2I_2]^-$.
 • MeOH first converted to MeI; MeI oxidatively adds to *cis*-$[Rh(CO)_2I_2]^-$.
 • Me group migrates to Rh-bound CO giving σ-bound C(O)Me; CO fills

vacant coordination site.
- Reductive elimination of MeC(O)I, then reaction with $H_2O$ gives $MeCO_2H$.
- Draw the cycle in Figure 26.4 in H&S to illustrate overall process.
- Tennessee-Eastman process is closely related to Monsanto process; converts $MeCO_2Me$ to $(MeCO)_2O$ using $cis$-$[Rh(CO)_2I_2]^-$ catalyst; role of Rh(I) catalyst is the same as in Monsanto process with MeI (formed along with $MeCO_2Li$ from $MeCO_2Me$) being converted to MeC(O)I; MeC(O)I reacts with $MeCO_2Li$ to yield $(MeCO)_2O$.
- Illustrate with a diagram how the secondary cycle in the Tennessee-Eastman process differs from that in the Monsanto process.

26.4    (a) If hydrogenation of the alkene can, in theory at least, lead to enantiomeric products, the alkene is prochiral. The alkenes to consider are **26.1-26.4.**

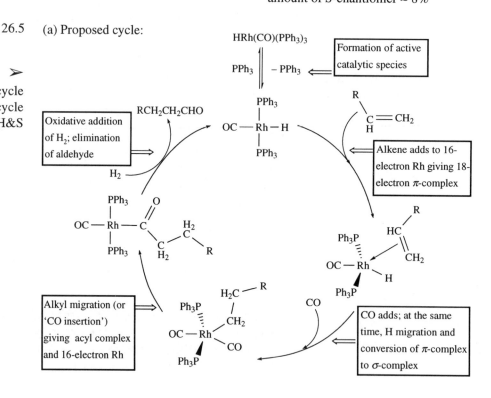

+ (E)-isomer        + (E)-isomer

**(26.1)**            **(26.2)**            **(26.3)**            **(26.4)**

The C atoms labelled * will *become* chiral *after* hydrogenation, e.g. **26.2** will become $PhMeCHCH_2Ph$. Therefore, **26.2** and **26.4** are prochiral.

(b)

$$\%ee = \left( \frac{|R - S|}{|R + S|} \right) \times 100 \qquad \therefore \frac{|R - S|}{|R + S|} = \frac{85}{100} = 0.85$$

The question states that $\%R > \%S$-enantiomer, so you can now proceed knowing that R > S. For amounts given in %:

R + S = 100                    R − S = 85

Combining these expressions gives:        amount of R-enantiomer ≈ 92%

amount of S-enantiomer ≈ 8%

26.5    (a) Proposed cycle:

Compare this proposed cycle with the inner catalytic cycle in Figure 26.6 in H&S

(b) Regioselectivity of hydroformylation of $RCH=CH_2$ is the selectivity towards $n$ or $i$-isomers; given by the $n:i$ ratio.

$$RCH_2CH_2CHO \text{ (}n\text{-isomer)} \xleftarrow{\text{CO/H}_2} RCH=CH_2 \xrightarrow{\text{CO/H}_2} R(CH_3)CHCHO \text{ (}i\text{-isomer)}$$

As temperature increases, $n:i$ ratio decreases, showing that there is a greater selectivity for the linear isomer at lower temperatures.

**26.6**

➤

Alkene isomerization: see answer 26.2

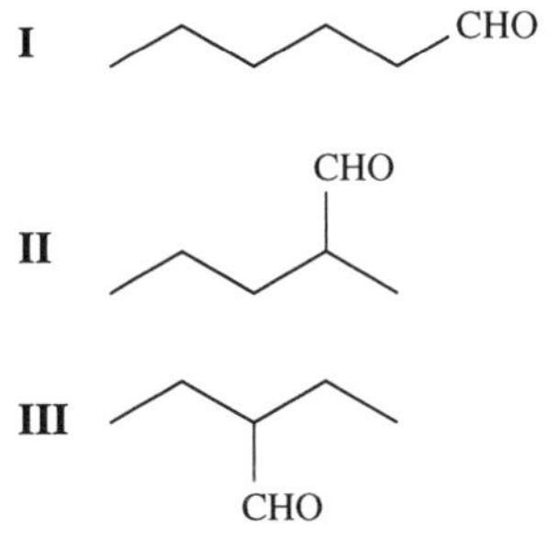

Hydroformylation of $CH_3CH_2CH=CHCH_3$ can lead to **I**, **II** and/or **III**. The active catalyst is $HCo(CO)_3$ which catalyses alkene isomerization *and* hydroformylation. Steric effects influence the product distribution. **I** is expected to be formed in the highest yield, and **III** in the lowest yield. Suggest that the 35:12:5 ratio corresponds to **I:II:III**. See equation 26.17 and accompanying discussion in H&S.

**I**

**II**

**III**

**26.7**

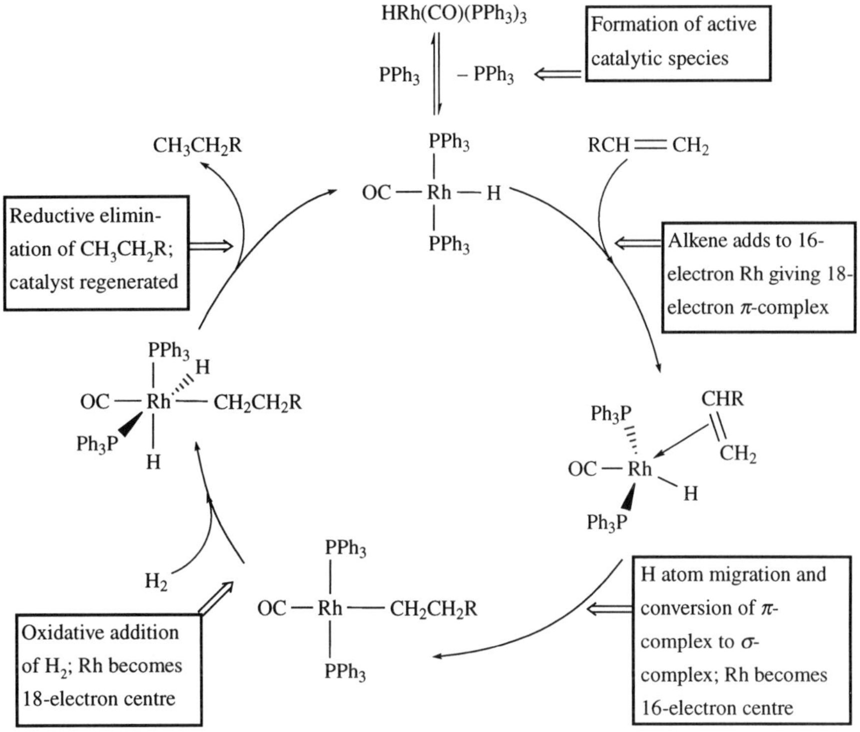

**26.8**

**(26.5)**     **(26.6)**

(a) Ligands **26.5** and **26.6** both coordinate through a *P*-donor. In the complexes $Fe(CO)_4(PPh_3)$ and $[Fe(CO)_4(\mathbf{26.6})]^+$, the fact that the IR spectral absorptions occur at very similar wavenumbers indicates that the amount of back donation into the CO $\pi^*$ orbitals is similar in each complex; there must be a similar charge distribution with respect to the Fe centre in each complex.

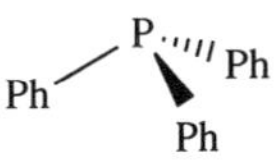

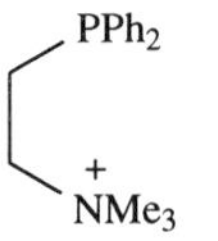

(b) The biphasic catalyst must be water-soluble. The choice of cation is critical – $Na^+$ will favour solubility of aqueous solution, whereas $[^nBu_4N]^+$ and $[Ph_4P]^+$ will increase the likelihood of the salt being soluble in an organic solvent. Therefore, the salt to choose is $Na[RuL_3]$.

**26.9**    Choose processes from those in Section 26.3 in H&S. Suggested plan for answer:
- Need for an organometallic complex containing a metal which can easily go from 16 to 18 to 16-electron centre; active catalyst must be coordinatively unsaturated, e.g. $HCo(CO)_3$, $[Rh(CO)_2I_2]^-$.

➤

Also see answers 26.3- 26.6

- An example of alkene hydrogenation, including a brief discussion of asymmetric hydrogenation and importance to drug industry.
- Monsanto process for making $MeCO_2H$; related manufacture of $(MeCO_2)_2O$.
- Hydroformylation and problems of regioselectivity and chemoselectivity.

**26.10**    (a)

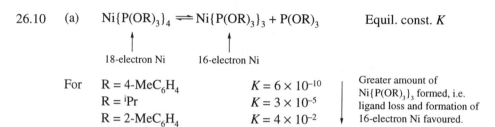

$$Ni\{P(OR)_3\}_4 \rightleftharpoons Ni\{P(OR)_3\}_3 + P(OR)_3 \qquad \text{Equil. const. } K$$

18-electron Ni        16-electron Ni

| For | | |
|---|---|---|
| $R = 4\text{-}MeC_6H_4$ | $K = 6 \times 10^{-10}$ | Greater amount of $Ni\{P(OR)_3\}_3$ formed, i.e. ligand loss and formation of 16-electron Ni favoured. |
| $R = {}^iPr$ | $K = 3 \times 10^{-5}$ | |
| $R = 2\text{-}MeC_6H_4$ | $K = 4 \times 10^{-2}$ | |

The active catalyst is the 16-electron species and is formed by loss of $P(OR)_3$. The data show that the dissociation step depends on the steric demands of R and is favoured when R is bulky.

(b) Catalytic cycle up to last 2 steps:

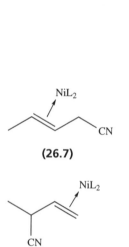

**(26.7)**

**(26.8)**

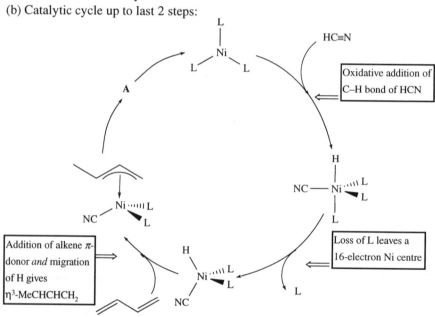

(c) The formation of **A** in the cycle above involves transfer of CN with **A** being either **26.7** or **26.7**. In the last step, addition of L releases the alkene and regenerates the active catalyst. For commercial purposes, linear alkene **26.7** is the target molecule.

**26.11**    (a) Cluster electron count in $H_2Os_3(CO)_{10} = (2 \times 1) + (3 \times 8) + (10 \times 2) = 46$

➤

Valence electron (ve) counts:
see Table 23.1, p. 186

For an $Os_3$ triangle, a 48 ve count is expected, and so $H_2Os_3(CO)_{10}$ is unsaturated. This can be represented by altering structure **26.15** in H&S to show an Os=Os double bond.

(b) The first step in the process is addition of the alkene (2 electron $\pi$-donor) to give a 48-electron $Os_3$-complex. The cluster-bound H atoms facilitate the isomerization as shown in the following cycle.

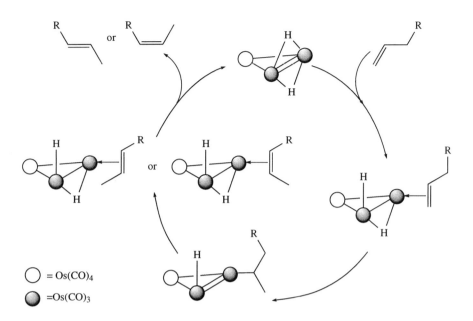

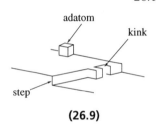

(26.9)

(26.10)

26.12 Ni has an fcc structure and so one could draw an atom arrangement as shown in Figure 5.1 (p. 43). In reality, the surface has imperfections such as those in **26.9**; also atoms missing in regions of flat surface creating 'holes'. Diagram **26.10** illustrates part of two layers of close-packed atoms, and shows a step. The 4 dark grey atoms define an $M_4$ 'butterfly' site; a triangular $M_3$ site is present on the top surface. Any 'flat' surface on clean Ni (assuming no adatoms or holes) is made up of triangular sites. In discrete cluster molecules, CO usually adopts terminal or bridging modes and such modes can also be adopted on a surface (**26.11**-**26.13**). At a step site, a CO ligand can bond in mode **26.14** and the C–O bond is weaker than in modes **26.11**-**26.13** – CO is 'activated'.

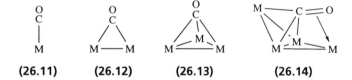

(26.11)   (26.12)   (26.13)   (26.14)

26.13 (a) Rh (a platinum-group metal) is rare and expensive; use of alumina support reduces cost. In the bulk metal, only the surface atoms are available for catalysis; small metal particles dispersed on γ-alumina (has a large surface area) gives a catalyst in which a high proportion of Rh atoms in the system are available. The γ-alumina support may play active role in the catalysis, e.g. in hydrocarbon reforming:
- Rh (or other platinum-group metal) catalyses alkane dehydrogenation;
- alkene isomerization is facilitated by the acidic γ-alumina surface;
- Rh catalyses conversion of isomerized alkene to more highly branched alkane.

(b) Oxidations in catalytic converter are not all catalysed by the same metal:

$2CO + O_2 \rightarrow 2CO_2$                catalysed by Pd and Pt

$C_nH_m + (n + {}^m/_4)O_2 \rightarrow nCO_2 + {}^m/_2H_2O$   catalysed by Pd and Pt

$2NO + 2CO \rightarrow 2CO_2 + N_2$          catalysed by Rh

$2NO + 2H_2 \rightarrow N_2 + 2H_2O$         catalysed by Rh

26.14          $2SO_2(g) + O_2(g) \rightleftharpoons 2SO_3(g)$          $\Delta_r H^o = -96$ kJ per mole of $SO_2$

(a) This is a gaseous equilibrium. In the forward reaction, 3 moles give 2 moles of gases, i.e. a *decrease* in pressure. By Le Chatelier's principle, increasing the external pressure encourages the forward reaction.

(b) By Le Chatelier, an increase in external temperature will be opposed; since the forward reaction is exothermic, the back reaction is endothermic and so is favoured.

For further details, see 'Production of $SO_3$ in the Contact process' in Section 26.6 in H&S

(c) Lower temperatures favour the forward reaction, but rate of $SO_3$ formation is slow (not useful commercially). Rate depends on temperature, but in the above equilibrium, the back reaction is favoured by an increase in temperature (not useful commercially). The rate is increased by using a $V_2O_5$ catalyst on $SiO_2$ carrier with $K_2SO_4$ promoter, passing reactants through several catalyst beds; action of catalyst:

$$SO_2 + V_2O_5 \rightleftharpoons 2VO_2 + SO_3$$

$$^1/_2 O_2 + 2VO_2 \rightarrow V_2O_5$$

26.15     (a)     $N_2(g) + 3H_2(g) \rightleftharpoons 2NH_3(g)$          $\Delta_r H^o = -92$ kJ per mole of $N_2$

The forward reaction is exothermic and proceeds with reduction in number of moles of gas; rate of formation of $NH_3$ is slow. As in answer 26.14, a higher temperature *decreases* yield of $NH_3$ and so is not a viable means of increasing the rate of formation of $NH_3$. Increasing the external pressure increases yield of $NH_3$ (application of Le Chatelier). Optimum conditions are 723 K and a high pressure in presence of $Fe_3O_4$/$K_2O/SiO_2/Al_2O_3$ catalyst – active catalyst is $\alpha$-Fe. Catalyst facilitates dissociation of $N_2$ and $H_2$; diatomics are adsorbed on catalyst surface and dissociate giving adsorbed N and H atoms which associate forming adsorbed NH, $NH_2$ and finally $NH_3$. $NH_3$ then desorbs. Rate for formation of $NH_3$ is sensitive to the metal catalyst.

$N_2(g)$

〜〜〜〜〜〜〜〜

↘

N(ad)     N(ad)
〜〜〜〜〜〜〜〜

**(26.15)**

(b) Rate determining step is adsorption of $N_2$ (**26.15**). V (early *d*-block metal) tends to retain adsorbed N atoms, blocking surface sites and hindering $NH_3$ formation. For Pt (late *d*-block metal), adsorption of $N_2$ has a high $\Delta G^{\ddagger}$ and reaction is slow. Fe is relatively cheap, but Os (one of Pt group metals) is rare and expensive.

26.16

For detailed structural data of the $MgCl_2$-supported catalyst, see B.L. Goodall (1986) *J. Chem. Ed.*, vol. 63, p. 191

(a) $MgCl_2$ has a layer structure which matches that of the active titanium chloride catalyst; $Ti^{4+}$ and $Mg^{2+}$ are similar sizes. Coordinatively unsaturated Ti must be present in surface sites – a vacant lattice site must be *cis* to the surface Cl (replaced by alkyl group during catalysis) as shown at the left-hand side of Figure 26.1.

(b) $Et_3Al$ (used with the $MgCl_2$ supported catalysts) and $Et_2AlCl$ are co-catalysts which alkylate the surface Ti (Figure 26.1).

(c) Part of the chain-growth (Cossee-Arlman mechanism) is shown in Figure 26.1;

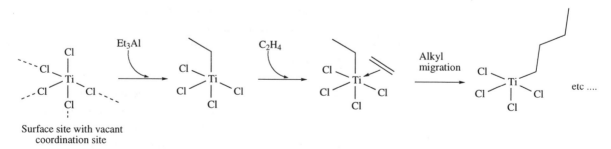

**Figure 26.1** Representation of alkene polymerization using an $MgCl_2$-supported $TiCl_4$ Ziegler-Natta catalyst.

surface alkylation, alkene addition *cis* to the alkyl group, followed by alkyl migration – then addition of the alkene again, alkyl migration and so on. Stereoregular products are formed – an essential feature of the commercial catalyst system.

26.17    Choose examples from Section 26.6 in H&S. Suggested plan for answer:
  • Heterogeneous catalyst is one which is in a different phase from the reaction it is catalysing, e.g. solid catalyst, gaseous reactants; examples: metal surfaces, alumina, silica, zeolites.
  • Fischer-Tropsch chain growth: $CO/H_2$ converted to hydrocarbons using Fe or Co catalyst; only commercially viable when oil stocks are limited/expensive; adsorbed CO and $H_2$ dissociate to give adsorbed atoms; adsorbed C and H combine to give CH and $CH_2$ ($H_2O$ desorbed) and chains build up from association on surface between adsorbed (ad) units:

$H(ad) + CH_2(ad) \rightarrow CH_3(ad)$
$CH_3(ad) + CH_2(ad) \rightarrow CH_3CH_2(ad)$
$CH_3CH_2(ad) + CH_2(ad) \rightarrow CH_3CH_2CH_2(ad)$    ....etc

Combination of alkyl group and H or two alkyl groups release an alkane; β-elimination from adsorbed alkyl group releases an alkene. Vinyl groups may also be involved as surface species.
  • Ziegler-Natta catalysts: for alkene polymerization with stereoregular products, e.g. isotactic polypropene (include a similar diagram as shown in answer 26.16).
  • Haber process for $NH_3$ production: see answer 26.15 for details.
  • Contact process for $SO_3$ production (first step in $H_2SO_4$ manufacture): see answer 26.14 for details.

26.18    Give a general comment about the structural properties of zeolites: aluminosilicates with macromolecular structures containing channels and pores of specific shapes and sizes depending on the zeolite.
(a) Separation of *n*- and *iso*-alkanes follows from the ability of a zeolite to be 'shape selective'; e.g. *n*-butane is a linear chain, but *iso*-butane is branched and more bulky. The pore size of 430 pm is selective in that zeolite 5A adsorbs *n*-alkanes but *iso*-alkanes are too sterically demanding to enter the pores.
(b) The Al sites in an aluminosilicate act as Brønsted acids:

and therefore zeolites are acid catalysts. The level of catalytic activity depends on the Al:Si ratio which varies with the zeolite.
Zeolite ZSM-5 ($Na_n[Al_nSi_{96-n}O_{192}] \approx 16H_2O$ where $n < 27$) is used commercially as an acid catalyst for isomerization of 1,3-xylene to 1,4-xylene — isomer selectivity presumably controlled by shape selectivity of ZSM-5 (see part (a)). Conversion of benzene to ethylbenzene requires an acid catalyst (compare use of $EtCl/AlCl_3$ in a Friedel-Crafts alkylation of $C_6H_6$). Adsorption of organic substrate onto surfaces within zeolite channels exposes substrate to highly active acid catalyst.

26.19    For full details, see Sections 26.5 and 26.6 in H&S. Suggested plan for answer:

- '3-way' catalytic converter is one which includes Rh/Pd/Pt in the catalyst to facilitate oxidation of hydrocarbons and CO *and* reduction of NO (see answer 26.13b for reactions).
- Metal particles supported on $Al_2O_3$ washcoat – gives high surface area; metal particle size important for optimal catalytic activity.
- High operating temperatures degrades metal particles and causes $Al_2O_3$ to undergo phase change reducing its surface area; group 2 oxides are added to stabilize the system against degradation.
- Light-off temperature is the minimum temperature at which the catalyst reaches a 50% efficiency; for the period of the cold start, the catalytic converter does not function unless a secondary source of heating is supplied.
- Air-fuel ratio has to be optimized: < 14.7:1 means too little $O_2$ for hydrocarbon and CO oxidation; > 14.7:1 means too much $O_2$ and competition for the $H_2$ which is needed to reduce NO.
- Storage of $O_2$ achieved using $CeO_2$ as the 'storage vessel': when $O_2$ is needed, reduce $CeO_2$ to $Ce_2O_3$; conversely, oxidize $Ce_2O_3$ to $CeO_2$ to store excess $O_2$.

# 27 Some aspects of solid state chemistry

27.1    (a) Schottky defect involves an atom or ion vacancy in a lattice; compound stoichiometry and electrical neutrality must be retained, e.g. in $CaCl_2$, $Ca^{2+}$ vacancies must be balanced by twice as many $Cl^-$ vacancies.
(b) Frenkel defect involves occupancy of a normally vacant lattice site (interstitial hole). AgBr has NaCl lattice and tetrahedral holes are large enough to accommodate an $Ag^+$ ion – $Ag^+$ migrates from its usual octahedral site to occupy a tetrahedral site. This does not alter stoichiometry.
(c) $Ag^+$ and $Cd^{2+}$ are similar sizes and so $Cd^{2+}$ can replace $Ag^+$ in AgCl without structural change. However, this replacement causes a charge imbalance which must be countered to retain overall electrical neutrality – achieved by having an additional $Ag^+$ vacancy for every $Cd^{2+}$ that enters the lattice: $2Ag^+/2Cl^-$ replaced by $Cd^{2+}/2Cl^-$.

27.2    Types of defect are varied: stoichiometry changes if sites are vacant or interstitial sites are occupied (cation or anion deficiencies or excesses). Example: metal deficiency in *d*-block metal oxides gives compounds such as $Fe_{1-x}O$ (*x* up to $\approx 0.1$) for Fe(II) oxide. To keep electrical neutrality, some Fe sites occupied by $Fe^{3+}$ ions – ability to form non-stoichiometric compound therefore relies on Fe having variable (but accessible) oxidation states; for a *d*-block metal, different ox. states usually involve loss of different numbers of electrons from the *same* set of *d*-orbitals. Defects may be *point defects* but in FeO, for example, there is evidence for regions in the lattice of clusters whose structures resemble $Fe_3O_4$. For metals with only one accessible oxidation state (e.g. groups 1 and 2), defects cannot arise by having oxidized or reduced metal sites (but see answer 27.1).

➢ See also answer 27.6

27.3    Consider the equation:

$$4Ni^{2+} + O_2(g) \rightleftharpoons 4Ni^{3+} + 2\square_+ + 2O^{2-} \qquad \square_+ = \text{vacant cation site}$$

For small deviations from stoichiometry, there are only a small number of vacant lattice sites and so for the bulk lattice, the concentrations of $Ni^{2+}$ and $O^{2-}$ are nearly constant. *K* for the above equilibrium can be approximated to:

$$K = \frac{[Ni^{3+}]^4[\square_+]^2}{p(O_2)}$$

The ratio of $\square_+$:$Ni^{3+} = 1:2$, and so $[\square_+] = \frac{1}{2}[Ni^{3+}]$ :     $K \propto \dfrac{[Ni^{3+}]^6}{p(O_2)}$

Since the electrical conductivity is proportional to $[Ni^{3+}]$ (see answer 27.5):

$$[Ni^{3+}] \propto \sqrt[6]{K \times p(O_2)}$$

and since *K* is a constant:     $[Ni^{3+}] \propto \sqrt[6]{p(O_2)}$

27.4    (a) Extrinsic defects arise by the presence of an impurity or dopant in a crystal (compare with extrinsic semiconductors where semiconduction arises from addition of a dopant). Intrinsic defects are inherent in a crystal (compare intrinsic semiconductors).
(b) Cubic $ZrO_2$ is used as a refractory material. On cooling pure $ZrO_2$, a phase change occurs from cubic to monoclinic at 1143 K – the material cracks. Adding

CaO to $ZrO_2$ stabilizes the cubic form; when CaO enters $ZrO_2$ lattice, $Ca^{2+}$ replace $Zr^{4+}$ ions and $O^{2-}$ vacancies are created; overall electrical neutrality retained: doped material has stoichiometry $Ca_xZr_{1-x}O_{2-x}$ and is anion-deficient.

(c) Solid solution can form between $Al_2O_3$ and $Cr_2O_3$ because (i) the stoichiometries of the solute and host are the same (cations have the same charge), (ii) the cationic radii are about the same.

27.5     Doping NiO with $Li_2O$ has to give an electrically neutral solid: introduction of $Li^+$ ions is accompanied by oxidation of an appropriate number of $Ni^{2+}$ to $Ni^{3+}$ to balance overall charge:

$$(1-x)Ni^{II}O + {}^x\!/_2Li_2O + {}^x\!/_4O_2 \rightarrow Li_xNi^{II}_{1-2x}Ni^{III}_xO$$

Positive holes: see Section 5.9 in H&S

The $Ni^{3+}$ ions can be regarded as positive holes into which electrons can move – electrons 'hop' from $Ni^{2+}$ to $Ni^{3+}$ resulting in an increase in electrical conductivity. As more $Li_2O$ is added, more $Ni^{3+}$ sites are formed in lattice ($x$ increases in above equation), providing a greater number of positive holes and, thus, an increase in conductivity.

27.6     (a) FeO exists as $Fe_{1-x}O$ ($0.04 < x < 0.11$) with some Fe sites occupied by $Fe^{3+}$ ions to retain electrical neutality. The lattice structure could be considered in terms of random point defects, but there is evidence for formation of domains which resemble $Fe_3O_4$; diagram **27.1** illustrates such clustered units.

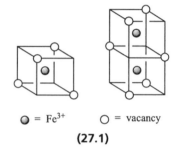

$\bullet$ = $Fe^{3+}$     $\circ$ = vacancy

**(27.1)**

(b) $UO_2$ exists as $UO_{2+x}$ with an excess of anions. Example of a $CaF_2$ type lattice with excess anions occupying interstitial sites. As well as the excess $O^{2-}$, $O^{2-}$ ions from 'normal' lattice positions are displaced into interstitial sites. Cluster domains are probably present in $UO_{2+x}$ rather than random point defects.

27.7     Diagram **27.2** shows the electrical circuit described in the question; AgI remains solid at 450 K. When current flows, the observation that one electrode gains in mass and one loses mass is explained by:

> *At the anode:*     $Ag \rightarrow Ag^+ + e^-$     electrode mass loss
> *At the cathode:*     $Ag^+ + e^- \rightarrow Ag$     electrode mass gain

The $Ag^+$ ions must be transported through the solid AgI, illustrating the property of an ion conductor, i.e. AgI conducts electricity by carrying $Ag^+$ ions rather than electrons.

Solid AgI
at 450 K

Ag anode     Ag cathode

**(27.2)**

27.8     At 300 K, NaCl is solid. The very low conductivity typifies an ionic solid which is an insulator. In the NaCl lattice, the ions lie in planes; planes perpendicular to one another have the same structure and so the conductivity is not direction dependent. For the other solids, conductivities are orders of magnitude higher than for NaCl indicating that they are electrical conductors.

Na β-alumina is $Na^+$ ion conductor; spinel-type layer structure with $Na^+$ ions in interlayer, channels; conducts only in the direction of these channels, so conductivity is direction dependent.

$Li_3N$ has a layer structure with $Li^+$ ions occupying sites within *and* between layers. A

Structure of Li$_3$N: see answer 10.12, p. 91

deficiency of Li$^+$ ions within layers leads to Li$^+$ ion conduction in a direction parallel to layers; Li$^+$ ions between the layers are not involved in conduction; conductivity is direction dependent.

**27.9**    Battery construction: Li and V$_6$O$_{13}$ electrodes separated by solid polymer electrolyte. Consider first the oxidation of Li to Li$^+$. Li$^+$ ions are then transported across the solid electrolyte to V$_6$O$_{13}$ where they are reversibly intercalated:

➤ See also Box 10.3 in H&S

$$x\text{Li}^+ + \text{V}_6\text{O}_{13} + xe^- \rightleftharpoons \text{Li}_x\text{V}_6\text{O}_{13}$$

Charging and discharging of the battery corresponds to Li$^+$ ions moving in opposite directions across electrolyte.

**27.10**    TiO, VO, MnO, FeO, CoO and NiO crystallize with NaCl lattices, but are non-stoichiometric, e.g. Fe$_{0.96}$O-FeO$_{0.89}$ and TiO$_{0.82}$-TiO$_{1.23}$. In TiO and VO, there is overlap of metal $t_{2g}$ orbitals giving partially occupied band (conduction band) and allowing electrical conduction. Crossing the $d$-block leads to increase in effective nuclear charge, with electrons tending to be localized on the metal centres. This rationalizes why MnO is an insulator at 298 K. FeO, CoO and NiO also show low conductivities

➤ See also answers 27.2 and 27.6

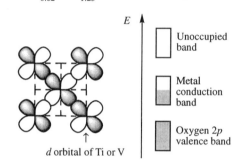

at 298 K. Raising the temperature increases the conductivities because of 'electron-hopping' from M$^{2+}$ to M$^{3+}$, the latter being present as a consequence of non-stoichiometry. In the presence of O$_2$, heating the oxide results in more M$^{2+}$ being oxidized to M$^{3+}$ – more positive holes lead to increased electrical conductivity.

**27.11**    (a) YBa$_2$Cu$_3$O$_7$ is high-temperature superconductor; consists of stacked perovskite-like units cells. Prototype perovskite is CaTiO$_3$ (see p. 47) – compare YBa$_2$Cu$_3$O$_7$ to CaTiO$_3$ by making replacements: Ba$^{2+}$ and Y$^{3+}$ for Ca$^{2+}$, and Cu$^{n+}$ for Ti$^{4+}$. Lattice is oxygen-deficient. Structure described in terms of CuO$_2$-BaO-CuO$_2$-Y-CuO$_2$-BaO-CuO$_2$ layers, with non-CuO$_2$ oxide layers isostructural with layers in NaCl. Combination of these relationships leads to the description given in the question.

➤ Look at Figures 27.5 and 27.6 in H&S

(b) MRI scanners use NbTi ($T_c$ = 9.5 K) multicore conductors. Although replacement by high-$T_c$ superconductors would be financially beneficial, problems must be overcome: (i) costs of cooling material to $T_c$; situation improves as higher values of $T_c$ are obtained; NbTi requires He coolant, but if $T_c$ > 77 K, liquid N$_2$ could be used; (ii) bulk cuprate superconductors lose their superconductivity after material has carried only limited amount of current; specialized fabrication techniques required.

**27.12**    (a) MgO doping of ZrO$_2$ parallels doping using CaO (refer to answer 27.4b).
(b) CaF$_2$ is doped with LaF$_3$ to form Ca$_{1-x}$La$_x$F$_{2+x}$ containing excess F$^-$ which occupy interstitial sites in the host lattice. Ca$_{1-x}$La$_x$F$_{2+x}$ is an F$^-$ ion conductor.
(c) Doping Si with B introduces electron deficient sites giving unoccupied level in band structure; band gap between this and occupied band beneath it is small, making thermal population of acceptor level possible. Positive holes left in valence band when electrons move to acceptor level act as charge carriers; gives p-type (extrinsic) semiconductor.

(d) Doping Si with As introduces electron rich sites; extra electrons occupy discrete level below conduction band with small band gap which allows thermal population of the conduction band – leads to n-type (extrinsic) semiconductor.

27.13    (a) $I^-$ oxidized to $I_2$; some V(V) reduced to V(IV); $Li^+$ ions intercalated in host vanadate:

$$xLiI + V_2O_5 \xrightarrow{\Delta} Li_xV_2O_5 + {}^x/_2I_2$$

(b) Formation of $[WO_4]^{2-}$ (no redox involved); $CaWO_4$ is the mineral scheelite:

$$CaO + WO_3 \xrightarrow{\Delta} CaWO_4$$

(c) Products are mixed metal oxides – do not contain discrete anions:

$$4SrO + Fe_2O_3 \xrightarrow{\Delta, O_2} 2Sr_2FeO_4 \ (or \ SrFeO_3)$$

27.14    (a) $BiCaVO_5$ contains Bi(III), Ca(II), V(V); reagents should have same ox. states:

$$Bi_2O_3 + 2CaO + V_2O_5 \rightarrow 2BiCaVO_5$$

(b) If $CuMo_2YO_8$ contains Mo(VI), then Cu is Cu(I) since Y has to be Y(III):

$$Cu_2O + 4MoO_3 + Y_2O_3 \rightarrow 2CuMo_2YO_8$$

(c) $Li_3InO_3$ contains $Li^+$ and In(III):

$$3Li_2O + In_2O_3 \rightarrow 2Li_3InO_3$$

(d) The only available ox. state for Y is Y(III), therefore $Ru_2Y_2O_7$ contains Ru(IV):

$$2RuO_2 + Y_2O_3 \rightarrow Ru_2Y_2O_7$$

27.15    Choose a CVD process from Section 27.6 in H&S, and give a sketch of the apparatus used. For uses in the semiconductor industry, include:
   • High-purity Si from $HSiCl_3$ and $H_2$, or thermal decomposition of $SiH_4$.
   • B-doped Si produced by depositing film of $\alpha$-BN on Si wafer and heating to facilitate diffusion of B atoms into Si.
   • Formation of high purity SiC from e.g. $SiCl_4$ and $NH_3$.
   • III-V semiconductors: e.g. GaAs from $Me_3Ga$ and $AsH_3$

27.16    (a) Wear-resistant coatings (e.g. for high-speed cutting tools) include $Al_2O_3$
➤    (deposited from $AlCl_3$, $CO_2$ and $H_2$), nitrides of Ti, Hf and Zr (from volatile metal

For more details for parts (a),    chloride, $H_2$ and $N_2$), and TiC (deposited using $TiCl_4$, $CH_4$ and $H_2$).
(b) and (d), see Section 27.6    (b) GaAs (intrinsic semiconductor) deposited from volatile $Me_3Ga$ and $AsH_3$.
in H&S    (c) Only light source for glass reflectors is car headlights; visibility depends on weather conditions. Energy source for LEDs includes sunlight; advantage of LEDs is that energy can be stored, and enhanced intensity of emission gives better visibility range.
(d) $BaTiO_3$ deposited using $Ti(O^iPr)_4$ and $Ba^{2+}$ β-ketonates; volatile complexes are difficult to find; F-substituents introduced to increase volatility, but films are contaminated with F. For $YBa_2Cu_3O_7$ formation studies have included β-ketonate complexes of $Ba^{2+}$, $Cu^{2+}$ and $Y^3$, but no results of commercial significance yet available.

# 28 The trace metals of life

28.1 (a) The general formula of an amino acid is shown in **28.1**. A peptide (or polypeptide) is produced by condensation of the $NH_2 + CO_2H$ groups of different amino acids (**28.2**). A peptide chain has an *N*-terminus ($NH_2$ group) and *C*-terminus ($CO_2H$ group).

**Table 28.1** Examples of naturally occurring α-amino acids.

| R (in **28.1**) | α-Amino acid |
| --- | --- |
| H | glycine |
| $CH_2CO_2H$ | L-aspartic acid |
| $CH_2SH$ | L-cysteine |
| $CH_2CH_2CO_2H$ | L-glutamic acid |
| $CH_2CHMe_2$ | L-leucine |
| $CH_2CH_2SMe$ | L-methionine |
| $CH_2OH$ | L-serine |

**(28.1)**     **(28.2)**

(b) 20 naturally occurring α-amino acids containing particular R groups (examples in Table 28.1); with the exception of glycine, all naturally occurring amino acids are chiral, and Nature is enantiomerically specific.

(c) A protein is a polypeptide with very high $M_r$; structures are complex; proteins classified as fibrous or globular. Prosthetic group in a protein is non-amino acid unit. In a metalloprotein, the prosthetic group binds a metal centre and this group is essential for biological activity of the protein, e.g. acts as redox-active centre or Lewis acid. Examples of metalloproteins: ferredoxins, haemoglobin, cytochrome *c*.

(d) If the metal ion is removed from a metalloprotein, the remaining organic molecule is the apoprotein.

(e) Haem is the prosthetic group in e.g. haemoglobin and myoglobin. Part of the co-ordination environment about Fe is a porphyrin group (**28.3**) with substituents which depend upon metalloprotein; axial ligand(s) have role in connecting haem to protein.

**(28.3)**

28.2 Refer to Section 28.2 in H&S for full details. Structures are complex, but answer should include representative examples of metal-binding sites. Points to discuss:
- Uptake of Fe into the blood as $Fe^{3+}$-containing transferrins; in this form, Fe is transported to protein 'storage vessel' called ferritin (e.g. in liver and spleen).
- Ferritin has very high $M_r$ ($\approx 445\,000$ for horse spleen apoferritin) and a complex structure – hollow shell accommodating $\leq 4500$ high-spin $Fe^{3+}$ as microcrystalline oxohydroxophosphate (e.g. of biomineralization). Iron probably enters cavity as $Fe^{2+}$, being oxidized once inside.
- Serum transferrin transports Fe to bone marrow; transferrin consists of single, high $M_r$ polypeptide chain containing 2 pockets in which $Fe^{3+}$ is coordinated by amino acid *O*- and *N*-donors; $[CO_3]^{2-}$ or $[HCO_3]^-$ must also be present.
- Aerobic microorganisms store and transport Fe as $Fe^{3+}$ using siderophores – *O*-donor ligands; complexes have very high stability constants.
- In mammals, thioneins contain soft *S*-donors and transport soft metal ions e.g. $Hg^{2+}$ and $Cd^{2+}$; act against toxic metals.

28.3 Model ligand $L^{2-}$ is anion of catechol **28.4**. The ligand $L'^{6-}$ (**28.5**, conjugate base of enterobactin) contains three catechol units and binds one $Fe^{3+}$. The coordination mode of **28.5** in $[FeL']^{3-}$ could be structurally modelled by using 3 catechol ligands

(28.4)

(28.5)

in the complex $[FeL]^{3-}$. $Fe^{3+}$ is $d^5$ whereas $Cr^{3+}$ is $d^3$ and kinetically inert. Use of $Cr^{3+}$ in place of $Fe^{3+}$ permits solution studies to be carried out – rates of substitution at $Cr^{3+}$ are slow.

**28.4**

Soft metal ions and ligands: see Table 6.9 in H&S

(a) Thioneins are sulfur-rich proteins with the protein chains folded into pockets, exhibiting soft $S$-donor binding sites. $Cd^{2+}$ is a soft metal ion. Stability constant of complex formed between soft metal ion and soft atom donor set is high. Other metal ions favoured by thioneins are $Cu^+$, $Zn^{2+}$ and $Hg^{2+}$.

(b) $[Cu_4(SPh)_6]^{2-}$ contains a cluster of $Cu^+$ centres with $[PhS]^-$ bridging ligands. Thionein pockets are rich in Cys residues and tend to bind multinuclear metal units. Thus, $[Cu_4(SPh)_6]^{2-}$ (a discrete complex, more readily studied by e.g. spectroscopic methods than a metalloprotein) represents a suitable model for the Cu-containing metalloprotein in yeast.

(c) The R group in the amino acid histidine is **28.6**. The heterocyclic ring is similar to that in imidazole (**28.7**) and trispyrazolylborate (**28.8**). Coordination complexes featuring **28.7** or **28.8** can be considered appropriate models for metalloproteins featuring His residues.

(28.6)    (28.7)

(28.8)

**28.5**

(28.9)

Fuller details are given in Section 28.3 in H&S; suggested answer plan is as follows:

(a) Single haem unit contains a functionalized porphyrin (protoporphyrin IX) which is an $N_4$-donor (**28.3**). In oxyhaemoglobin, iron is bound as $Fe^{3+}$ with axial His residue from protein chain and axial $O_2$ unit (**28.9**); deoxyhaemoglobin contains $Fe^{2+}$ and binding of $O_2$ is accompanied by electron transfer oxidizing $Fe^{2+}$ to $Fe^{3+}$, and reducing $O_2$ to $[O_2]^-$. Low-spin $Fe^{3+}$ ($d^5$) and $[O_2]^-$ each has an unpaired electron – antiferromagnetic coupling renders oxyhaemoglobin diamagnetic.

(b) 'Picket fence' porphyrins are used in model complexes; they have bulky substituents that prevent formation of $[O_2]^{2-}$ bridged species – this occurs in less sterically hindered complexes such as $[Fe(tpp)]$ ($H_2tpp$ = tetraphenylporphyrin):

$$2[Fe^{II}(tpp)] + O_2 \rightarrow [(tpp)Fe^{III}\text{--}O\text{--}O\text{--}Fe^{III}(tpp)]$$

(c) Haemoglobin is tetrameric and the 4 haem units communicate with each other via changes in the conformation of the protein chains; once one $O_2$ is bound, the affinity of the remaining Fe sites for $O_2$ increases dramatically, and similarly, release of the first $O_2$ triggers release of the other three.

(d) Change from high-spin $Fe^{2+}$ (octahedral $d^6$) to low-spin $Fe^{3+}$ (octahedral $d^5$) which is antiferromagnetically coupled to $[O_2]^-$: deoxy- to oxy-form is a change from paramagnetic to diamagnetic species.

28.6    More details for this answer are found in Section 28.3 in H&S; include the following points plus comments on model compounds:

(a) In mammals, myoglobin binds $O_2$ in the same way as haemoglobin (see answer 28.5 for details) *except* for the fact that myoglobin is monomeric and shows no cooperativity. Supporting evidence comes from model/structural studies using picket fence porphyrins and from magnetic data.

(b) Haemerythrin is a *non*-haem Fe-containing $O_2$ carrier in marine invertebrates. Active site is dinuclear (crystallographic data for both deoxy- and oxy-forms); magnetic data show antiferromagnetically coupled Fe(II) centres in the deoxyform. $O_2$ binding involves coordinatively unsaturated Fe(II) and $\mu$-OH:

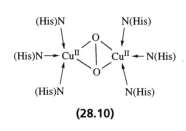

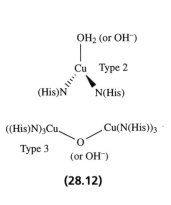

**(28.10)**

(c) Haemocyanins are Cu-containing $O_2$ carriers in e.g. arthropods; *no* haem unit. Colourless Cu(I) deoxy-form turns blue on $O_2$ uptake as Cu(I) oxidized to Cu(II). Active site contains 2 Cu(I)(His)$_3$ units in close proximity (Cu---Cu = 354 pm, non-bonded) and oxy-form contains an Cu(II)–$[O_2]^{2-}$–Cu(II) unit (**28.10**) in which Cu(II) centres are antiferromagnetically coupled (superexchange). Evidence for structures and bonding modes: magnetic data, Raman spectroscopic data for $v$(O–O), and crystallographic data for both deoxy- and oxy-forms.

28.7    For additional details, see Section 28.4 in H&S. Points to include:

- Cu centres in blue copper proteins fall into 3 classes with spectroscopic (as well as structural) distinctions; proteins contain at least one Type 1 centre.
- Function: redox centres utilizing $Cu^{2+}/Cu^+$ couple.
- Type 1 centre exhibits intense absorption in electronic spectrum with $\lambda_{max} \approx$ 600 nm ($\varepsilon_{max} \approx 100 \times$ that of $Cu^{2+}$(aq)) assigned to charge transfer from Cys ligand to $Cu^{2+}$; EPR spectrum shows narrow hyperfine splitting ($Cu^{2+}$ is $d^9$).
- Electronic and EPR spectra of Type 2 centre typical of simple $Cu^{2+}$ coordination complexes.
- Type 3 centre exhibits absorption with $\lambda_{max} \approx$ 330 nm; consists of 2 $Cu^{2+}$ centres, antiferromagnetically coupled giving a diamagnetic system. The $Cu_2$-unit acts as 2-electron transfer centre – involved in $O_2$ reduction.
- Protein crystallographic data provide structural information: Type 1 centre in plastocyanin is 4-coordinate (**28.11**) (see answer 28.8 for more detail); ascorbate oxidase (catalyses $O_2$ reduction to $H_2O$) contains one Type 1 centre (as **28.11**), one Type 2 and one Type 3 centre – Type 2 and Type 3 Cu atoms form $Cu_3$ unit (**28.12**); combination of Type 1, 2 and 3 centres in ascorbate oxidase facilitates electron transfer from organic substrate (at Type 1 Cu) and reduction of $O_2$ (at Type 2/3 Cu site).

**Type 1**
$$N(His)$$
$$|$$
$$(His)N—Cu^{\prime\prime\prime\prime}S(Cys)$$
$$S(Met)$$

**(28.11)**

$$OH_2 \text{ (or } OH^-)$$
$$|$$
$$Cu \quad \text{Type 2}$$
$$(His)N \quad N(His)$$

$$((His)N)_3Cu \quad Cu(N(His))_3$$
$$O$$
**Type 3** $\quad$ (or $OH^-$)

**(28.12)**

28.8    Plastocyanin is a blue copper protein containing Type 1 Cu (see answer 28.7) . His, Cys and Met residues on protein backbone provide coordination site **28.11**;

Cu(II) and Cu(I) prefer 4- and 3-coordination respectively: although **28.11** is 4-coordinate, Cu–S(Met) > Cu–S(Cys), so provides a compromise environment for Cu(I) and Cu(II) – rapid electron transfer is possible. Reduction of Cu(II) to Cu(I) accompanied by lengthening of Cu–N and Cu–S bonds. The fact that in **28.11** one Cu–S bond is long implies tending towards 3-coordination, i.e. favouring Cu(I); supported by high reduction potential for plastocyanin (+370 mV at pH 7).

28.9    Ascorbate oxidase is a blue copper protein (crystallographic data available) and contains Type 1, 2 and 3 centres:
   - Type 1 Cu (see **28.11**) is remote from other 3 Cu sites but able to communicate with them via protein chain; coordination is by a Cys, Met and two His residues (see answers 28.7 and 28.8 for more details).
   - Type 2 centre (single Cu) and Type 3 centre ($Cu_2$ unit with metal atoms antiferromagnetically coupled through $\mu$-O or $\mu$-OH) form a $(Cu)(Cu_2)$ unit (see **28.12**); all protein residues are His.
   - Function: reduction of $O_2$ to $H_2O$ coupled with oxidation of organic substrate (a phenol):

$$O_2 + 4H^+ + 4e^- \rightleftharpoons 2H_2O \qquad\qquad 4RH + O_2 \rightarrow 4R^{\bullet} + 2H_2O$$

   - All Cu centres involved in electron transfer via Cu(II)/Cu(I) couple; Type 1 Cu involved in electron transfer from organic substrate; $O_2$ reduction occurs at Type 2/3 site.

28.10    (a) Blue colour arises from $Cu^{2+}$ and represents the oxidized form of the blue copper protein; reduced form contains $Cu^+$ and is colourless.
(b) [4Fe-4S] ferredoxin (**28.13**) contains $Fe^{3+}/Fe^{2+}$ centres; 1-electron redox process involves $Fe_4$ unit – not localized at one Fe centre but can be represented as:

$$2Fe(III).2Fe(II) + e^- \rightleftharpoons Fe(III).3Fe(II)$$

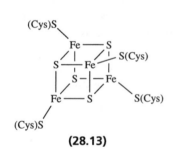

**(28.13)**

Cluster is held in a pocket of the protein chain; changes in conformation of metal-binding pocket alter the Fe coordination environment and affect reduction potential.
(c) $O_2$ acts as a $\pi$-acceptor when it binds to Fe(II) in deoxyhaemoglobin to form the Fe(III) oxy-complex. CO is also a $\pi$-acceptor ligand and therefore competes for Fe(II) in the same binding site as $O_2$. Once bound, CO blocks the haem site preventing $O_2$ coordination. [CN]$^-$ is also a $\pi$-acceptor ligand, but favours higher oxidation state metal centres than CO; [CN]$^-$ binds to Fe(III) in cytochromes (involved in electron transfer processes).

28.11    For full details, see 'The mitrochondrial electron-transfer chain' in Section 28.4 in H&S. Points to include:
   - Mitrochondrial electron-transfer chain is means of transferring electrons in living cells.
   - Draw out the chain shown in Figure 28.12 in H&S.
   - Emphasize range of reduction potentials that must be covered by biological systems at pH 7: –414 mV for $H^+$ reduction to $H_2$, to +815 mV for $O_2$ reduction to $H_2O$; each member of the chain operates within a very small potential range – hence need for a chain of electron transfer mediators.
   - Electron transfer involving a metal centre in metalloprotein is a 1-electron process; transfer involving organic substrate is usually a 2-electron step.

• Quinones play a vital role in overcoming the 1-electron/2-electron mismatch since they can undergo both 1- and 2-electron processes:

$$O=\text{◯}=O \;\xrightleftharpoons[1e^-]{H^\bullet}\; O=\text{◯}\text{—OH} \;\xrightleftharpoons[1e^-]{H^\bullet}\; HO\text{—◯—}OH$$

Quinone                Semiquinone                Hydroquinone (quinol)

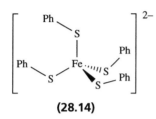

**(28.14)**

**28.12** (a) $[Fe(SPh)_4]^{2-}$ (**28.14**) contains tetrahedral Fe(II) with $[PhS]^-$ ligands. Rubredoxin contains a tetrahedral $Fe\{S(Cys)\}_4$-site with the S(Cys) residues attached to the protein chain. Thus, **28.14** is a good structural model. For the model compound, observed values of $\mu_{eff}$ are 5.05 and 5.85 $\mu_B$ for reduced and oxidized forms. These compare well with $\mu$(spin-only) for tetrahedral high-spin $Fe^{2+}$ ($d^6$) and $Fe^{3+}$ ($d^5$):

$$\mu(\text{spin-only}) \text{ high-spin } Fe^{2+} = \sqrt{n(n+2)} = \sqrt{4(4+2)} = 4.90 \ \mu_B$$

$$\mu(\text{spin-only}) \text{ high-spin } Fe^{3+} = \sqrt{n(n+2)} = \sqrt{5(5+2)} = 5.92 \ \mu_B$$

(b) Spinach ferredoxin is a [2Fe-2S] system with an $Fe_2(\mu\text{-}S)_2\{S(Cys)\}_4$ core; the model complex **28.15** is structurally related to this, with Ph substituents replacing Cys residues. While **28.15** models the active site, it cannot provide information about the effects of the protein chain (e.g. its conformation).

**(28.15)**

(c) Nitrogenase contains two types of clusters which are active sites: (i) Cys bridged double [4Fe-4S] site ('P-cluster'); (ii) FeMo-cofactor. The discrete complex in the question models half of FeMo-cofactor (Figure 28.1), but in the model, each Fe is 4-coordinate whereas the active site involves coordinatively unsaturated Fe atoms which are considered to be the sites of addition of $N_2$. For the model, Mössbauer data consistent with delocalization of charge, i.e. 2.67 is an average oxidation state of Fe(II).2Fe(III), a model that is appropriate for the active site.

**28.13** Although charge is not localized, redox reactions in [4Fe-4S] protein summarized as:

$$4Fe(III) \underset{\mathbf{A}}{\rightleftharpoons} 3Fe(III).Fe(II) \underset{\mathbf{B}}{\rightleftharpoons} 2Fe(III).2Fe(II) \underset{\mathbf{C}}{\rightleftharpoons} Fe(III).3Fe(II) \underset{\mathbf{D}}{\rightleftharpoons} 4Fe(II)$$

Each of steps **A** to **D** is a 1-electron reduction or oxidation.
(a) Under physiological conditions, couples **B** and **C** are accessed; $E'$ values must fall within appropriate limits at pH 7 (see answer 28.11); typically, $E'$ for **C** $\approx -350$ to $-450$ mV, and for **B**, $\approx +350$ mV.

**Figure 28.1** Model compound (left-hand diagram) for part of the FeMo cofactor in nitrogenase; the part modelled is indicated with the dashed circle in the right-hand diagram. The identity of atom X is uncertain.

(b) HIPIP (high potential protein) corresponds to couple **B** – highest oxidized state of the $Fe_4$ unit under physiological conditions.

See Figure 28.12 in H&S

(c) See part (a) for typical $E'$ values. Mitrochondrial chain ranges from $-414$ mV for $H^+$ reduction to $H_2$, to $+815$ mV for $O_2$ reduction to $H_2O$. The [4Fe-4S] ferredoxins are involved in electron transfer at the most negative potential end of the chain, while HIPIP operates further along the chain (more +ve potential). Combining **B** and **C** to give a single 2-electron reduction or oxidation does not occur.

28.14    Active site of [2Fe-2S] ferredoxin is represented in **28.16** and that in Rieske protein in **28.17**; structural difference is coordination by His (**28.17**) rather than Cys residues. At physiological pH, $E'$ values of [2Fe-2S] ferredoxins are negative (depends on protein: $\approx -300$ to $-400$ mV), whereas Rieske protein has a positive potential ($+290$ mV for protein isolated from spinach). Their roles in the mitrochondrial chain (see answer 28.11) are therefore different: electron transfer in Rieske protein is coupled to oxidation of plastoquinol to plastosemiquinone which releases $H^+$; electron transfer in ferredoxins is utilized at the most negative potential end of the chain.

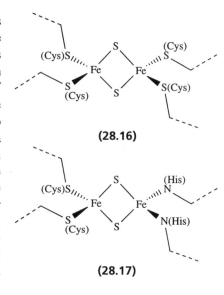

**(28.16)**

**(28.17)**

28.15    (a) Haem unit contains $Fe^{2+}$ or $Fe^{3+}$ coordinated within $N_4$-donor set of porphyrin group (see **28.1**). Deoxymyoglobin contains protoporphyrin IX and binds Fe(II); does not become Fe(III) until $O_2$ bound in oxymyoglobin. Cytochrome $c$ is an electron transfer centre, and Fe(II) or Fe(III) is present depending on redox state. Deoxymyoglobin contains 5-coordinate Fe(II) with axial site occupied by His residue (**28.18**) from protein chain. In cytochrome $c$, Fe is 6-coordinate and haem unit is bound to the protein chain through axial His and Met residues, (**28.19**), and through 2 Met residues covalently linked to the porphyrin ring (refer to Figure 28.18 in H&S).

(b) Cytochrome $c$ is a 1-electron transfer centre, utilizing $Fe^{3+}/Fe^{2+}$ redox couple. In the mitrochondrial electron-transfer chain (see answer 28.11) cytochrome $c$ accepts an electron from cytochrome $c_1$ and transfers it to cytochrome $c$ oxidase:

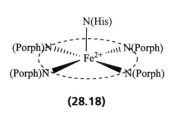

**(28.18)**

**(28.19)**

Electron transfer between these metalloproteins is proposed to occur by a mechanism that involves tunnelling through the edge of the haem unit. Electrons so transferred are finally used in the 4-electron reduction of $O_2$ to $H_2O$.

28.16    (a) Cytochrome $c$ oxidase is the terminal member of mitrochondrial electron transfer chain and catalyses reduction of $O_2$ to $H_2O$ (electron transfer is coupled to $H^+$ pumping):

$$O_2 + 4H^+ + 4e^- \rightleftharpoons 2H_2O$$

(b) Four active sites (structurally confirmed) are $Cu_A$, $Cu_B$, haem $a$ and haem $a_3$:
- $Cu_A$ contains a $Cu_2\{\mu\text{-}S(Cys)\}_2$ site;
- haem $a$ contains 6-coordinate Fe with axial His residues;
- $Cu_B$ is 3-coordinate (3 His residues) with Cu atom facing (but not bonded to) Fe in haem $a_3$;
- haem $a_3$ contains 5-coordinate Fe with the vacant coordination site facing $Cu_B$ (Fe----Cu = 450 pm); haem $a_3$ and $Cu_B$ are antiferromagnetically coupled.

Proposed operation:
- Electron transfer involves $Cu_A$ and haem $a$; electrons are transferred from cytochrome $c$ (see scheme in answer 28.15) to $Cu_A$ and are passed to haem $a$.
- Haem $a_3$ and $Cu_B$ together are the $O_2$ binding site – at this site, $O_2$ is reduced to $H_2O$ and the site is also involved in pumping $H^+$ across the mitochondrial inner membrane.
- The 4 sites work together even though $Cu_A$ is $\approx$ 2000 pm away from haem $a$ and haem $a_3$, and haem $a$ is 900 pm distant from haem $a_3$; a hydrogen-bonded network involving protein chain residues, haem propanoate substituents, and a $Cu_A$-bound His residue are thought to provide an electron-transfer route between $Cu_A$ and haem $a$.

28.17   (a) Haemoglobin utilizes haem-Fe for $O_2$ binding; haem unit provides a versatile 5- to 6-coordinate binding site with the $N_4$-donors of the porphyrin ring allowing the $Fe^{2+}$ or $Fe^{3+}$ centre to lie out of or in their coordination plane. Fe in cytochromes undergo less marked structural changes than Fe in haemoglobin; the coordination sphere (e.g. 6-coordinate in cytochrome $c$) tolerates reversible $Fe^{3+}$ to $Fe^{2+}$ change as an electron is transferred.

(b) See answer 28.16b for details; emphasize the fact that metal centres present have to be able to transfer electrons, bind $O_2$ *and* pump $H^+$.

(c) 5-Coordinate $Fe^{2+}$ is present in deoxy-form of haemoglobin; vacant site is filled by $O_2$; binding is concomitant with 1-electron transfer to give $Fe^{3+}$ and coordinated $[O_2]^-$; this is reversible, with the 5-coordinate 'rest state' being reformed when $O_2$ is released. Cytochrome $c$ does not need to bind another ligand (although it does when $[CN]^-$ poisons the system) since its role is one of electron transfer with the 6-coordinate Fe centre reversibly going between $Fe^{2+}$ and $Fe^{3+}$.

(d) NP1 is a haem protein in salivary glands of blood-sucking insect *Rhodnius prolixus*. Fe(III) in NP1 binds NO 10× more tightly at the pH of the insect's saliva than at the pH of the insect's victim. Once insect saliva is released into the victim, NO is released causing expansion of the blood vessels and inhibiting blood clotting.

28.18   Full details are give in Section 28.5 in H&S. Suggested plan for answer:
- $Zn^{2+}$ is a hard metal centre; coordinated by hard *N*- and *O*-donors; highly polarizing and able to act as a Lewis acid.
- Metalloenzymes containing $Zn^{2+}$ include carbonic anhydrase II (CAII) and carboxypeptidase A (CPA).
- CAII catalyses reversible hydration of $CO_2$:

$$H_2O + CO_2 \rightleftharpoons H^+ + [HCO_3]^-$$

In active site, $Zn^{2+}$ ion bound by three protein His residues and $[OH]^-$ (**28.20**); $[OH]^-$ is hydrogen bonded to Glu and Thr residues. $CO_2$ enters a hydrophobic pocket and is converted to $[HCO_3]^-$ which is released; $H_2O$ enters and is deprotonated giving back the active site.

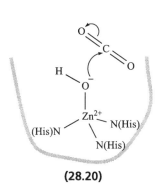

(**28.20**)

- CPA catalyses cleavage of peptide link in polypeptide:

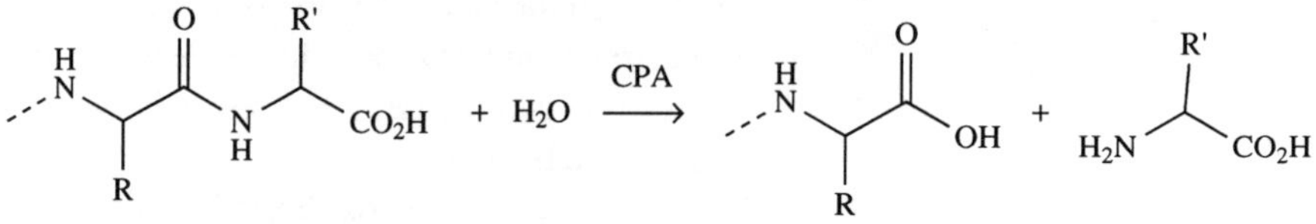

Specific cleavage at *C*-terminus and selective for *C*-terminal amino acid with large aliphatic or Ph substituent. In active site, $Zn^{2+}$ bound by *O,O'*-Glu, 2 His residues and $H_2O$. Use Figure 28.23 in H&S to describe the mechanism of catalysed peptide link cleavage, emphasizing not only the Lewis acidity of $Zn^{2+}$, but also the cooperation between protein residues and the role of hydrogen bonding.

- The answer could also be extended to include carboxypeptidase G2 in which the active site contains 2 $Zn^{2+}$ centres.

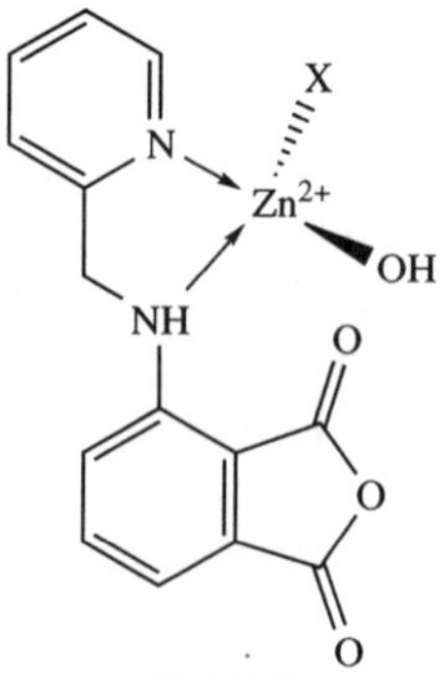

28.19 Rate of hydrolysis depends on concentrations of acid anhydride, $Zn^{2+}$ *and* [OH]$^-$, showing that all are involved in rate determining step. Tetrahedral environment for $Zn^{2+}$ is likely; propose initial complex **28.21** where X could be $H_2O$ or an *O*-donor from the acid anhydride. A plausible mechanism involves attack by coordinated [OH]$^-$ at carbonyl C atom followed by ring opening:

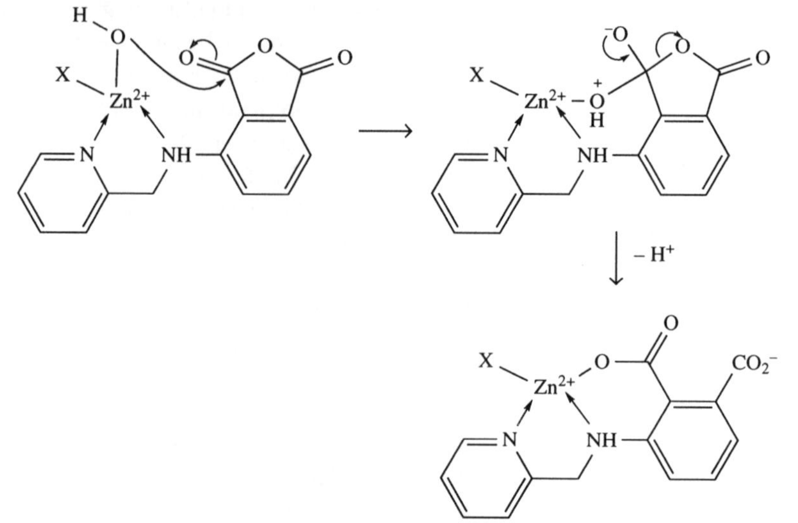

**(28.21)**

Note that the degree of protonation depends on pH.

28.20 Full details are found at the end of Section 28.5 in H&S. Points to include:
- Carbonic anhydrase contains $Zn^{2+}$, $d^{10}$ ion; spectroscopic and magnetic studies (common techniques for investigating metalloproteins or model compounds) are not appropriate for a $d^{10}$ ion – filled *d* level means diamagnetic complexes in all geometries and no '*d-d*' electronic transitions. Therefore need to substitute $Zn^{2+}$ by a metal ion that can be a spectroscopic and magnetic probe, *but* substitution must not perturb the coordination environment.
- $Co^{2+}$ substitution is suitable: $Co^{2+}$ is $d^7$ and so gives electronic spectroscopic and magnetic data; ionic radii of $Co^{2+}$ and $Zn^{2+}$ are similar; $Co^{2+}$ and $Zn^{2+}$ can be accommodated within similar coordination geometries; replacement of $Zn^{2+}$ in a protein by $Co^{2+}$ often has only small effects on protein conformation.